Space Science and Technologies

Series Editor

Peijian Ye, China Academy of Space Technology, Beijing, China

Space Science and Technologies publishes a host of recent advances and achievements in the field – quickly and informally. It covers a wide range of disciplines and specialties, with a focus on three main aspects: key theories, basic implementation methods, and practical engineering applications. It includes, but is not limited to, theoretical and applied overall system design, subsystem design, major space-vehicle supporting technologies, and the management of related engineering implementations.

Within the scopes of the series are monographs, professional books or graduate textbooks, edited volumes, and reference works purposely devoted to support education in related areas at the graduate and post-graduate levels.

More information about this series at http://www.springer.com/series/16385

Yaobing Wang

Space Robotics

北京理工大学出版社
BEIJING INSTITUTE OF TECHNOLOGY PRESS

Springer

Yaobing Wang
Beijing Institute of Spacecraft
System Engineering
CAST
Beijing, China

ISSN 2730-6410 ISSN 2730-6429 (electronic)
Space Science and Technologies
ISBN 978-981-15-4901-4 ISBN 978-981-15-4902-1 (eBook)
https://doi.org/10.1007/978-981-15-4902-1

Jointly published with Beijing Institute of Technology Press
The print edition is not for sale in China (Mainland). Customers from China (Mainland) please order the print book from: Beijing Institute of Technology Press.

This Springer imprint is published by the registered company Springer Nature Singapore Pte Ltd.
The registered company address is: 152 Beach Road, #21-01/04 Gateway East, Singapore 189721, Singapore

Series Editor's Preface

China's space technology and science research have earned a place in the world, but have not been compiled into a series of systematic publications yet. In 2018, the series Space Science and Technology edited mainly by me and co-authored by the leading figures in China's space industry was published in China, when China Academy of Space Technology was celebrating the 50th anniversary of its founding. This collection contains 23 volumes in Chinese, only 10 of which have been selected, recreated, and translated into English. In addition, each English volume has been recreated at the suggestion of Springer, by deleting the contents similar to Springer's existing publications and adding the contents that are internationally advanced and even leading, and bear both Chinese characteristics and worldwide universality. This series fully reflects the knowledge and engineering experience recently accumulated by Chinese scientists and engineers in space technology and science research.

As the Editor-in-Chief of this series, I always insist that this collection must be of high quality, either in Chinese version or English version. First, the contents of this series must be condensed and sublimated based on the combination of theory and practice, so as to provide both theoretical value and engineering guidance. Second, the relationships between past knowledge and state of the art and between other people's work and our own new findings should be properly balanced in the book contents to ensure the knowledge systematicness and continuity and to highlight new achievements and insights. Each volume intends to introduce the readers to something new. Third, the English version should be customized for international exposure and play a solid supporting role for China to contribute to the world's space field.

This collection consists of 10 volumes, including *Spacecraft Thermal Control Technologies*, *Spacecraft Power System Technologies*, *Spacecraft Electromagnetic Compatibility Technologies*, *Technologies for Spacecraft Antennas Engineering Design*, *Satellite Navigation Systems and Technologies*, *Satellite Remote Sensing Technologies*, *Spacecraft Autonomous Navigation Technologies Based on Multi-source Information Fusion*, *Technologies for Deep Space Exploration*, *Space Robotics*, and *Manned Spacecraft Technologies*.

Space robot is a basic tool for human beings to explore and utilize the space. While summarizing many years' research achievements related to robot theory and technology, this volume, namely *Space Robotics*, refines and integrates the practical engineering experience and product development rules for space robots into its main contents. This volume has the following features:

1. Starting from the particularity of space environment and application, the book discusses the theory and method of space robot design. The purpose is to provide the basic concepts and theories related to space robots, and introduce the basic methods and steps of engineering implementation of space robots.
2. Based on the engineering practice of several space robot projects, the book emphasizes the combination of theory and application. In addition to the design theory and method of those space robots suitable for engineering applications, it also presents various engineering examples of several space robots.

The publication of this series adds a new member to the international family of space technology and science publications, and intends to play an important role in promoting academic exchanges and space business cooperation. It provides comprehensive, authentic, and rich information for international space scientists and engineers, enterprises, and institutions as well as government sectors to have a deeper understanding of China's space industry. Of course, I believe that this series will also be of great reference value to researchers, engineers, graduate students, and university students in the related fields.

Peijian Ye
Academician, Chinese Academy of Sciences
Beijing, China

Preface

This book was written by the frontline experts and engineers who are engaged in the research of space robotics. It is oriented to the scientific and technological personnel occupied in the relevant work or the senior undergraduates and post-graduates in some colleges and universities. As a summary and extraction of several space robot projects, this book has a high engineering reference value.

This book presents the basic concepts and theories of space robots, as well as the basic implementation methods and steps of space robot projects. To especially highlight the combination of theory and application, this book uses much ink in elaborating on three space-robot design examples oriented to the engineering applications. In order to fully and accurately understand the contents of this book, the readers had better have basic knowledge of calculus, matrices and vectors, mechanics, and control.

The book is divided into four parts in 16 chapters. The first part mainly addresses the basic robot theories, including the kinematics, dynamics, and motion planning and force control of space robots, especially the dynamics and control of floating-based space robots. The second part mainly introduces the engineering design method of space robots, focusing on how to carry out product design and verification while giving full consideration to the special working environment (vacuum, high and low temperatures, irradiation, etc.) and special design constraints (launch environment, verification difficulty, etc.) for space robots. The third part mainly provides several engineering design examples of space robots and briefly describes the process from task requirement formulation to product planning. At the end of this book, the existing space robot products around the world are summarized, and future technological development is prospected.

The completion of this book benefits from the joint efforts of the experts, scholars, and technicians devoted to the research of space robots, who are working for the Beijing Institute of Spacecraft System Engineering. Chapters 1, 15, and 16 were completed by Wang Yaobing, Wang Youyu, Zhang Dawei, and Li Daming; Chap. 2 by Pan Dong and Li Jianfei; Chap. 3 by Zhang Xiaodong, Pan Dong, and Liu Xin; Chap. 4 by Liu Xin and Zhang Xiaodong; Chap. 5 by Li Jianfei; Chap. 6 by Wang Youyu, Hou Liang, and Wu Shuang; Chap. 7 by Wang Kang, Yang Xu,

Lin Yuncheng, Zhang Wenming, and Li Delun; Chap. 8 by Liu Xin, Zhang Xiaodong, and Hou Liang; Chap. 9 by Chen Lei; Chap. 10 by Wu Zhihong; Chap. 11 by Zhang Dawei; Chap. 12 by Wang Youyu, Hu Chengwei, and Tang Zixin; Chap. 13 by Pan Dong, Yuan Baofeng, and Liu Yafang; and Chap. 14 by Ma Ruqi, Jiang Shuiqing, and Du Xiaodong. All the chapters were collected and organized by Wang Yaobing and Zhang Dawei. The researchers Hu Chengwei and Gao Sheng provided detailed and in-depth technical guidance in the compilation of this book. The Academician Yang Mengfei, together with the researchers Chen Liemin, Gao Yaonan, Zhang Bainan, Wang Dayi, and Li Zhi, reviewed the book and put forward specific suggestions on revision. As the Editor-in-Chief of the whole series, the Academician Ye Peijian planned and arranged the direction and contents of this book. Here, we would like to express our heartfelt thanks to the above experts for their guidance and help!

The compilation of this book also received great support from the leaders of the China Academy of Space Technology and the Beijing Institute of Spacecraft System Engineering. We acknowledge their help on this book!

Beijing, China
December 2019

Yaobing Wang

Contents

About the Author

Yaobing Wang, Ph.D., is research professor of Beijing Institute of Spacecraft System Engineering, and Director of Beijing Key Laboratory of Intelligent Space Robotic Systems Technology and Applications. His research interests focus on space robotics and spacecraft structure and mechanisms.

Chapter 1
Introduction

1.1 Definition, Characteristics, and Classification of Space Robots

1.1.1 Definition of Space Robot

Nowadays, the term "Space Robotics" is very familiar to those working in the field of robotics [1], but there is still no universally accepted definition of a "space robot". Lin Yiming et al. defined space robots in their work "The Current Status and Analysis of Space Robots" as "a class of special robots that perform tasks in space such as the support for the construction and operation of space stations, the satellite assembly and service, and the planetary surface exploration and testing [2]".

In the Chinese Institute of Electronics (CIE) standard "General Technical Requirements for Space Robotics" (T/CIE045-2017), a space robot is defined as "the robot that is applied to the space beyond the Earth's atmosphere (including the space inside and outside spacecraft, and celestial bodies)". By combining the above definitions, we think that the main difference between space robots and other robots stems from the particularity of their operating environment. Therefore, in this book, the space robot will be defined as a kind of special robot used in space beyond the Earth's atmosphere.

1.1.2 Characteristics of Space Robots

Space robots are special robots used in space. Due to the particularity of their application environment, space robots have the following characteristics compared to terrestrial robots:

(1) Special working environment. Space robots work in extraterrestrial space. For the space robots usually working on orbit, such conditions as ultra-vacuum, high

Y. Wang, *Space Robotics*, Space Science and Technologies,
https://doi.org/10.1007/978-981-15-4902-1_1

and low temperatures, strong radiation, microgravity, and complex illumination need to be considered. For the space robots that perform planetary exploration tasks, special topography, temperature, special atmosphere, gravel, dust, gravity, and other factors should be considered additionally. In addition, all static force, vibration, noise, shock, and other loads within the mission profile also need to be considered for space robots.

(2) Multi-facet design constraints. In addition to meeting the requirements of ground launch and space working environment, space robots should be designed under the consideration of the resource constraints on configuration, mass, power consumption, installation space, communication, and field of view, as well as the constraints on related interfaces, functions, and performance.
(3) High-reliability requirements. Space robots need to work in space for a long time, basically without any maintenance during work. This requires them to have high reliability under the onboard limited resources.
(4) Diversified tasks. Space robots usually undertake a variety of tasks, such as the identification, measurement, capture, handling, installation, assembly, and replacement of space targets, and sample collection and processing. The manipulated objects include flight products, cabins, modules, equipment, instruments, special tools, and space debris. Therefore, in the design of space robots, it is necessary to take into account the special requirements of different missions and different objects.
(5) Complex system composition. A space robot is a complex space system involving multiple disciplines, such as materials, mechanics, machinery, electronics, thermal control, optics, and control theory. In terms of composition, a space robot includes not only a mechanical system consisting of multiple joints and end effectors, but also a perceptual system consisting of cameras and sensors, a control system consisting of an arm controller and several joint controllers, and a human–robot interaction system composed of an instruction generating module and a telemetry feedback module.
(6) Difficult ground verification. A space robot is a kind of special robot designed for the space environment. It is difficult to carry out physical test verification of all operating conditions of space robots directly under ground gravity conditions. And the coupling effect of vacuum, microgravity or low gravity, high and low temperatures, and other environmental conditions during on-orbit operation is difficult to simulate on the ground, which brings great problems to the comprehensiveness and sufficiency of space-robotics ground verification.

Based on the above characteristics of space robots, special attention should be paid to the following problems their design and verification:

(1) Multidisciplinary design and optimization of space robot systems;
(2) Design of space environment adaptability, including the design and selection of raw materials and components, the design of anti-cold welding and lubrication of moving parts, the thermal matching design of parts, and the design of the adaptability of parts and systems to force, heat, magnetism, and irradiation;
(3) Reliability and safety design for long-term services in harsh environments;

(4) Space robot teleoperation under large delay conditions and its autonomous planning and control algorithm design;
(5) Modification and verification of simulation models;
(6) Adequacy and coverage of ground test verification.

1.1.3 Classification of Space Robots

According to different principles, space robots can be classified in many ways. According to the characteristics of missions and environmental factors, space robots can be divided into on-orbit operation robots and planetary surface exploration robots. The former includes those robots performing on-orbit monitoring, assembly, maintenance, and upgrade tasks for the spacecraft, and assisting astronauts in completing on-orbit operations. The latter includes those robots performing planetary surface exploration, sample collection and scientific experiment tasks, and assisting astronauts in completing planetary exploration missions. All space robots that have been verified by flight can be included in these two types. With reference to this classification, this book divides space robots into on-orbit operation robots and planetary exploration robots.

1.1.3.1 On-Orbit Operation Robots

The on-orbit operation robots mainly refer to the space robots that perform various operation tasks in the microgravity orbit environment. They include, but are not limited to, free-flying robots, intra- and extra-vehicular operating robots in manned space station/space laboratory, and operating robots in unmanned space service stations. These robots are used to provide a variety of on-orbit operation services, such as target capture, target transfer and release, on-orbit assembly, on-orbit services (fuel refueling, module replacement, etc.), on-orbit manufacturing, and assisting astronauts in extra-vehicular activities (EVAs). The characteristics of their working environment include vacuum, strong radiation, microgravity, and alternating temperature.

1.1.3.2 Planetary Exploration Robots

Planetary exploration robots mainly refer to the space robots that perform tasks on the moon, planets, and asteroids and other celestial bodies. They include, but are not limited to, unmanned/manned exploration robots, planetary surface construction robots. These robots usually have a wheeled or legged mobile system, and are typically equipped with manipulators. The tasks they perform mainly involve moving and operating, including planetary surface roaming, exploration in extreme areas, sample collection, scientific experiments, and base construction on planet surfaces, as well as joint exploration with astronauts. The characteristics of their working environment mainly include vacuum or special atmosphere, strong radiation, sand dust, special geological conditions, gravity, and special temperature environment.

1.2 Basic Composition and Main Research Areas of Space Robots

1.2.1 Basic Composition of Space Robots

Considering the complete loop of space robot mission, the ground control system is usually included in the space robot system. In this case, space robot systems can be divided into the space portion and the ground portion in general. The space portion refers to the part that performs tasks on orbit or on the celestial bodies. It is the main part of the space robot, which covers the main functional units of the space robot (such as perceptual unit, control unit, and execution unit) and the space teleoperation unit (such as the intra-vehicular teleoperation system). The ground portion refers to the functional units deployed on the ground to perform the space-oriented tasks such as operational command input, feedback information reception, and operational status monitoring. It is also called the ground teleoperation system of space robots.

According to its function, a space robot system can be divided into different units. The typical division is as follows:

(1) Mechanical system: It is a system that supports a space robot to perform a specified action, consisting of the mechanical components designed in specific forms according to the requirements of the mission. It usually includes hold-down and release mechanisms, joints that provide motion function for the robot, and end effectors that perform specific operations, as well as the wheels and their steering and drive units that support the movement of the space robot.
(2) Perceptual system: It is a system that supports a space robot to sense the working environment, the operational object, and its own state. It is composed of various types of sensors, typically including imaging devices that acquire visual information, force/tactile sensors that acquire force/tactile information, and information processing units.
(3) Control system: It is a system that supports the space robot to complete analysis, decision-making, planning, and motion control. It is usually composed of controllers (consisting of a processing chip and peripheral circuits), processing modules (exchanger), etc.
(4) Power system: It is a system that supports the space robot to obtain external power sources and perform power distribution according to the power requirements of each component. It usually includes power supply modules, power distribution modules, and wire harnesses.
(5) Thermal control system: It is a system that supports the space robot to maintain the temperature of each component and device within the allowable temperature

range. It typically includes a passive thermal control subsystem composed of multilayer insulation and thermal control coatings and an active thermal control subsystem consisting of temperature sensors, heaters, etc.

(6) Human–robot interaction system: It is a system that supports operators to perform various mode controls on space robots. It usually includes a human–robot interaction interface device and a space robot state feedback display module. The commonly used human–robot interaction mode is a teleoperation mode. The human–robot interaction system in this mode is usually called a teleoperation system, which typically includes a feedback module, a calculation module, an input module, and a simulation module.

(7) Communication system: It is a system that supports space robots to interact with ground teleoperation systems or spacecraft systems and other space robots. It includes the communication links required for information interaction between space and ground.

In the mission profile of a space robot, if there is a period of time when the space robot operates as an independent spacecraft, for example, as a free-flying robot or a planetary surface rover to perform independent missions, the space robot generally has all of the above 7 systems. If the space robot only performs space tasks as a payload of the spacecraft, then in actual product design, it is not necessarily designed with all the systems described above. For example, a space robot system assigned to a space station usually communicates with other spacecraft by means of a communication system of the space station, so it does not have its own communication system. The space robot system with high autonomy can independently perform space operation tasks without using a teleoperation system, so it doesn't need a teleoperation system. In addition, for space robot systems that may perform relatively simple tasks or functions, some systems can be merged. For example, the thermal control system is incorporated into the mechanical system, and the power system into the control system.

This book highlights the main systems of space robots, i.e., mechanical system, perceptual system, control system, and human–robot interaction system (mainly used for teleoperation).

1.2.2 Main Research Areas of Space Robotics [3]

The main research areas of space robotics are related to the composition and application of space robots. The theories, methods, and engineering implementation technologies involved are all those that the technicians should learn and study, specifically including

(1) Space robot kinematics and dynamics;
(2) Space robot task planning and motion planning, including task decomposition, path planning, and trajectory planning;
(3) Space robot control, including robot motion control and force control;

(4) Space robot system design;
(5) Space robot key subsystems design, such as the design of mechanical, control, perceptual, and interaction subsystems;
(6) Space robot software;
(7) Space environment adaptability of space robots;
(8) Simulation and test verification of space robots;
(9) Human–robot collaboration and multi-robot collaboration;
(10) Space robot autonomy and intelligence.

References

1. B. Siciliano, O. Khatib, *Handbook of Robotics* (Springer, New York, 2007)
2. Y. Lin, D. Li, Y. Wang et al., Current status and analysis of space robot. Spacecraft Eng. **24**(5), 1–7 (2015)
3. Y. Xiong, *Robotics* (Mechanical Industry Press, Beijing, 1993)

Chapter 2
Kinematics and Dynamics of Space Robots

Kinematics and dynamics are the basis of analyzing the characteristics and control of space robots. In this chapter, the kinematics equations of a typical space robot system are given, including a description of topological relationships and the kinematics modeling process, which can be used for robot dynamics analysis and path planning. Based on analytical mechanics and vector mechanics, the dynamic models for rigid space robots and flexible space robots are established in this chapter, which can be used for robot dynamics analysis and control algorithm design.

2.1 Topology Mathematical Description

A space robot is a typical multi-body system with a certain topology formed by multiple links connected by hinges. In order to model and analyze the multi-body system, first, the system structure should be simplified, abstracted, and then described in a mathematical language such that the connection relationship of the multi-body system can be accurately and uniquely represented. Figure 2.1 shows a typical multi-body system topology.

According to the graph theory, the topological relationship of the multi-body system is mathematically described by introducing the path vector B and the correlation matrix S. Taking the n-degree-of-freedom (DOF) multi-body system composed of $n + 1$ bodies (where body 0 is the base) as an example, the path vector B is used to find the body in the system internally associated with the body $i\,(i = 1, \ldots, n)$, and the element B_i is the labeled number of the body internally associated with the body i. The correlation matrix S is used to find the body externally associated with the body i. The row number and column number of the element S_{ij} correspond to the labels of the body and the joint, respectively. The values of the element $S_{ij}\,(i, j = 1, \ldots, n)$ are defined as

Y. Wang, *Space Robotics*, Space Science and Technologies,
https://doi.org/10.1007/978-981-15-4902-1_2

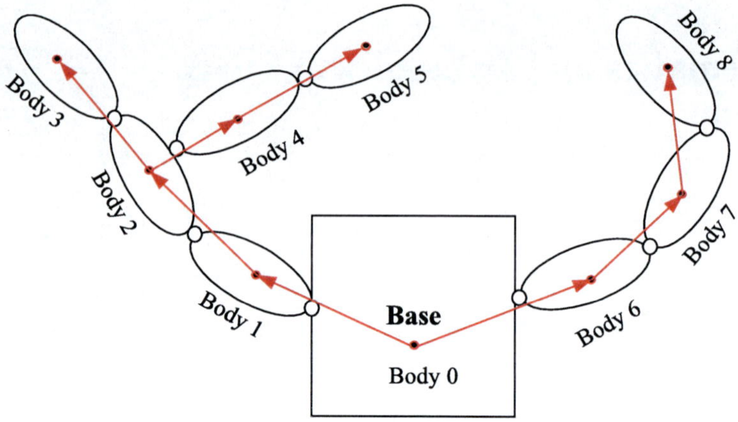

Fig. 2.1 Multi-body system topology

$$S_{ij} = \begin{cases} 1 & (i = B_j) \\ -1 & (i = j) \\ 0 & (i \neq B_j, j) \end{cases}$$

To check whether the body i is connected to the base, it can be described by the vector S_{0j} ($j = 1, \ldots, n$), and defined as

$$S_{0j} = \begin{cases} 1 & (0 = B_j) \\ 0 & (0 \neq B_j) \end{cases}$$

To determine whether the body i is the end, it can be described by the vector S_{ej} ($j = 1, \ldots, n$), and defined as

$$S_{ej} = \begin{cases} 1 & \text{(body } j \text{ is the end body)} \\ 0 & \text{(body } j \text{ is not the end body)} \end{cases}$$

Take the system in Fig. 2.1 as an example. The topology can be described as

$$B = [0, 1, 2, 2, 4, 0, 6, 7],\ S_0 = [1, 0, 0, 0, 0, 1, 0, 0],\ S_e = [0, 0, 1, 0, 1, 0, 0, 1]$$

$$S = \begin{bmatrix} -1 & 1 & 0 & 0 & 0 & 0 & 0 & 0 \\ 0 & -1 & 1 & 1 & 0 & 0 & 0 & 0 \\ 0 & 0 & -1 & 0 & 0 & 0 & 0 & 0 \\ 0 & 0 & 0 & -1 & 1 & 0 & 0 & 0 \\ 0 & 0 & 0 & 0 & -1 & 0 & 0 & 0 \\ 0 & 0 & 0 & 0 & 0 & -1 & 1 & 0 \\ 0 & 0 & 0 & 0 & 0 & 0 & -1 & 1 \\ 0 & 0 & 0 & 0 & 0 & 0 & 0 & -1 \end{bmatrix}.$$

2.2 Definition of Coordinate and Coordinate Transformation

The reference coordinate is the basis for position/orientation measurement and motion control of space robots. The coordinates commonly used in a space robot system include inertial coordinate, base coordinate, joint coordinate, and end coordinate. In this section, the coordinate of the space robot system is defined, and the transformation relationships between the coordinate systems are represented by the direction cosine matrix, thus laying a foundation for subsequent kinematics and dynamics calculations.

Figure 2.2 shows a chain-type space robot system consisting of a base body and n links, which are connected by rotating joints, and an end effector is connected to the

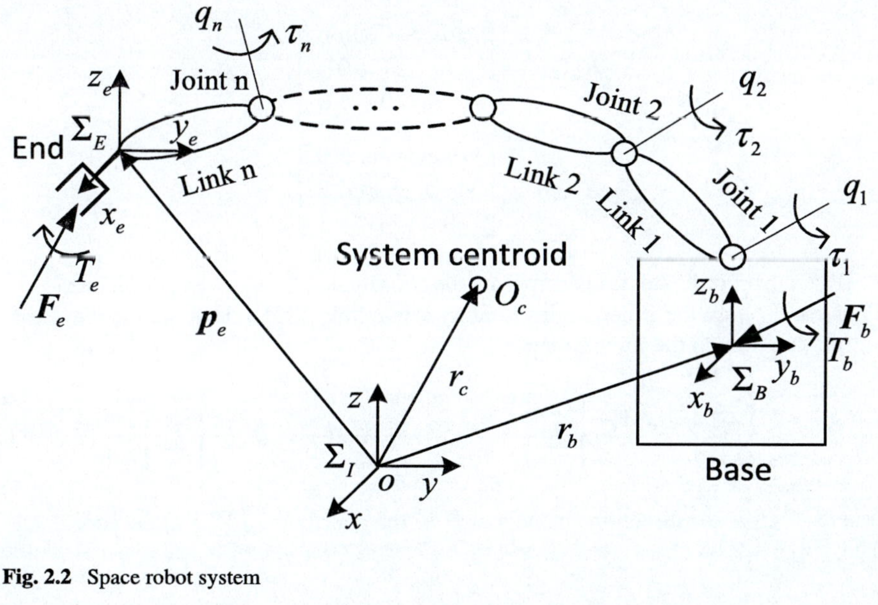

Fig. 2.2 Space robot system

link n. The coordinates of the space robot system is defined as follows: for the inertia coordinate $\{\Sigma_I\}$, the J2000 coordinate is generally selected as the inertial coordinate $\{\Sigma_I\}$; for the base coordinate system $\{\Sigma_B\}$, the origin is located on the base centroid, and the three coordinate axes are the three main inertia axes of the spacecraft; for the end coordinate system $\{\Sigma_E\}$, the origin is located on the end effector operation center, and the three-axis directions are determined according to the operational tasks and the direction of the target coordinate. The coordinates of the two joint coordinate pairs are defined as follows: for the outboard frame $\left\{\Sigma_{\hat{i}}\right\}$ connecting with link $i-1$, the origin is located at the intersection of the joint i rotating axis and the link axis, and the z-axis is the direction of the joint axis, and the x-axis points to the joint $i+1$; for the inboard frame $\{\Sigma_i\}$ $(i = 1, \ldots, n)$ connecting with link i, when the joint coordinate is zero, it coincides with $\left\{\Sigma_{\hat{i}}\right\}$; when in motion, it translates along the z-axis relative to $\left\{\Sigma_{\hat{i}}\right\}$ or rotates. In the process of deriving the dynamic equation, for convenience of description, a centroid body coordinate $\{\Sigma_{cmi}\}$ and a centroid inertial coordinate $\{\Sigma_{Oi}\}$ are defined as auxiliaries. The origin of the centroid body coordinate $\{\Sigma_{cmi}\}$ is at the centroid of the link i, and the direction is consistent with $\{\Sigma_i\}$; the origin of the centroid inertial coordinate $\{\Sigma_{Oi}\}$ is at the centroid of the link i, and its direction is the same as the inertial coordinate.

The relationship between the coordinates is determined by the direction cosine matrix, and the coordinate transformation can be completed by three rotations around the x, y, and z axes between any two coordinates. The angle rotated by three rotations is called the Euler angle (θ, φ, ϕ), and the basic rotation matrix is

$$\boldsymbol{R}_x(\theta) = \begin{bmatrix} 1 & 0 & 0 \\ 0 & \cos\theta & -\sin\theta \\ 0 & \sin\theta & \cos\theta \end{bmatrix},$$

$$\boldsymbol{R}_y(\varphi) = \begin{bmatrix} \cos\varphi & 0 & \sin\varphi \\ 0 & 1 & 0 \\ -\sin\varphi & 0 & \cos\varphi \end{bmatrix},$$

$$\boldsymbol{R}_z(\phi) = \begin{bmatrix} \cos\phi & -\sin\phi & 0 \\ \sin\phi & \cos\phi & 0 \\ 0 & 0 & 1 \end{bmatrix}.$$

The joint coordinates transform from the coordinate $\{\Sigma_{i-1}\}$ to the coordinate $\{\Sigma_i\}$; θ_i, φ_i, and ϕ_i are the Euler angles rotating around the XYZ axis, and q_i is the angle of rotation around the joint i, then

$$\left\{\sum_{i-1}\right\} = {}^{i-1}\boldsymbol{C}_i\left\{\sum_{i}\right\} = \boldsymbol{R}_x(\theta_i)\boldsymbol{R}_y(\varphi_i)\boldsymbol{R}_z(\phi_i + q_i)\left\{\sum_{i}\right\}, \tag{2.1}$$

where ${}^{i-1}\boldsymbol{C}_i$ is the direction cosine matrix of the coordinate $\left\{\sum_i\right\}$ to the coordinate $\left\{\sum_{i-1}\right\}$.

The base of the space robot system is in a free state; its orientation in the inertial coordinate can be expressed as

$$\left\{\sum_{I}\right\} = {}^{I}\boldsymbol{C}_0\left\{\sum_{0}\right\} = [\boldsymbol{R}_x(\theta_0)\boldsymbol{R}_y(\varphi_0)\boldsymbol{R}_z(\phi_0)]\left\{\sum_{0}\right\}, \tag{2.2}$$

where ${}^{I}\boldsymbol{C}_0$ is the direction cosine matrix of the robot base coordinate to the inertial coordinate, θ_0, φ_0, and ϕ_0 are the roll, pitch, and yaw angles of the base, respectively.

2.3 Space Robot Kinematics [1]

The forward kinematics problem of space robot is to determine the end position and orientation of the robot when the configuration parameters and joint angle of the robot are known. In other words, if the joint angle at each moment is known, the end position and orientation at any time can be obtained through the positive kinematics equation.

However, if the end of the robot is to be placed in a desired position and orientation, the angles of all joints must be known. This is called the inverse kinematics analysis of the robot. The essence of the inverse kinematics analysis is to find the inverse solution of the positive equation of motion, that is, the joint angles can be obtained with the known end position and orientation. Inverse kinematics is relatively important for robots and is the basis for robot motion planning and control.

2.3.1 Position and Speed of Each Body

The parameters of the robot links are shown in Fig. 2.3. The parameters are defined as follows: $oxyz$ is the inertial coordinate; o_i, $\hat{o}_i$ are the joint points of the joint i on link $i-1$ and link i, respectively. When it is a rotating joint, the two coincide; $\boldsymbol{l}_i$ is the vector of the joint i pointing to the joint $i+1$, o_{cm}^{i} is the centroid of the link i, c_i is the vector of the joint i to the centroid of the link i, $\boldsymbol{b}_i$ is the vector of the centroid of link i to the joint $\boldsymbol{i}+1$, $\boldsymbol{R}_i$ is the position vector of the joint i in the inertial coordinate, and $\boldsymbol{r}_i$ is the position vector of the centroid of the link i in the inertial coordinate.

According to the relationships in the figure, the recursive relationship of the link centroid position is:

$$\boldsymbol{r}_i = \boldsymbol{r}_{i-1} + \boldsymbol{b}_{i-1} + \boldsymbol{c}_i, \tag{2.3}$$

where $\boldsymbol{b}_{i-1}$ and $\boldsymbol{c}_i$ can be represented by the geometry of the link $\boldsymbol{i}-1$ and the link i and the orientation transformation matrix of the joint coordinate to the inertial

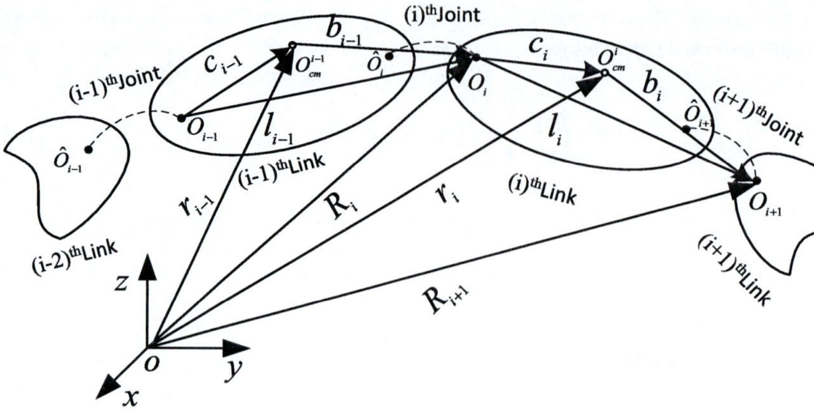

Fig. 2.3 Relationships of robot links

coordinate, $\boldsymbol{b}_{i-1} = {}^I\boldsymbol{C}_{i-1}{}^{i-1}\boldsymbol{b}_{i-1}$, $\boldsymbol{c}_i = {}^I\boldsymbol{C}_i{}^i\boldsymbol{c}_i$. ${}^{i-1}\boldsymbol{b}_{i-1}$, is the vector representation of the distances between the centroids of the link $i-1$ to the joint i in the coordinates of joint $i-1$; and ${}^i\boldsymbol{c}_i$ is the vector representation of the distances between the joint i to the centroids of the link i in the coordinates of the joint i. Then Eq. (2.3) changes to:

$$\boldsymbol{r}_i = \boldsymbol{r}_{i-1} + {}^I\boldsymbol{C}_{i-1}{}^{i-1}\boldsymbol{b}_{i-1} + {}^I\boldsymbol{C}_i{}^i\boldsymbol{c}_i. \tag{2.4}$$

It can be seen with Figs. 2.2 and 2.3 that when r_b, q_i, ${}^{i-1}\boldsymbol{b}_{i-1}$, and ${}^i\boldsymbol{c}_i$ are known, the position of any link under the inertial coordinate can be recursed by Eq. (2.4).

Differentiating Eq. (2.3) with respect to time yields:

$$\dot{\boldsymbol{r}}_i = \dot{\boldsymbol{r}}_{i-1} + \dot{b}_{i-1} + \dot{c}_i. \tag{2.5}$$

From the absolute derivative and relative derivative theorems of the variable vector, we have

$$\dot{\boldsymbol{b}}_{i-1} = \frac{\tilde{\mathrm{d}}\boldsymbol{b}_{i-1}}{\mathrm{dt}} + \omega_{i-1} \times \boldsymbol{b}_{i-1} \tag{2.6}$$

$$\dot{\boldsymbol{c}}_i = \frac{\tilde{\mathrm{d}}\boldsymbol{c}_i}{\mathrm{dt}} + \omega_i \times \boldsymbol{c}_i, \tag{2.7}$$

where $\frac{\tilde{\mathrm{d}}\boldsymbol{b}_{i-1}}{\mathrm{dt}}$ is the relative derivative of the vector b_{i-1}, b_{i-1} and c_i are constant values in the rigid-coupling coordinate, and the relative derivative is 0. ω_{i-1} and ω_i are the angular velocities of the link $\boldsymbol{i}-1$ and the link i, respectively, and $\boldsymbol{v}_i = \dot{\boldsymbol{r}}_i$ is the centroid velocity of the link i.

Then the recursive relationship of the centroid of the link is

$$\boldsymbol{v}_i = \boldsymbol{v}_{i-1} + \boldsymbol{\omega}_{i-1} \times \boldsymbol{b}_{i-1} + \boldsymbol{\omega}_i \times \boldsymbol{c}_i. \tag{2.8}$$

The rotation matrix ${}^{I}\boldsymbol{C}_m$ between any coordinate $\{\sum_m\}$ and the inertial coordinate is a function of time t. At each instant t, there is a rotation axis (represented by its unit vector k) and an angle q, such that ${}^{I}\boldsymbol{C}_m = \boldsymbol{C}_k(q)$. The change rate of ${}^{I}\dot{\boldsymbol{C}}_m$ caused by the change rate of the angle q is

$${}^{I}\dot{\boldsymbol{C}}_m = \frac{\mathrm{d}}{\mathrm{d}t}\boldsymbol{C}_k(q) = \left[\frac{\partial}{\partial q}\boldsymbol{C}_k(q)\right]\frac{\mathrm{d}q}{\mathrm{d}t} = \dot{q}\frac{\partial}{\partial q}\boldsymbol{C}_k(q). \tag{2.9}$$

From Euler equation, we get

$$\frac{\partial}{\partial q}\boldsymbol{C}_k(q) = \boldsymbol{k} \times \boldsymbol{C}_k(q) = \boldsymbol{k} \times {}^{I}\boldsymbol{C}_m. \tag{2.10}$$

Substituting Eq. (2.10) into Eq. (2.9), we have

$${}^{I}\dot{\boldsymbol{C}}_m = \dot{q}\boldsymbol{k} \times {}^{I}\boldsymbol{C}_m = \boldsymbol{\omega}_m \times {}^{I}\boldsymbol{C}_m, \tag{2.11}$$

where $\boldsymbol{\omega}_m = \dot{q}\boldsymbol{k}$ is the expression of the angular velocity of the coordinate $\{\Sigma_m\}$ relative to the inertial coordinate in the inertial system.

According to the properties of Eqs. (2.9), (2.10), and (2.11), the recursive relationship of the centroid angular velocity of the serial robot link connected by the revolute joint in Fig. 2.3 is obtained. ${}^{I}\boldsymbol{C}_i$ is the rotation matrix of the link $\boldsymbol{i}$ relative to the inertial system, then:

$${}^{I}\dot{\boldsymbol{C}}_i = \sum_{j=0}^{i}\left[\dot{q}_j\frac{\partial}{\partial q_j}{}^{I}\boldsymbol{C}_i\right] \tag{2.12}$$

$$\frac{\partial}{\partial q_j}{}^{I}\boldsymbol{C}_i = {}^{I}\boldsymbol{C}_{j-1}\left[\frac{\mathrm{d}}{\mathrm{d}q_j}{}^{j-1}\boldsymbol{C}_j(q_j)\right]{}^{j}\boldsymbol{C}_i. \tag{2.13}$$

From Eq. (2.10), Eq. (2.13) can be transformed into

$$\frac{\partial}{\partial q_j}{}^{I}\boldsymbol{C}_i = \boldsymbol{k}_j \times {}^{I}\boldsymbol{C}_i. \tag{2.14}$$

Substituting Eq. (2.14) into Eq. (2.12) gives

$${}^{I}\dot{\boldsymbol{C}}_i = \sum_{j=0}^{i}\dot{q}_j\boldsymbol{k}_j \times {}^{I}\boldsymbol{C}_i = \omega_i \times {}^{I}\boldsymbol{C}_i. \tag{2.15}$$

Therefore, the recursive relationship of the centroid velocity of the link i is obtained as

$$\boldsymbol{\omega}_i = \sum_{j=0}^{i} \dot{q}_j \boldsymbol{k}_j = \boldsymbol{\omega}_{i-1} + \dot{q}_i \boldsymbol{k}_i. \tag{2.16}$$

Simultaneously, from Eqs. (2.8), (2.16), when the base centroid velocity $\boldsymbol{v}_b$ and the angular velocity $\boldsymbol{\omega}_b$ are known, the centroid velocity and angular velocity of each link of the free-floating space robot in the inertial system can be recursively calculated.

$${}^I\boldsymbol{v}_i = {}^I\boldsymbol{v}_{i-1} + {}^I\boldsymbol{\omega}_{i-1} \times {}^I\boldsymbol{b}_{i-1} + {}^I\boldsymbol{\omega}_i \times {}^I\boldsymbol{c}_i \tag{2.17}$$

$${}^I\boldsymbol{\omega}_i = {}^I\boldsymbol{\omega}_{i-1} + {}^I\boldsymbol{C}_i {}^i\boldsymbol{k}_i \dot{q}_i. \tag{2.18}$$

By deriving Eqs. (2.17) and (2.18), the recursive relationship of the link acceleration is obtained as

$$\begin{aligned} {}^I\dot{\boldsymbol{v}}_i = {}^I\dot{\boldsymbol{v}}_{i-1} + {}^I\dot{\boldsymbol{\omega}}_{i-1} \times {}^I\boldsymbol{b}_{i-1} + {}^I\boldsymbol{\omega}_{i-1} \times ({}^I\boldsymbol{\omega}_{i-1} \times {}^I\boldsymbol{b}_{i-1}) \\ + {}^I\dot{\boldsymbol{\omega}}_i \times {}^I\boldsymbol{c}_i + {}^I\boldsymbol{\omega}_i \times ({}^I\boldsymbol{\omega}_i \times {}^I\boldsymbol{c}_i) \end{aligned} \tag{2.19}$$

$${}^I\dot{\boldsymbol{\omega}}_i = {}^I\dot{\boldsymbol{\omega}}_{i-1} + {}^I\boldsymbol{\omega}_i \times ({}^I\boldsymbol{C}_i {}^i\boldsymbol{k}_i \dot{q}_i) + {}^I\boldsymbol{C}_i {}^i\boldsymbol{k}_i \ddot{q}_i. \tag{2.20}$$

2.3.2 *End Position and Speed*

When the robot is performing the operation tasks, the condition of the end in the operation space should be acquired. Since it is difficult to directly measure the position and speed of the robot end, they are generally derived from the joint angles and angular velocities of the robot.

Figure 2.4 shows an n-DOF floating-based robot system, where $\boldsymbol{r}_b$ is the position vector of the base centroid in the inertial space, $\boldsymbol{l}_0$ is the vector of the base centroid pointing to the robot mounting point, $\boldsymbol{l}_i (i = 1, \ldots, n)$ is the vector between two adjacent joints, the low joint directing to the high joint, $\boldsymbol{R}_i (i = 1, \ldots, n)$ is the position vector of the joints of the robot in the inertial system, and $\boldsymbol{x}_e$ is the robot end position vector.

According to the connection relationship in Fig. 2.4, the end position can be expressed as

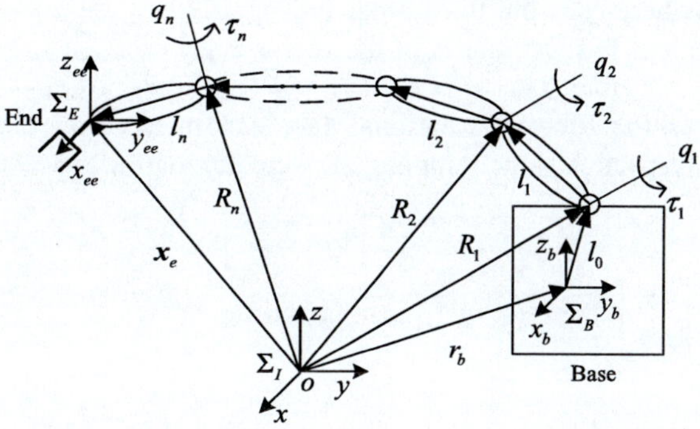

Fig. 2.4 End kinematics of the robot

$$\boldsymbol{x}_e = \boldsymbol{r}_b + \boldsymbol{l}_0 + \sum_{i=1}^{n} \boldsymbol{l}_i. \tag{2.21}$$

By deriving Eq. (2.21), we have

$$\dot{\boldsymbol{x}}_e = \boldsymbol{v}_b + \boldsymbol{\omega}_b \times \boldsymbol{l}_0 + \sum_{i=1}^{n} \boldsymbol{\omega}_i \times \boldsymbol{l}_i, \tag{2.22}$$

where $\boldsymbol{v_b}$ and $\boldsymbol{\omega_b}$ are the centroid velocity and angular velocity of the base, respectively, $\boldsymbol{\omega_i}$ is the angular velocity of the link $\boldsymbol{i}$.

According to Eq. (2.16):

$$\boldsymbol{\omega}_i = \sum_{j=0}^{i} \dot{q}_j \boldsymbol{k}_j = \boldsymbol{\omega}_b + \sum_{j=1}^{i} \dot{\boldsymbol{q}}_j \boldsymbol{k}_j. \tag{2.23}$$

Substituting Eq. (2.23) into Eq. (2.22), the recursive relationship of the robot end velocity is given as

$$\boldsymbol{v}_e = \boldsymbol{v}_b + \boldsymbol{\omega}_b \times (\boldsymbol{x}_e - \boldsymbol{r}_b) + \sum_{j=1}^{n} \dot{\boldsymbol{q}}_j \boldsymbol{k}_j \times (\boldsymbol{x}_e - \boldsymbol{R}_j) \tag{2.24}$$

$$\boldsymbol{\omega}_e = \boldsymbol{\omega}_b + \sum_{j=1}^{n} \dot{\boldsymbol{q}}_j \boldsymbol{k}_j. \tag{2.25}$$

Equations (2.24) and (2.25) are expressed in matrix form:

$$\begin{bmatrix} \boldsymbol{v}_e \\ \boldsymbol{\omega}_e \end{bmatrix} = \begin{bmatrix} \mathbf{I}_3 & \boldsymbol{x}_{eb}^{\times\,T} \\ 0 & \mathbf{I}_3 \end{bmatrix} \begin{bmatrix} \boldsymbol{v}_b \\ \boldsymbol{\omega}_b \end{bmatrix} + \begin{bmatrix} \boldsymbol{J}_{Te} \\ \boldsymbol{J}_{Re} \end{bmatrix} \dot{\boldsymbol{q}}, \tag{2.26}$$

where I_3 is the third-order unit matrix, $\boldsymbol{x}_{eb} = \boldsymbol{x}_e - \boldsymbol{x}_b$ and $\boldsymbol{x}_{eb}^{\times}\boldsymbol{x}_{eb}^{\times}$ represent the cross-product matrix of the vector $\boldsymbol{x}_{eb}$.

Define $\boldsymbol{J}_{Te} = [\boldsymbol{k}_1 \times (\boldsymbol{x}_e - \boldsymbol{R}_1), \ldots, \boldsymbol{k}_n \times (\boldsymbol{x}_e - \boldsymbol{R}_n)], \boldsymbol{J}_{Re} = [\boldsymbol{k}_1, \boldsymbol{k}_2, \ldots, \boldsymbol{k}_n]$ as the robot translational Jacobian matrix and the rotational Jacobian matrix, respectively, the joint angular velocity matrix is $\dot{\boldsymbol{q}} = [\dot{\boldsymbol{q}}_1, \dot{\boldsymbol{q}}_2, \ldots, \dot{\boldsymbol{q}}_n]$.

Make $\boldsymbol{J}_b \in R^{6\times 6} \equiv \begin{bmatrix} I_3 & \boldsymbol{x}_{eb}^{\times\,T} \\ 0 & I_3 \end{bmatrix}$, $\boldsymbol{J}_m \in R^{6\times n} \equiv \begin{bmatrix} \boldsymbol{J}_{Te} \\ \boldsymbol{J}_{Re} \end{bmatrix}$, $\dot{\boldsymbol{x}}_e = \begin{bmatrix} \boldsymbol{v}_e \\ \boldsymbol{\omega}_e \end{bmatrix}$, $\dot{\boldsymbol{x}}_b = \begin{bmatrix} \boldsymbol{v}_b \\ \boldsymbol{\omega}_b \end{bmatrix}$, then Eq. (2.26) can be written as

$$\dot{\boldsymbol{x}}_e = \boldsymbol{J}_b\dot{\boldsymbol{x}}_b + \boldsymbol{J}_m\dot{\boldsymbol{q}}. \tag{2.27}$$

The physical meanings of each item in the above equation are very clear. The first item on the right represents the end speed of the robot caused by the speed of the base, and the second item represents the end speed caused by the angular velocity of the joints.

By deriving Eq. (2.27), the end acceleration is obtained as

$$\ddot{\boldsymbol{x}}_e = \boldsymbol{J}_b\ddot{\boldsymbol{x}}_b + \dot{\boldsymbol{J}}_b\dot{\boldsymbol{x}}_b + \boldsymbol{J}_m\ddot{\boldsymbol{q}} + \dot{\boldsymbol{J}}_m\dot{\boldsymbol{q}}. \tag{2.28}$$

2.4 Space Rigid Robot Dynamic Equations [2]

A space robot system is a complex system with multiple DOFs, strong coupling, and nonlinearity, as well as flexible arm booms and joints. In some missions, the flexibility of the arm booms and joints cannot be neglected; a flexible multi-body dynamic modeling method should be adopted. For some small-sized space robots, or those with slow motion speed and lower motion accuracy, a multi-rigid-body system modeling method is adopted. The most commonly used modeling methods for rigid robot systems are the Lagrange method and the Newton–Euler method. The Lagrange method takes the robot system as an indivisible whole, and obtains the system's differential equations by selecting generalized coordinates. The Newton–Euler modeling method takes each rigid body as the reference object, and establishes its dynamic equations based on Newton's second law and Euler's equation. In addition, the dynamic model of the whole system can be simultaneously obtained under the consideration of the constraints and forces between different rigid bodies. Both modeling methods have their advantages and disadvantages, and can be applied to different situations.

The Lagrange and Newton–Euler modeling methods study the problem from the perspectives of analytical mechanics and vector mechanics, respectively, but are consistent in essence. The equations established by the Lagrange method have good mathematical properties, which are very conducive to the design of theoretical analysis and control algorithms. The equations established by the Newton–Euler method have obvious physical meanings, and the computational efficiency is far superior to the Lagrange equation when the DOFs increase. Generally, inverse dynamics problems are calculated by the Newton–Euler method, while positive dynamics problems are first solved by the Newton–Euler inverse dynamics algorithm, combined with the special value method to solve the numerical solution of the matrix in the Lagrange equation, then the acceleration is obtained by solving the Lagrange equation. In recent years, some researchers have directly established control algorithms based on the Newton–Euler equation, for example, Zhu Wenhong proposed the Virtual Decomposition Control (VDC) method, which inherited the characteristics of the Newton–Euler dynamics algorithm.

2.4.1 Dynamic Equations of Space Rigid Robots Established by Lagrange Method

Taking the space robot system in Sect. 2.2 as the object, its dynamic model is established. The Lagrange function is defined as $L = T - V$, where T is the kinetic energy of the system and V is the potential energy of the system. The Lagrange equation can be written as

$$\frac{\mathrm{d}}{\mathrm{d}t}\left(\frac{\partial L}{\partial \dot{\boldsymbol{\varphi}}}\right) - \frac{\partial L}{\partial \boldsymbol{\varphi}} = \boldsymbol{Q}, \tag{2.29}$$

where $\boldsymbol{\varphi} = \begin{bmatrix} \boldsymbol{x}_b^T \\ \boldsymbol{q}^T \end{bmatrix}$ is the generalized coordinate of the selected space robot system, $\boldsymbol{Q}$ is the generalized force corresponding to the nonconservative active force system.

According to Eqs. (2.17) and (2.18), the kinetic energy of the robot link $\boldsymbol{i}$ relative to the inertial system is obtained as

$$T_i = \frac{1}{2} m_i \boldsymbol{v}_i^T \cdot \boldsymbol{v}_i + \frac{1}{2} \boldsymbol{\omega}_i^T \cdot \boldsymbol{I}_i \cdot \boldsymbol{\omega}_i = \frac{1}{2} \begin{bmatrix} \boldsymbol{v}_i \\ \boldsymbol{\omega}_i \end{bmatrix}^T \begin{bmatrix} m_i \boldsymbol{I}_3 & \boldsymbol{0} \\ \boldsymbol{0} & \boldsymbol{I}_i \end{bmatrix} \begin{bmatrix} \boldsymbol{v}_i \\ \boldsymbol{\omega}_i \end{bmatrix} \tag{2.30}$$

where $\boldsymbol{m}_i$ is the mass of the link $\boldsymbol{i}$ and $\boldsymbol{I}_i$ is the inertial matrix of the link $\boldsymbol{i}$ with respect to the centroid.

According to Eq. (2.26):

$$\begin{bmatrix} \boldsymbol{v}_i \\ \boldsymbol{\omega}_i \end{bmatrix} = \begin{bmatrix} I_3 & \boldsymbol{r}_{ib}^{\times T} \\ \boldsymbol{0} & I_3 \end{bmatrix} \dot{\boldsymbol{x}}_b + \begin{bmatrix} \boldsymbol{J}_{Ti} \\ \boldsymbol{J}_{Ri} \end{bmatrix} \dot{\boldsymbol{q}}, \tag{2.31}$$

where $\boldsymbol{r}_{ib} = \boldsymbol{r}_i - \boldsymbol{r}_b$. Substituting Eq. (2.31) into Eq. (2.30), the kinetic energy of the robot system is

$$T = \sum_{i=0}^{n} T_i = \frac{1}{2}\begin{bmatrix} \dot{\boldsymbol{x}}_b \\ \dot{\boldsymbol{q}} \end{bmatrix}^T \begin{bmatrix} \boldsymbol{H}_b & \boldsymbol{H}_{bm} \\ \boldsymbol{H}_{bm}^T & \boldsymbol{H}_m \end{bmatrix} \begin{bmatrix} \dot{\boldsymbol{x}}_b \\ \dot{\boldsymbol{q}} \end{bmatrix} = \frac{1}{2}\dot{\boldsymbol{\varphi}}^T \boldsymbol{H}(\boldsymbol{\varphi})\boldsymbol{\varphi}, \tag{2.32}$$

where $\boldsymbol{H}(\boldsymbol{\varphi})$ is the generalized inertial matrix of the system, $\boldsymbol{H}_b$ is the inertia array caused by the base, $\boldsymbol{H}_m$ is the inertia array caused by the robot, and $\boldsymbol{H}_{bm}$ is the coupling term between the base and the robot. The specific forms are as follows:

$$\boldsymbol{H}_b \in R^{6\times 6} \equiv \begin{bmatrix} \boldsymbol{H}_v & \boldsymbol{H}_{vw} \\ \boldsymbol{H}_{vw}^T & \boldsymbol{H}_w \end{bmatrix} \tag{2.33}$$

$$\boldsymbol{H}_v \in R^{3\times 3} \equiv \boldsymbol{M}\boldsymbol{I}_3 \tag{2.34}$$

$$\boldsymbol{H}_w \in R^{3\times 3} \equiv \sum_{i=1}^{n} \left(\boldsymbol{I}_i + m_i \boldsymbol{r}_{ib}^{\times\ T} \boldsymbol{r}_{ib}^{\times}\right) + \boldsymbol{I}_0 \tag{2.35}$$

$$\boldsymbol{H}_{vw} \in R^{3\times 3} \equiv \mathrm{M}\boldsymbol{r}_{ib}^{\times T} \tag{2.36}$$

$$\mathrm{M} = \sum_{i=0}^{n} (m_i) \tag{2.37}$$

$$\boldsymbol{r}_{cb} \in R^3 \equiv \boldsymbol{r}_c - \boldsymbol{r}_b \tag{2.38}$$

$$\boldsymbol{H}_m \in R^{n\times n} \equiv \sum_{i=1}^{n} \left(\boldsymbol{J}_{Ri}^T \boldsymbol{I}_i \boldsymbol{J}_{Ri} + m_i \boldsymbol{J}_{Ti}^T \boldsymbol{J}_{Ti}\right) \tag{2.39}$$

$$\boldsymbol{H}_{bm} \in R^{6\times n} \equiv \begin{bmatrix} \boldsymbol{H}_{vq} \\ \boldsymbol{H}_{wq} \end{bmatrix} \tag{2.40}$$

$$\boldsymbol{H}_{vq} \in R^{3\times n} \equiv \sum_{i=1}^{n} m_i \boldsymbol{J}_{Ti} \tag{2.41}$$

$$\boldsymbol{H}_{wq} \in R^{3\times n} \equiv \sum_{i=1}^{n} \left(\boldsymbol{I}_i \boldsymbol{J}_{Ri} + m_i \boldsymbol{r}_{ib}^{\times} \boldsymbol{J}_{Ti}\right). \tag{2.42}$$

According to the forms of the robot end Jacobian matrix, the Jacobian matrix of any link $\boldsymbol{i}$ can obtained as

$$\boldsymbol{J}_{Ti} \in R^{3\times n} \equiv [\boldsymbol{k}_1 \times (\boldsymbol{r}_i - \boldsymbol{R}_1), \boldsymbol{k}_2 \times (\boldsymbol{r}_i - \boldsymbol{R}_2), \ldots, \boldsymbol{k}_i \times (\boldsymbol{r}_i - \boldsymbol{R}_i), 0, \ldots, 0] \tag{2.43}$$

$$\boldsymbol{J}_{Ri} \in R^{3\times n} \equiv [\boldsymbol{k}_1, \boldsymbol{k}_2, \ldots, \boldsymbol{k}_i, 0, \ldots, 0]. \tag{2.44}$$

When the space robot system is in a microgravity environment, the potential energy caused by gravity is negligible. Substituting Eq. (2.32) into Eq. (2.29):

$$\boldsymbol{H}\ddot{\boldsymbol{\varphi}} + \dot{\boldsymbol{H}}\dot{\boldsymbol{\varphi}} - \frac{1}{2}\dot{\boldsymbol{\varphi}}^T \frac{\partial \boldsymbol{H}}{\partial \boldsymbol{\varphi}}\dot{\boldsymbol{\varphi}} = \boldsymbol{Q}. \tag{2.45}$$

Let $\dot{\boldsymbol{H}}\dot{\boldsymbol{\varphi}} - \frac{1}{2}\dot{\boldsymbol{\varphi}}^T \frac{\partial \boldsymbol{H}}{\partial \boldsymbol{\varphi}}\dot{\boldsymbol{\varphi}} = \boldsymbol{C}(\boldsymbol{\varphi}, \dot{\boldsymbol{\varphi}})\dot{\boldsymbol{\varphi}}$, then Eq. (2.45) can be written as

$$\boldsymbol{H}(\boldsymbol{\varphi})\ddot{\boldsymbol{\varphi}} + \boldsymbol{C}(\boldsymbol{\varphi}, \dot{\boldsymbol{\varphi}})\dot{\boldsymbol{\varphi}} = \boldsymbol{Q}. \tag{2.46}$$

Equation (2.46) is the generalized form of the space robot dynamic model. In the equation, $\boldsymbol{H}(\boldsymbol{\varphi})\ddot{\boldsymbol{\varphi}}$ is the inertial force term proportional to the generalized acceleration, and each element in $\boldsymbol{C}(\boldsymbol{\varphi}, \dot{\boldsymbol{\varphi}})\dot{\boldsymbol{\varphi}}$ is the secondary type of generalized speed, including centrifugal force and Coriolis force.

According to the external force of the space robot system, it can be determined that the generalized force Q mainly includes three parts: the driving force/torque $\boldsymbol{F}_b$ on the base body during orientation control, the robot joint driving torque $\boldsymbol{\tau}_m$, and the force/torque $\boldsymbol{F}_e$ at the end during task operations. These generalized forces can be expressed as

$$\boldsymbol{Q} = \begin{bmatrix} \boldsymbol{F}_b \\ \boldsymbol{\tau}_m \end{bmatrix} + \begin{bmatrix} \boldsymbol{J}_b^T \\ \boldsymbol{J}_m^T \end{bmatrix} \boldsymbol{F}_e \tag{2.47}$$

$$\boldsymbol{J}_b \in R^{6\times 6} \equiv \begin{bmatrix} I_3 & \boldsymbol{x}_{eb}^{\times\ T} \\ 0 & I_3 \end{bmatrix} \tag{2.48}$$

$$\boldsymbol{J}_m \in R^{6\times n} \equiv \begin{bmatrix} \boldsymbol{J}_{Te} \\ \boldsymbol{J}_{Re} \end{bmatrix}. \tag{2.49}$$

Let $\boldsymbol{C}(\boldsymbol{\varphi}, \dot{\boldsymbol{\varphi}})\dot{\boldsymbol{\varphi}} = [\boldsymbol{c}_b, \boldsymbol{c}_m]^T$, then Eq. (2.46) can be expressed as

$$\begin{bmatrix} \boldsymbol{H}_b & \boldsymbol{H}_{bm} \\ \boldsymbol{H}_{bm}^T & \boldsymbol{H}_m \end{bmatrix} \begin{bmatrix} \ddot{\boldsymbol{x}}_b \\ \ddot{\boldsymbol{q}} \end{bmatrix} + \begin{bmatrix} \boldsymbol{c}_b \\ \boldsymbol{c}_m \end{bmatrix} = \begin{bmatrix} \boldsymbol{F}_b \\ \boldsymbol{\tau}_m \end{bmatrix} + \begin{bmatrix} \boldsymbol{J}_b^T \\ \boldsymbol{J}_m^T \end{bmatrix} \boldsymbol{F}_e. \tag{2.50}$$

2.4.2 *Dynamic Equation of Space Rigid Robot Established by Newton–Euler Method*

(1) Speed relationships between coordinates

As shown in Fig. 2.5, the speeds of two space coordinates $\{\boldsymbol{A}\}$ and $\{\boldsymbol{B}\}$ under the inertial system $\{\boldsymbol{I}\}$ are v_A and v_B, the angular velocities are ω_A and ω_B, respectively, the velocity of $\{\boldsymbol{B}\}$ relative to $\{\boldsymbol{A}\}$ is v_B^A, the angular velocity is ω_B^A; a six-dimensional velocity matrix V is defined as a combination of velocity $\boldsymbol{v}$ and angular velocity $\boldsymbol{\omega}$, i.e.:

$$\boldsymbol{V}_A = \begin{bmatrix} \boldsymbol{v}_A \\ \boldsymbol{\omega}_A \end{bmatrix}, \boldsymbol{V}_B = \begin{bmatrix} \boldsymbol{v}_B \\ \boldsymbol{\omega}_B \end{bmatrix}, \quad \boldsymbol{V}_B^A = \begin{bmatrix} \boldsymbol{v}_B^A \\ \boldsymbol{\omega}_B^A \end{bmatrix}. \tag{2.51}$$

From the velocity synthesis theorem of rigid bodies, the following relations hold:

$$\begin{aligned} \boldsymbol{v}_B &= \boldsymbol{v}_A + \boldsymbol{\omega}_A \times \boldsymbol{p}_B^A + \boldsymbol{v}_B^A \\ \boldsymbol{\omega}_B &= \boldsymbol{\omega}_A + \boldsymbol{\omega}_B^A, \end{aligned} \tag{2.52}$$

where $\boldsymbol{p}_B^A$ is the vector pointing from the $\{\boldsymbol{A}\}$ origin to the $\{\boldsymbol{B}\}$ origin in the inertial system. Transforming to the coordinate $\{\boldsymbol{B}\}$, both sides of the equation are multiplied by ${}^B\boldsymbol{C}_I$. For any rotation matrix $\boldsymbol{C}$ and vectors $\boldsymbol{a}$ and $\boldsymbol{b}$, it holds that $\boldsymbol{C}(\boldsymbol{a} \times \boldsymbol{b}) = (\boldsymbol{C}\boldsymbol{a}) \times (\boldsymbol{C}\boldsymbol{b})$, then

$$\begin{aligned} {}^B\boldsymbol{v}_B &= {}^B\boldsymbol{C}_A^A\boldsymbol{v}_A + {}^B\boldsymbol{C}_A({}^A\boldsymbol{\omega}_A \times {}^A\boldsymbol{p}_B^A) + {}^B\boldsymbol{C}_A^A\boldsymbol{v}_B^A \\ {}^B\boldsymbol{\omega}_B &= {}^B\boldsymbol{C}_A^A\boldsymbol{\omega}_A + {}^B\boldsymbol{C}_A^A\boldsymbol{\omega}_B^A, \end{aligned} \tag{2.53}$$

where ${}^B\boldsymbol{v}_B$, ${}^B\boldsymbol{\omega}_B$ are the speed, angular velocity of coordinate $\{\boldsymbol{B}\}$ relative to the inertial system expressed in the coordinate $\{\boldsymbol{B}\}$; ${}^A\boldsymbol{v}_B^A$, ${}^A\boldsymbol{\omega}_B^A$ are the speed, angular velocity of $\{\boldsymbol{B}\}$ relative to $\{\boldsymbol{A}\}$ expressed in the coordinate $\{\boldsymbol{A}\}$; $({}^A\boldsymbol{p}_B^A)$ is the vector of the $\{\boldsymbol{A}\}$ origin pointing to the $\{\boldsymbol{B}\}$ origin represented in the coordinate $\{\boldsymbol{A}\}$; ${}^B\boldsymbol{C}_A$ is the rotation matrix of $\{\boldsymbol{A}\}$ relative to $\{\boldsymbol{B}\}$; and the meaning of the following symbolic can be analogized accordingly. Equation (2.53) can be rewritten as

$${}^B\boldsymbol{V}_B = {}^B\boldsymbol{T}_A^A\boldsymbol{V}_A + {}^B\boldsymbol{V}_B^A, \tag{2.54}$$

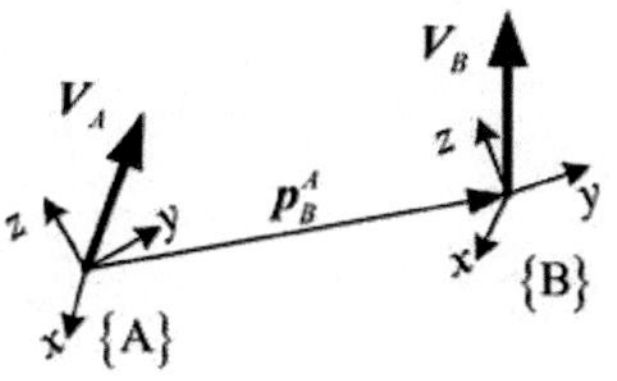

Fig. 2.5 Speed relationships between coordinates

where

$$ {}^{B}\boldsymbol{T}_{A}=\begin{bmatrix} {}^{B}\boldsymbol{C}_{A} & -{}^{B}\boldsymbol{C}_{A}\left({}^{A}\boldsymbol{p}_{\boldsymbol{B}}^{A}\right)^{\times} \\ \mathbf{O}_{3} & {}^{A}\boldsymbol{C}_{B} \end{bmatrix} \tag{2.55} $$

is called the Six-dimensional Velocity Transformation Matrix from $\{\boldsymbol{B}\}$ to $\{\boldsymbol{A}\}$, where O_3 is a 3-dimensional zero-square matrix, $\left({}^{A}\boldsymbol{p}_{\boldsymbol{B}}^{A}\right)^{\times}$ is ${}^{A}\boldsymbol{p}_{\boldsymbol{B}}^{A}$ the cross-multiplication matrix of ${}^{A}\boldsymbol{p}_{\boldsymbol{B}}^{A}$. The six-dimensional velocity transformation matrix is transitive, that is, for any three coordinates $\{\boldsymbol{A}\}$, $\{\boldsymbol{B}\}$, $\{\boldsymbol{C}\}$, ${}^{A}\boldsymbol{T}_{C} = {}^{A}\boldsymbol{T}_{\boldsymbol{B}}^{B}\boldsymbol{T}_{C}$ holds.

Differentiating Eq. (2.54) with respect to time, the acceleration relationship holds as

$$ {}^{B}\mathring{\boldsymbol{V}}_{B} = {}^{B}\boldsymbol{T}_{A}\,{}^{A}\mathring{\boldsymbol{V}}_{A} + {}^{B}\dot{\boldsymbol{T}}_{A}\,{}^{A}\boldsymbol{V}_{A} + {}^{B}\mathring{\boldsymbol{V}}_{B}^{A}, \tag{2.56} $$

where the first term on the right side of the equation is the convective acceleration, the second term contains partial Coriolis acceleration, and the third term is the relative acceleration. It should be specially noted that ${}^{B}\mathring{\boldsymbol{V}}_{B} = \frac{\tilde{d}}{dt}\left({}^{B}\boldsymbol{V}_{B}\right)$ is the relative derivative of $\boldsymbol{V}_{B}$ in $\{\boldsymbol{B}\}$, not equal to ${}^{B}\dot{\boldsymbol{V}}_{B}$, the coordinate of $\dot{\boldsymbol{V}}_{B}$ in $\{\boldsymbol{B}\}$ ${}^{B}\dot{\boldsymbol{V}}_{B}$. Consider that ${}^{B}\dot{\boldsymbol{C}}_{A} = \left({}^{A}\boldsymbol{\omega}_{\boldsymbol{B}}^{A}\right)^{\times}\,{}^{B}\boldsymbol{C}_{A}$, then

$$ {}^{B}\dot{\boldsymbol{T}}_{A} = \begin{bmatrix} -\left({}^{A}\boldsymbol{\omega}_{\boldsymbol{B}}^{A}\right)^{\times B}\boldsymbol{C}_{A} & \left({}^{A}\boldsymbol{\omega}_{\boldsymbol{B}}^{A}\right)^{\times B}\boldsymbol{C}_{A}\left({}^{A}\boldsymbol{p}_{\boldsymbol{B}}^{A}\right)^{\times} - {}^{B}\boldsymbol{C}_{A}\left({}^{A}\boldsymbol{v}_{\boldsymbol{B}}^{A}\right)^{\times} \\ \mathbf{O}_{3} & -\left({}^{A}\boldsymbol{\omega}_{\boldsymbol{B}}^{A}\right)^{\times B}\boldsymbol{C}_{A} \end{bmatrix}. \tag{2.57} $$

There are several special situations:

(a) If $\{A\}$, $\{B\}$ are two fixed coordinate frames on the rigid body, then

$$ {}^{B}\boldsymbol{V}_{B} = {}^{B}\boldsymbol{T}_{A}^{A}\boldsymbol{V}_{A} \tag{2.58} $$

$$ {}^{B}\mathring{\boldsymbol{V}}_{B} = {}^{B}\boldsymbol{T}_{A}^{A}\mathring{\boldsymbol{V}}_{A}. \tag{2.59} $$

In addition, it is proved by the translational theorem of force that the forces at the two coordinates $\{\boldsymbol{A}\}$ and $\{\boldsymbol{B}\}$ have the following equivalent relations:

$$ {}^{A}\boldsymbol{F}_{A} = {}^{B}\boldsymbol{T}_{A}^{T\,B}\boldsymbol{F}_{B}. \tag{2.60} $$

(b) If $\{\boldsymbol{A}\}$, $\{\boldsymbol{B}\}$ are the coordinates that rotate around a standard axis (e.g., around the z-axis), then

$$ {}^{B}\boldsymbol{V}_{B} = {}^{B}\boldsymbol{T}_{A}^{A}\boldsymbol{V}_{A} + \boldsymbol{\omega}\cdot z_{6} \tag{2.61} $$

$$ {}^{B}\overset{\circ}{V}_{B} = {}^{B}T_{A}\,{}^{A}\overset{\circ}{V}_{A} + {}^{B}\dot{T}_{A}\,{}^{A}V_{A} + \dot{\nu}\cdot z_{3}, \tag{2.62} $$

where

$$ {}^{A}T_{B} = \begin{bmatrix} {}^{A}C_{B} & O_{3} \\ O_{3} & {}^{A}C_{B} \end{bmatrix} \tag{2.63} $$

$$ {}^{B}\dot{T}_{A} = \begin{bmatrix} -\left({}^{A}\omega_{B}^{A}\right)^{\times}\,{}^{A}C_{B} & O_{3} \\ O_{3} & -\left({}^{A}\omega_{B}^{A}\right)^{\times}\,{}^{A}C_{B} \end{bmatrix}, \tag{2.64} $$

where ω is the angular velocity of the fixed axis, $z_{6} = \begin{bmatrix} 0 & 0 & 0 & 0 & 0 & 1 \end{bmatrix}^{T}$.

(c) If $\{A\}$, $\{B\}$ are coordinates that move relatively along a standard axis (e.g., along the z-axis), then

$$ {}^{B}V_{B} = {}^{B}T_{A}^{A}V_{A} + \nu\cdot z_{3} \tag{2.65} $$

$$ {}^{B}\overset{\circ}{V}_{B} = {}^{B}T_{A}^{A}\overset{\circ}{V}_{A} + {}^{B}\dot{T}_{A}^{A}V_{A} + \dot{\nu}\cdot z_{3}, \tag{2.66} $$

where

$$ {}^{B}T_{A} = \begin{bmatrix} I_{3} & -\left({}^{A}p_{B}^{A}\right)^{\times} \\ O_{3} & I_{3} \end{bmatrix} \tag{2.67} $$

$$ {}^{B}\dot{T}_{A} = \begin{bmatrix} O_{3} & -\left({}^{A}v_{B}^{A}\right)^{\times} \\ O_{3} & O_{3} \end{bmatrix}, \tag{2.68} $$

ν is the relative velocity, $z_{3} = \begin{bmatrix} 0 & 0 & 1 & 0 & 0 & 0 \end{bmatrix}^{T}$.

(2) Dynamic equations of a single rigid body

For the rigid body i, applying the Newton–Euler equation in the centroid inertial system $\{\Sigma_{Oi}\}$ gives

$$ \begin{bmatrix} m_{i}I_{3} & \\ & I_{oi}(t) \end{bmatrix}\dot{V}_{Oi} + \begin{bmatrix} m_{i}g \\ \omega_{i}^{\times}I_{oi}(t)\omega_{i} \end{bmatrix} = F_{oi}^{*} \tag{2.69} $$

where I_{3} is a 3-dimensional unit matrix; m_{i} is the mass of the rigid body i; I_{Oi} is the representation of the inertial matrix of the rigid body i in the centroid inertial system (it is noted that its value will change with the rotation of the rigid body), $I_{Oi}(t) = {}^{oi}C_{i}\,{}^{i}I_{i}\left({}^{oi}C_{i}\right)^{T}$; ${}^{i}I_{i}$ is the inertial matrix in the body-fixed coordinate system $\{\Sigma_{i}\}$, which is time-invariant; $\dot{V}_{Oi}$ is the centroid acceleration and angular acceleration in the inertial system, $\dot{\boldsymbol{V}}_{oi} = \begin{bmatrix} \dot{\boldsymbol{v}}_{Oi} \\ \dot{\boldsymbol{\omega}}_i \end{bmatrix}$; $\dot{\boldsymbol{v}}_{Oi}$ is the centroid velocity of the rigid body $\boldsymbol{i}$ in the inertial system; $\boldsymbol{\omega}_i$ is the angular velocity of the rigid body $\boldsymbol{i}$ in the inertial system; $\boldsymbol{g}$ is the gravitational acceleration; $\boldsymbol{F}^*_{Oi}$ is the joint force and moment of the rigid body $\boldsymbol{i}$ expressed in the centroid inertial system.

Using the speed and force transformation relationship (2.58), (2.60), we have

$$\boldsymbol{F}^*_i = \left({}^{cmi}\boldsymbol{T}_i\right)^T \boldsymbol{F}^*_{cmi} = \left({}^{cmi}\boldsymbol{T}_i\right)^T \begin{bmatrix} {}^{cmi}\boldsymbol{C}_{oi} & \\ & {}^{cmi}\boldsymbol{C}_{oi} \end{bmatrix} \boldsymbol{F}^*_{oi} \tag{2.70}$$

$$\boldsymbol{V}_{oi} = \begin{bmatrix} {}^{Oi}\boldsymbol{C}_{cmi} & \\ & {}^{oi}\boldsymbol{C}_{cmi} \end{bmatrix} {}^{cmi}\boldsymbol{T}_i \boldsymbol{V}_i, \tag{2.71}$$

where $\boldsymbol{F}^*_i$ is the joint force of the rigid body $\boldsymbol{i}$ expressed in the coordinate system $\{\Sigma_i\}$.

Differentiating Eq. (2.71) with respect to time yields

$$\dot{V}_{oi} = \begin{bmatrix} \omega_i^{\times 0i}\,{}^{0i}\boldsymbol{C}_{cmi} & \\ & \omega_i^{\times 0i}\boldsymbol{C}_{cmi} \end{bmatrix} {}^{cmi}\boldsymbol{T}_i \boldsymbol{V}_i + \begin{bmatrix} {}^{Oi}\boldsymbol{C}_{cmi} & \\ & {}^{Ici}\boldsymbol{C}_{cmi} \end{bmatrix} {}^{cmi}\boldsymbol{T}_i \mathring{\boldsymbol{V}}_i. \tag{2.72}$$

Substituting (2.70), (2.71), and (2.72) into (2.69) yields

$$\boldsymbol{M}_i \cdot \mathring{\boldsymbol{V}}_i + \boldsymbol{C}_i \cdot \boldsymbol{V}_i + \boldsymbol{G}_i = \boldsymbol{F}^*_i, \tag{2.73}$$

where

$$\boldsymbol{M}_i = \begin{bmatrix} m_i \boldsymbol{I}_{3\times3} & -m_i\left({}^i\boldsymbol{p}^i_{ci}\right)^\times \\ m_i\left({}^i\boldsymbol{p}^i_{ci}\right)^\times & {}^i\boldsymbol{I}_i - m_i\left({}^i\boldsymbol{p}^i_{ci}\right)^\times\left({}^i\boldsymbol{p}^i_{ci}\right)^\times \end{bmatrix} \tag{2.74}$$

$$\boldsymbol{C}_i = \begin{bmatrix} m_i(\boldsymbol{\omega}_i)^\times & -m_i(\boldsymbol{\omega}_i)^\times\left({}^i\boldsymbol{p}^i_{ci}\right)^\times \\ m_i\left({}^i\boldsymbol{p}^i_{ci}\right)^\times(\boldsymbol{\omega}_i)^\times & (\boldsymbol{\omega}_i)^{\times i}\boldsymbol{I}_i + {}^i\boldsymbol{I}_i(\boldsymbol{\omega}_i)^\times - m_i\left({}^i\boldsymbol{p}^i_{ci}\right)^\times(\boldsymbol{\omega}_i)^\times\left({}^i\boldsymbol{p}^i_{ci}\right)^\times \end{bmatrix} \tag{2.75}$$

$$\boldsymbol{G}_i = \begin{bmatrix} m_i{}^i\boldsymbol{C}_{ci}\boldsymbol{g} \\ m_i\left({}^i\boldsymbol{p}^i_{ci}\right)^{\times i}\boldsymbol{C}_{ci}\boldsymbol{g} \end{bmatrix} \tag{2.76}$$

are the inertial matrix of the rigid body $\boldsymbol{i}$, the Coriolis force and the centrifugal force matrix, and the gravity term vector, respectively. Note that the inertial matrix is no more symmetric, but it is easy to prove that it is still positive definite. Equation (2.73) is still a linear function of a set of inertial parameters, i.e., for the selected inertial parameter vector $\boldsymbol{\theta}_1 \in \mathbb{R}^s (s \leq 13)$, the following relations hold:

$$\boldsymbol{M}_i \cdot \mathring{\boldsymbol{V}}_i + \boldsymbol{C}_i \cdot \boldsymbol{V}_i + \boldsymbol{G}_i = \boldsymbol{Y}_i \boldsymbol{\theta}_i, \tag{2.77}$$

where $\boldsymbol{Y}_i$ is the regression matrix for the inertial parameters $\boldsymbol{\theta}_i$. For different rigid bodies, the forms of $\boldsymbol{M}_i$, $\boldsymbol{C}_i$, and $\boldsymbol{G}_i$ are unified, and $\boldsymbol{Y}_i$ is identical, which is not necessary to deduce again.

(3) Dynamic equations of space robot systems

The velocity and acceleration of every coordinate can be obtained from the kinematic relationship between the coordinates. To simplify writing, for the following velocities and forces in the dynamic equation for a single rigid body, if the left superscript and the right subscript are the same, they can be omitted as long as no confusion is caused. Let the velocity of the outboard frame of the base $\{\Sigma_{\hat{1}}\}$ be $\boldsymbol{V}_b$, the joint (angle) displacement vector be $\boldsymbol{q}$, and $\boldsymbol{q}_i$ be the $\boldsymbol{i}$th element of $\boldsymbol{q}$, then the velocity and acceleration of each coordinate can be obtained by forward recursion.

The velocity relationship ($i = 1, 2, \ldots, n$) is

$$V_{\hat{1}} = \boldsymbol{V}_b \tag{2.78}$$

$$\boldsymbol{V}_i = {}^i\boldsymbol{T}_{\hat{i}}\boldsymbol{V}_{\hat{i}} + (\sigma_i z_3 + \bar{\sigma}_i z_6)\dot{q}_i \tag{2.79}$$

$$\boldsymbol{V}_{\widehat{i+1}} = {}^{\widehat{i+1}}\boldsymbol{T}_i\boldsymbol{V}_i. \tag{2.80}$$

The acceleration relationship ($i = 0, 1, 2, \ldots, n$) is

$$\mathring{\boldsymbol{V}}_{\hat{1}} = \dot{\boldsymbol{V}}_b \tag{2.81}$$

$$\mathring{\boldsymbol{V}}_i =^i \boldsymbol{T}_{\hat{i}}\mathring{V}_{\hat{i}} + {}^i\dot{\boldsymbol{T}}_{\hat{i}}V_{\hat{i}} + (\sigma_i z_3 + \bar{\sigma}_i z_6)\ddot{q}_i \tag{2.82}$$

$$\mathring{\boldsymbol{V}}_{\widehat{i+1}} = {}^{\widehat{i+1}}\boldsymbol{T}_i\,\mathring{\boldsymbol{V}}_i \tag{2.83}$$

where σ_i represents the type of joint $\boldsymbol{i}$, if it is a translational joint, $\sigma_i = 1$; if it is a rotating joint, $\sigma_i = 0$. In the task coordinate, the force at the robot end is written as ${}^e\boldsymbol{F}_e$, then the force and moment at each joint can be obtained by reverse recursion.

The force and moment relationship ($i = n, n-1, \ldots, 1$) is

$$\boldsymbol{F}_{\widehat{n+1}} = {}^e\boldsymbol{F}_e \tag{2.84}$$

$$\boldsymbol{F}_i^* = \boldsymbol{M}_i \cdot \mathring{\boldsymbol{V}}_i + \boldsymbol{C}_i \cdot \boldsymbol{V}_i + \boldsymbol{G}_i \tag{2.85}$$

$$\boldsymbol{F}_i = \boldsymbol{F}_i^* + \left({}^{\widehat{i+1}}\boldsymbol{T}_i\right)^T \boldsymbol{F}_{\widehat{i+1}} \tag{2.86}$$

$$\boldsymbol{F}_{\hat{i}} = \left({}^i\boldsymbol{T}_{\hat{i}}\right)^T \boldsymbol{F}_i. \tag{2.87}$$

The output of the joint is

$$\tau_i = (\sigma_i z_3 + \bar{\sigma}_i z_6)^T \boldsymbol{F}_i. \tag{2.88}$$

Considering that the base is free-floating, the dynamics system expressed by the above equation is actually an under-actuated system. For the closeness of the equations, six additional constraint equations are required:

$$\boldsymbol{F}_0 = \boldsymbol{F}_0^* + \left({}^{\hat{1}}\boldsymbol{T}_0\right)^T \boldsymbol{F}_{\hat{1}} = O \tag{2.89}$$

Thus, Eqs. (2.78)–(2.89) simultaneously form the Newton–Euler dynamic equations of the system. For robot systems operating in a microgravity environment, the above gravity terms $\boldsymbol{G}_i$ are all zeros.

2.4.3 Situations Without Considering Base Floating

For some space robot systems whose base effects can be ignored, the base coordinate can be regarded as the inertial system to simplify the space robot dynamic model, and transform it into a fixed-base system. The Lagrange dynamic equation of a fixed base robot can be written as:

$$\boldsymbol{H}(\boldsymbol{q})\ddot{\boldsymbol{q}} + \boldsymbol{C}(\boldsymbol{q}, \dot{\boldsymbol{q}})\dot{\boldsymbol{q}} + \boldsymbol{G}(\boldsymbol{q}) = \boldsymbol{\tau} - \boldsymbol{J}^T \boldsymbol{F}_e. \tag{2.90}$$

Considering the nonredundant non-singular case, because $\dot{\boldsymbol{x}} = \boldsymbol{J}\dot{\boldsymbol{q}}, \ddot{\boldsymbol{x}} = \boldsymbol{J}\dot{\boldsymbol{q}} + \dot{\boldsymbol{J}}\dot{\boldsymbol{q}}$, then

$$\dot{\boldsymbol{q}} = \boldsymbol{J}^{-1}\dot{\boldsymbol{x}} \tag{2.91}$$

$$\ddot{\boldsymbol{q}} = \boldsymbol{J}^{-1}\ddot{\boldsymbol{x}} - \boldsymbol{J}^{-1}\dot{\boldsymbol{J}}\boldsymbol{J}^{-1}\dot{\boldsymbol{x}}. \tag{2.92}$$

Substituting Eqs. (2.91) and (2.92) into (2.90) and multiplying by $\boldsymbol{J}^{-T}$ yields

$$\begin{aligned} &\boldsymbol{J}^{-T}\boldsymbol{H}(\boldsymbol{q})\boldsymbol{J}^{-1}\ddot{\boldsymbol{x}} - \boldsymbol{J}^{-T}\boldsymbol{H}(\boldsymbol{q})\boldsymbol{J}^{-1}\dot{\boldsymbol{J}}\boldsymbol{J}^{-1}\dot{\boldsymbol{x}} + \boldsymbol{J}^{-T}\boldsymbol{C}(\boldsymbol{q}, \dot{\boldsymbol{q}})\boldsymbol{J}^{-1}\dot{\boldsymbol{x}} + \boldsymbol{J}^{-T}\boldsymbol{G}(\boldsymbol{q}) \\ &= \boldsymbol{J}^{-T}\boldsymbol{\tau} - \boldsymbol{F}_e \end{aligned} \tag{2.93}$$

It can be proven that $-\boldsymbol{J}^{-1}\dot{\boldsymbol{J}}\boldsymbol{J}^{-1} = \frac{\mathrm{d}}{\mathrm{d}t}(\boldsymbol{J}^{-1})$. If we denote

$$\overline{\boldsymbol{H}} = \boldsymbol{J}^{-T}\boldsymbol{H}\boldsymbol{J}^{-1}, \quad \overline{\boldsymbol{C}} = \boldsymbol{J}^{-T}\boldsymbol{C}\boldsymbol{J}^{-1} + \boldsymbol{J}^{-1}\boldsymbol{H}\frac{d}{dt}(\boldsymbol{J}^{-1}), \quad \overline{\boldsymbol{G}} = \boldsymbol{J}^{-T}\boldsymbol{G}, \tag{2.94}$$

the dynamic equation can be rewritten as

$$\overline{H}(q)\ddot{x} + \overline{C}(q, \dot{q})\dot{x} + \overline{G}(q) = J^{-T}\tau - F_e, \tag{2.95}$$

which is a form of Cartesian space dynamic equation.

Because the Newton–Euler equation takes a subsystem as the research object, the local change of the model only affects the local part of the dynamic equation, which turns out to be modular. If the base is fixed, without the constraint of (2.89), then the Eqs. (2.78) and (2.81) become

$$V_{\hat{1}} = O_6 \tag{2.96}$$

$$\overset{\circ}{V}_{\hat{1}} = O_6. \tag{2.97}$$

2.5 Space Flexible Robot Dynamic Equations [3, 4]

The space robot has the characteristics: long links, lightweight, and large payloads. Elastic deformation and end vibration can be easily generated in the flexible arm boom, which has a great impact on the fast and stable operation of the robot. The dynamics of flexible robots is the basis for dynamics analysis, robot controller design, and robot vibration suppression. The keystone of flexible robot dynamics is to effectively describe the deformation of flexible arm booms, for which the Assumed Modes Method is the most widely used and most effective. The following is an application of the assumed modes method to determine the flexibility of the arm boom and a study of the dynamics of a space flexible robot system based on the Lagrange method.

2.5.1 *Flexible Body Deformation and Kinetic Energy*

The deformation of the flexible body is described based on the assumed modes method, as shown in Fig. 2.6, where $\{\Sigma_I\}$ is the inertial coordinate, with the proximal joint of each flexible body as the origin, the tangential direction of the flexible body as the x-axis, and the direction of the joint axis as the z-axis, then the body-fixed coordinate system of each body $o_i x_i y_i (i = 1, \ldots, n)$ is established. The angle of each joint is $\varphi_i (i = 1, \ldots, n)$, and the position vector of each joint in the inertial system is $R_i (i = 1, \ldots, n)$.

Point $\boldsymbol{p}$ is any point on the flexible body $\boldsymbol{i}$. According to the figure, the position vector of point $\boldsymbol{p}$ in the inertial coordinate can be expressed as

$$\boldsymbol{r}_{ip} = \boldsymbol{R}_i + {}^I\boldsymbol{C}_i \boldsymbol{u}'_i = \boldsymbol{R}_i + {}^i\boldsymbol{C}_i \left(\boldsymbol{u}_i + \boldsymbol{u}_{if}\right) \tag{2.98}$$

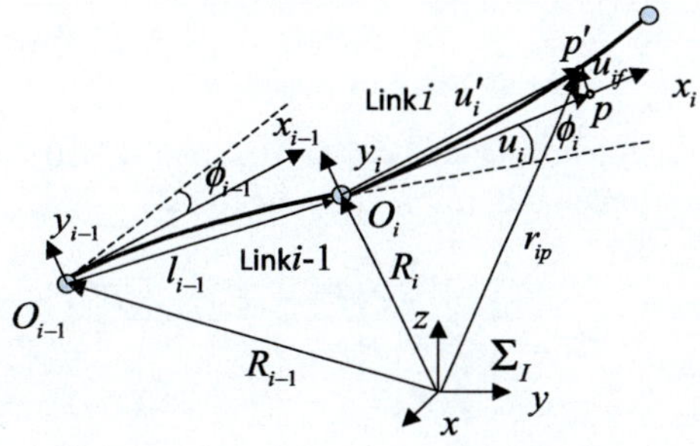

Fig. 2.6 Description of flexible bodies

where $\boldsymbol{u}'_i$ is the position vector of point p in the follow-up coordinate $o_i x_i y_i$, ${}^I\boldsymbol{C}_i$ is the transformation matrix of the follow-up coordinate to the inertial coordinate $\boldsymbol{\Sigma}_I$, $\boldsymbol{u}_i$ is the position vector of point p in the follow-up coordinate system $o_i x_i y_i$ before deformation of the flexible body, it is a constant value, and $\boldsymbol{u}_{if}$ is a deformation vector in the follow-up coordinate at the flexible body point p.

According to the assumed modes method

$$\boldsymbol{u}_{if} = \boldsymbol{\Phi}_i \boldsymbol{q}_{if} \tag{2.99}$$

where $\boldsymbol{\Phi}_i$ is the mode matrix that satisfies the requirements, and $\boldsymbol{q}_{if}$ is the generalized coordinate of the deformation (mode coordinates).

Then, Eq. (2.98) can be written as

$$\boldsymbol{r}_{ip} = \boldsymbol{R}_i + {}^I\boldsymbol{C}_i \boldsymbol{u}'_i = \boldsymbol{R}_i + {}^I\boldsymbol{C}_i\left(\boldsymbol{u}_i + \boldsymbol{\Phi}_i \boldsymbol{q}_{if}\right). \tag{2.100}$$

The velocity vector of point p in the inertial system is

$$\dot{\boldsymbol{r}}_{ip} = \dot{\boldsymbol{R}}_i + {}^I\dot{\boldsymbol{C}}_i \boldsymbol{u}'_i + {}^I\boldsymbol{C}_i \dot{\boldsymbol{u}}'_i = \dot{\boldsymbol{R}}_i + {}^I\dot{\boldsymbol{C}}_i \boldsymbol{u}'_i + {}^I\boldsymbol{C}_i \boldsymbol{\Phi}_i \dot{\boldsymbol{q}}_{if}. \tag{2.101}$$

Select $\boldsymbol{R}_i$, $\boldsymbol{P}_i$, and $\boldsymbol{q}_{if}$ as the position generalized coordinate, the orientation generalized coordinate, and the deformation generalized coordinate that describe the follow-up coordinate $o_i x_i y_i$, where $\boldsymbol{P}_i = [\boldsymbol{P}_0 \boldsymbol{P}_1 \boldsymbol{P}_2 \boldsymbol{P}_3]^T$ is the attitude quaternion, and satisfies $\boldsymbol{P}_i \boldsymbol{P}_i^T = 1$.

According to the attitude quaternion and the Euler axis angle relationship.

$$\boldsymbol{p} = \begin{bmatrix} \boldsymbol{p}_0 \\ \boldsymbol{p}_r \end{bmatrix} = [\boldsymbol{p}_0, \boldsymbol{p}_1, \boldsymbol{p}_2, \boldsymbol{p}_3]^T = \left[\cos\frac{\theta}{2}, \boldsymbol{e}_x \sin\frac{\theta}{2}, \boldsymbol{e}_y \sin\frac{\theta}{2}, \boldsymbol{e}_z \sin\frac{\theta}{2}\right]^T, \tag{2.102}$$

where $\boldsymbol{p}_r$, θ are the Euler axis vector and angle, respectively.

The rotation matrix expressed by the Euler axis angle is

$$\boldsymbol{C} = \left[\boldsymbol{I} + \boldsymbol{p_r}\sin\theta + 2\boldsymbol{p}_r^2\sin^2\frac{\theta}{2}\right]. \tag{2.103}$$

According to Eqs. (2.102) and (2.103), the rotation matrix represented by the attitude quaternion is obtained as

$$^{I}\boldsymbol{C}_i = \begin{bmatrix} 1-2\left(\boldsymbol{P}_2^2+\boldsymbol{P}_3^2\right) & 2(\boldsymbol{P}_1\boldsymbol{P}_2-\boldsymbol{P}_0\boldsymbol{P}_3) & 2(\boldsymbol{P}_1\boldsymbol{P}_3+\boldsymbol{P}_0\boldsymbol{P}_2) \\ 2(\boldsymbol{P}_1\boldsymbol{P}_2+\boldsymbol{P}_0\boldsymbol{P}_3) & 1-2\left(\boldsymbol{P}_1^2+\boldsymbol{P}_3^2\right) & 2(\boldsymbol{P}_2\boldsymbol{P}_3-\boldsymbol{P}_0\boldsymbol{P}_1) \\ 2(\boldsymbol{P}_1\boldsymbol{P}_2-\boldsymbol{P}_0\boldsymbol{P}_2) & 2(\boldsymbol{P}_2\boldsymbol{P}_3+\boldsymbol{P}_0\boldsymbol{P}_1) & 1-2\left(\boldsymbol{P}_1^2+\boldsymbol{P}_2^2\right) \end{bmatrix} \tag{2.104}$$

Equation (2.104) is changed into a two-matrix multiplication form as

$$^{I}\boldsymbol{C}_i = \boldsymbol{C}\hat{\boldsymbol{C}}^{\boldsymbol{T}} \tag{2.105}$$

$$\boldsymbol{C} = \begin{bmatrix} -\boldsymbol{P}_1 & \boldsymbol{P}_0 & -\boldsymbol{P}_3 & \boldsymbol{P}_2 \\ -\boldsymbol{P}_2 & \boldsymbol{P}_3 & \boldsymbol{P}_0 & -\boldsymbol{P}_1 \\ -\boldsymbol{P}_3 & -\boldsymbol{P}_2 & \boldsymbol{P}_1 & \boldsymbol{P}_0 \end{bmatrix} \tag{2.106}$$

$$\hat{\boldsymbol{C}} = \begin{bmatrix} -\boldsymbol{P}_1 & \boldsymbol{P}_0 & \boldsymbol{P}_3 & -\boldsymbol{P}_2 \\ -\boldsymbol{P}_2 & -\boldsymbol{P}_3 & \boldsymbol{P}_0 & \boldsymbol{P}_1 \\ -\boldsymbol{P}_3 & \boldsymbol{P}_2 & -\boldsymbol{P}_1 & \boldsymbol{P}_0 \end{bmatrix}. \tag{2.107}$$

The angular velocity cross-multiplication matrix $\tilde{\omega}_i$ in the following coordinate is defined as follows:

$$\tilde{\omega}_i = {}^{I}\boldsymbol{C}_i^{T}\,{}^{I}\dot{\boldsymbol{C}}_i. \tag{2.108}$$

Then, we have the following relationship:

$$^{I}\dot{\boldsymbol{C}}_i\boldsymbol{u}_i' = {}^{I}\boldsymbol{C}_i\tilde{\omega}_i\boldsymbol{u}_i'. \tag{2.109}$$

Simultaneously, from Eqs. (2.104), (2.105), and (2.108), we get

$$\omega_i = 2\hat{\boldsymbol{C}}\dot{\boldsymbol{P}}_i. \tag{2.110}$$

Substituting (2.110) into (2.109):

$$^{I}\dot{\boldsymbol{C}}_i\boldsymbol{u}_i' = -2^{I}\boldsymbol{C}_i\tilde{\boldsymbol{u}}_i'\hat{\boldsymbol{C}}\dot{\boldsymbol{P}}_i. \tag{2.111}$$

Equation (2.101) can be expressed as a matrix form:

$$\dot{\boldsymbol{r}}_{\boldsymbol{ip}} = \left[\boldsymbol{I}\ \boldsymbol{D}\ {}^{I}\boldsymbol{C}_i\boldsymbol{\Phi}_i\right]\left[\dot{\boldsymbol{R}}_i\ \dot{\boldsymbol{P}}_i\ \dot{\boldsymbol{q}}_{if}\right]^{T} \tag{2.112}$$

$$\boldsymbol{D} = -2^I\boldsymbol{C}_i\tilde{\boldsymbol{u}}_i'\hat{\boldsymbol{C}}. \tag{2.113}$$

The kinetic energy of the flexible body $\boldsymbol{i}$ is

$$T_i = \frac{1}{2}\int\limits_V \rho\dot{\boldsymbol{r}}_{iP}^T\dot{\boldsymbol{r}}_{iP}dV, \tag{2.114}$$

where $\boldsymbol{\rho}$ is the density of the flexible body. Substitute Eq. (2.112) into (2.114):

$$\boldsymbol{T}_i = \frac{1}{2}\dot{\boldsymbol{q}}_i^T\boldsymbol{H}_i\dot{\boldsymbol{q}}_i, \tag{2.115}$$

where $\boldsymbol{q}_i = \begin{bmatrix} \boldsymbol{R}_i^T & \boldsymbol{P}_i^T & \boldsymbol{q}_{if}^T \end{bmatrix}^T$ is the generalized coordinates, $\boldsymbol{H}_i$ is the mass matrix, then we have

$$\boldsymbol{H}_i = \int\limits_V \rho\begin{bmatrix} \boldsymbol{I} \\ \boldsymbol{D}^T \\ ({}^I\boldsymbol{C}_i\boldsymbol{\Phi}_i)^T \end{bmatrix}[\boldsymbol{I}\,\boldsymbol{D}\,{}^I\boldsymbol{C}_i\boldsymbol{\Phi}_i]dV = \int\limits_V \rho\begin{bmatrix} \boldsymbol{I} & \boldsymbol{D} & {}^I\boldsymbol{C}_i\boldsymbol{\Phi}_i \\ & \boldsymbol{D}^T\boldsymbol{D} & \boldsymbol{D}^{T\,I}\boldsymbol{C}_i\boldsymbol{\Phi}_i \\ \cdots & & \boldsymbol{\Phi}_i^T\boldsymbol{\Phi}_i \end{bmatrix}dV, \tag{2.116}$$

and the mass matrix $\boldsymbol{H}_i$ is a symmetric matrix. Divide and define each sub-equation as

$$\boldsymbol{H} = \begin{bmatrix} H_{tt} & H_{tr} & H_{tf} \\ & H_{rr} & H_{rf} \\ \cdots & & H_{ff} \end{bmatrix}. \tag{2.117}$$

Then Eq. (2.115) can be written as a matrix form

$$\boldsymbol{T}_i = \frac{1}{2}\dot{\boldsymbol{q}}_i{}^T\boldsymbol{H}_i\dot{\boldsymbol{q}}_i = \frac{1}{2}\begin{bmatrix} \dot{\boldsymbol{R}}_i^T & \dot{\boldsymbol{P}}_i^T & \dot{\boldsymbol{q}}_{if}^T \end{bmatrix}\begin{bmatrix} H_{tt} & H_{tr} & H_{tf} \\ & H_{rr} & H_{rf} \\ \cdots & & H_{ff} \end{bmatrix}\begin{bmatrix} \dot{\boldsymbol{R}}_i \\ \dot{\boldsymbol{P}}_i \\ \dot{\boldsymbol{q}}_{if} \end{bmatrix}. \tag{2.118}$$

According to Eqs. (2.116), (2.118), the specific expressions and the meanings of the sub-equations in Eq. (2.117) are as follows:

$H_{tt} = \int_V \rho\boldsymbol{I}dV$: The quality characteristics of translation of a flexible body;

$H_{tr} = \int_V \rho\boldsymbol{D}dV$: The quality characteristics of the translational and rotational coupling of a flexible body;

$H_{tf} = \int_V \boldsymbol{\rho}^I\boldsymbol{C}_i\boldsymbol{\Phi}_i dV$: The quality characteristics of translational and flexible vibration coupling of a flexible body;

$H_{rr} = \int_V \rho\boldsymbol{D}^T\boldsymbol{D}dV$: The rotation quality characteristics of a flexible body;

$H_{rf} = \int_V \rho\boldsymbol{D}^{T\,I}\boldsymbol{C}_i\boldsymbol{\Phi}_i dV$: The quality characteristics of rotational and flexible vibration coupling of a flexible body;

$H_{ff} = \int_V \rho \boldsymbol{\Phi}_i^T \boldsymbol{\Phi}_i dV$: The quality characteristics of flexible vibration of a flexible body.

By further processing of the above equations, we can obtain the display form of the kinetic energy of the flexible body $\boldsymbol{i}$ with respect to the body mass, the rotation inertia, the flexible vibration mode matrix and the generalized coordinates, and then solve it by programming.

2.5.2 Elastic Force and Generalized Force of the Flexible Body

The flexible body also has the elastic potential energy generated by strain. For the convenience of derivation, the elastic potential energy is presented in the form of generalized elastic force, and the elastic deformation of the flexible body is assumed. Based on the theory of elastic mechanics, the expression of the virtual work of the internal force caused by the elastic deformation of the flexible body is

$$
\begin{aligned}
\delta W_i &= -\int_V \sigma^T \delta\varepsilon dV = -\int_V (E\varepsilon)^T \delta\varepsilon dV \\
&= -\int_V \boldsymbol{q}_{if}^T (\boldsymbol{D}^* \boldsymbol{\Phi}_i)^T E^T (\boldsymbol{D}^* \boldsymbol{\Phi}_i) \delta \boldsymbol{q}_{ij} dV \\
&= -\boldsymbol{q}_{if}^T \left[\int_V (\boldsymbol{D}^* \boldsymbol{\Phi}_i)^T E^T (\boldsymbol{D}^* \boldsymbol{\Phi}_i) dV \right] \delta \boldsymbol{q}_{jf} \\
&= -\boldsymbol{q}_{if}^T \boldsymbol{K}_f \delta \boldsymbol{q}_{if}
\end{aligned}
\tag{2.119}
$$

where σ, ε, and E are the elastic stress, strain, and elastic modulus of the elastic body, respectively, $\sigma = E\varepsilon, \boldsymbol{\varepsilon} = \boldsymbol{D}^* \boldsymbol{u}_{if} = \boldsymbol{D}^* (\boldsymbol{\Phi}_i \boldsymbol{q}_{if})$, $\boldsymbol{D}^*$ is the differential operator matrix.

Extend Eq. (2.119) to the matrix form of generalized coordinates as

$$
\delta W_i = \left[\boldsymbol{R}_i^T \boldsymbol{P}_i^T \boldsymbol{q}_{if}^T \right] \begin{bmatrix} 0 & 0 & 0 \\ 0 & 0 & 0 \\ 0 & 0 & \boldsymbol{K}_{ff} \end{bmatrix} \begin{bmatrix} \delta \boldsymbol{R}_i \\ \delta \boldsymbol{P}_i \\ \delta \boldsymbol{q}_{if} \end{bmatrix} = -\boldsymbol{q}_i^T \boldsymbol{K}_i \delta \boldsymbol{q}_i. \tag{2.120}
$$

Assume that the resultant force acting on the point p on the flexible body $\boldsymbol{i}$ is $\boldsymbol{F}_i$, then the virtual work done by $\boldsymbol{F}_i$ is

$$
\delta W = \boldsymbol{F}_i \delta \boldsymbol{r}_{ip} = \boldsymbol{F}_i \left(\sum_{j=1}^{n} \frac{\partial \boldsymbol{r}_{ip}}{\partial q_j} \delta q_j \right)
$$

$$= \sum_{j=1}^{n} \left(\boldsymbol{F}_i \frac{\partial \boldsymbol{r}_{ip}}{\partial q_j} \right) \delta q_j \triangleq \sum_{j=1}^{n} Q_j \delta q_j \tag{2.121}$$

where Q_j is a generalized force corresponding to the generalized coordinate q_j, and n is the number of generalized coordinates. Then the generalized force of the flexible body $\boldsymbol{i}$ can be expressed as

$$\boldsymbol{Q}_i = \left[\boldsymbol{Q}_{ti}, \boldsymbol{Q}_{ri}, \boldsymbol{Q}_{fi} \right]^T = \sum_{k=1}^{N} \boldsymbol{F}_k \frac{\partial \boldsymbol{r}_k}{\partial \boldsymbol{q}_i}. \tag{2.122}$$

2.5.3 Interbody Constraint Equation of the Flexible Body

The bodies of a flexible multi-body system are connected with each other by certain mechanical constraints. These constraints of each single body should be described in mathematical form, then the dynamic equations of every single body in the system are assembled together to obtain the dynamic equation of the multi-body system. The constraint equation of the holonomic constraint is a function of the generalized coordinate $\boldsymbol{q}$ and time $\boldsymbol{t}$ of the multi-body system. The general form is

$$\boldsymbol{C}(\boldsymbol{q}, t) = 0. \tag{2.123}$$

Assume that each body is a single-DOF rotary joint connection; there are altogether five constraint equations between the two bodies, three of which are position constraint equations, two are attitude constraint equations.

Assume that the adjacent two bodies $\boldsymbol{i},\boldsymbol{j}$ are connected by $\boldsymbol{r}^{ij}$ (as shown in Fig. 2.7), and then the position constraint equation is

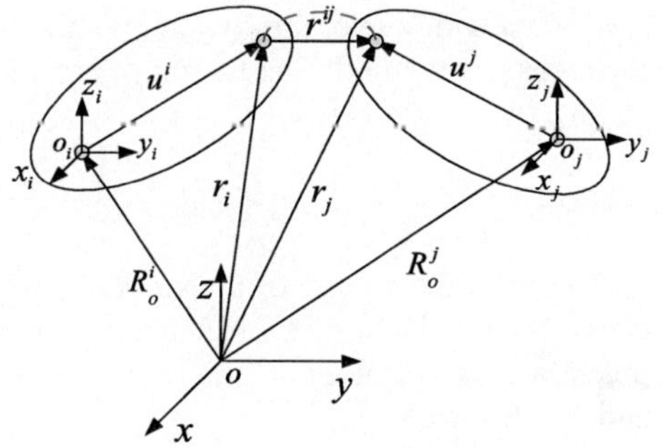

Fig. 2.7 Position constraint relationship

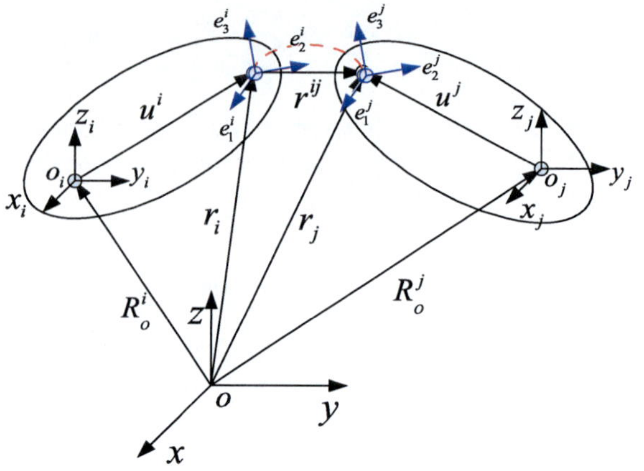

Fig. 2.8 Attitude constraint relationship

$$\boldsymbol{C}_p = \left(\boldsymbol{R}_o^j + \boldsymbol{C}^j\left(\boldsymbol{u}_O^{\prime j} + \boldsymbol{u}_f^j\right)\right) - \left(\boldsymbol{R}_O^i + \boldsymbol{C}^i\left(\boldsymbol{u}_O^{\prime i} + \boldsymbol{u}_f^i\right)\right) = 0. \tag{2.124}$$

The unit direction vectors of the hinge point coordinates of the two bodies $\boldsymbol{i}, \boldsymbol{j}$ are $\left[e_1^i\ e_2^i\ e_3^i\right]^T, \left[e_1^j\ e_2^j\ e_3^j\right]^T$, respectively (as shown in Fig. 2.8), $\boldsymbol{e_3}$ is the direction of the joint rotation axis, then the attitude constraint equation is

$$\boldsymbol{C}_r = \begin{bmatrix} \left(e_3^i\right)^T e_1^j \\ \left(e_3^i\right)^T e_2^j \end{bmatrix} = 0. \tag{2.125}$$

2.5.4 Dynamic Model of the Flexible Multi-Body System

Based on the kinetic energy, elastic force, and generalized force of the flexible body obtained above, the dynamic equation of the flexible body $\boldsymbol{i}$ is obtained by the Lagrange method. The Lagrange equation is

$$\frac{\mathrm{d}}{\mathrm{d}t}\left(\frac{\partial L}{\partial \dot{q}}\right) - \frac{\partial L}{\partial q} = \boldsymbol{Q}. \tag{2.126}$$

The potential energy of the space flexible multi-body system only considers the elastic potential energy caused by the deformation of the flexible body, ignoring the potential energy caused by gravity, thus $L = T - W$; substitute Eqs. (2.118), (2.120), and (2.122), we have

$$\boldsymbol{H}_i(\boldsymbol{q}_i)\ddot{\boldsymbol{q}}_i + \boldsymbol{C}(\boldsymbol{q}_i; \dot{\boldsymbol{q}}_i)\dot{\boldsymbol{q}}_i + \boldsymbol{K}_i\boldsymbol{q}_i = \boldsymbol{Q}_i \tag{2.127}$$

where $C(q_i, \dot{q}_i) = \frac{dH_i}{dt}\dot{q}_i - \frac{1}{2}\frac{\partial H_i}{\partial q_i}\ddot{q}_i$ is a velocity nonlinear term, the first term is the Coriolis force, and the second term is the centrifugal force.

For a multi-body system with n DOFs, two bodies are connected by a rotating joint, and the dynamic equations of n motion bodies are assembled by the constraint equation. The system dynamic equations with multipliers are obtained as follows:

$$\begin{gathered} H(q)\ddot{q} + C(q, \dot{q})\dot{q} + Kq + C_q^T\lambda = Q \\ C_i(q, t) = 0(i = 1, \ldots, k), \end{gathered} \tag{2.128}$$

where

$$H = \text{diag}[H_1, H_2, \ldots, H_n]$$

$$q = \left[q_1^T q_2^T, \ldots, q_n^T\right]^T$$

$$K = \text{diag}[K_1, K_2, \ldots, K_n]$$

$$C_q = \left[C_{1q}^T, C_{2q}^T, \ldots, C_{kq}^T\right]^T$$

$$\lambda = [\lambda_1\lambda_2, \ldots, \lambda_k]^T$$

$$Q = \left[Q_{F_1}^T Q_{F_2}^T, \ldots, Q_{F_n}^T\right]^T.$$

Equation (2.128) is a typical differential-algebraic equations. Although there are a large number of equations, it is a highly sparse matrix, suitable for efficient solution, and applicable to both holonomic and non-holonomic constraints. It can be converted to a closed dynamic equation by introducing additional acceleration constraint equations and solved.

Assume that the constraint equation $C_i(q, t)$ is continuously derivable to the second order of time t and is linearly independent of each other, then the acceleration constraint equation is

$$C_q\ddot{q} = \gamma \tag{2.129}$$

$$\gamma = -C_{tt} - 2C_{qt}\dot{q} - \left(C_q\dot{q}\right)_q\dot{q}, \tag{2.130}$$

where C_x is the partial derivation of the constraint equation to the independent variable x, γ is the right term of the acceleration constraint equation, and Eqs. (2.128), (2.129), and (2.130) are simultaneous to obtain the closed multi-body system dynamic equation as

$$\begin{bmatrix} \boldsymbol{H} & \boldsymbol{C}_q^T \\ \boldsymbol{C}_q & 0 \end{bmatrix} \begin{bmatrix} \ddot{\boldsymbol{q}} \\ \boldsymbol{\lambda} \end{bmatrix} = \begin{bmatrix} \boldsymbol{Q} - \boldsymbol{C}(\boldsymbol{q}, \dot{\boldsymbol{q}})\dot{\boldsymbol{q}} - \boldsymbol{K}\boldsymbol{q} \\ \boldsymbol{\gamma} \end{bmatrix} = \begin{bmatrix} \boldsymbol{Q}^* \\ \boldsymbol{\gamma} \end{bmatrix}. \tag{2.131}$$

And it can be proved that if the coefficient matrix of the above equations is a non-singular matrix, the unique solution of the independent variables $\ddot{\boldsymbol{q}}, \boldsymbol{\lambda}$ can be obtained, and then $\dot{\boldsymbol{q}}, \boldsymbol{q}$ can be obtained by integrating the step size.

2.5.5 Dynamics of Space Flexible Robot

Based on the dynamic Eq. (2.128) of flexible multi-body systems, the dynamic modeling of the space flexible robot system (Fig. 2.2) is established. The dynamic model of the space flexible robot system is as follows:

$$\begin{bmatrix} \boldsymbol{H}_b & \boldsymbol{H}_{bm} & \boldsymbol{H}_{bf} \\ \boldsymbol{H}_{bm}^T & \boldsymbol{H}_m & \boldsymbol{H}_{mf} \\ \boldsymbol{H}_{bf}^T & \boldsymbol{H}_{mf}^T & \boldsymbol{H}_f \end{bmatrix} \begin{bmatrix} \ddot{\boldsymbol{x}}_b \\ \ddot{\boldsymbol{q}} \\ \ddot{\boldsymbol{q}}_f \end{bmatrix} + \begin{bmatrix} \boldsymbol{c}_b \\ \boldsymbol{c}_m \\ \boldsymbol{c}_f \end{bmatrix} + \begin{bmatrix} 0 \\ 0 \\ \boldsymbol{K}_{ff}\boldsymbol{q}_f \end{bmatrix} = \begin{bmatrix} \boldsymbol{F}_b \\ \boldsymbol{\tau}_m \\ 0 \end{bmatrix} + \begin{bmatrix} \boldsymbol{J}_b^T \\ \boldsymbol{J}_m^T \\ 0 \end{bmatrix} \boldsymbol{F}_e \tag{2.132}$$

where $\boldsymbol{H}_b$ is the inertial matrix of the base, $\boldsymbol{H}_m$ is the inertial matrix of the robot system, $\boldsymbol{H}_f$ is the inertial matrix corresponding to the flexible mode, $\boldsymbol{H}_{bm}$ is the coupled inertial matrix of the base and the robot, $\boldsymbol{H}_{bf}$ is the coupled inertial matrix of the base and the flexible mode, and $\boldsymbol{H}_{mf}$ is the coupled inertial matrix of the robot and flexible mode; $\boldsymbol{x}_b$ is the base position/orientation, $\boldsymbol{q}$ is the joint angle of the robot, $\boldsymbol{q}_f$ is the flexible arm mode coordinate; $\boldsymbol{c}_b$, $\boldsymbol{c}_m$, $\boldsymbol{c}_f$ are the nonlinear terms corresponding to the base, robot, and mode coordinate velocity, respectively; $\boldsymbol{K}_{ff}$ is the mode rigidness matrix, $\boldsymbol{F}_b$, $\boldsymbol{\tau}_m$ are the base force/torque and the robot joint control torque, $\boldsymbol{F}_e$ is the external force/torque acting on the end of the robot; $\boldsymbol{J}_b$ is the Jacobian matrix of the robot end relative to the base, and $\boldsymbol{J}_m$ is the Jacobian matrix of the end relative to the robot joint.

References

1. H. Wei, *Robot Dynamics and Control* (Higher Education Press, Beijing, 2005)
2. Y. Liu, *Multi-Rigid-Body System Dynamics* (Higher Education Press, Beijing, 1989)
3. W. Huang, C. Shao, *Multi-flexible Body System Dynamics* (Science Press, Beijing, 1996)
4. D.A. Turcic, A. Midha, Dynamic analysis of elastic mechanism systems, part I: applications. J. Dyn. Syst. Meas. Control **106**(4), 249–254 (1984)

Chapter 3
Motion Planning of Space Robot

For different types of space robots, motion planning has different meanings. Motion planning of space robots usually refers to the process of generating the desired motion trajectory in the robot joint space or Cartesian space according to the mission target. Motion planning of the space-moving robot generally refers to generating the motion path of the robot; and the key of motion planning for a leg-based walking robot is how to design an effective gait to ensure the walking stability, speed, and efficiency of the robot.

3.1 Motion Planning for Space Robot

3.1.1 Description of the Planning Problem

The trajectory usually refers to the position, velocity, and acceleration of the robot during its whole movement. The trajectory planning of a robot is to study the trajectory generation method in joint space or Cartesian space based on the robot kinematics and dynamics as well as the task requirements. This section introduces the trajectory generation method of robot motion in joint space or Cartesian space.

The motion of a robot is usually described as the motion of the tool coordinate system $\{T\}$ relative to the base coordinate system $\{S\}$. For the "Picking/Placing" operation robot, the main task focuses on the initial state and the target state, i.e., the initial value of the tool coordinate system $\{T_0\}$ and its target value $\{T_g\}$. This motion is called point-to-point (PTP) motion.

For other operations, not only the starting and ending points state of the robot but also sufficient intermediate points state or the complete curve equation must be defined. This kind of motion planning is called contour motion planning or continuous-path (continuous-path motion) planning.

Y. Wang, *Space Robotics*, Space Science and Technologies,
https://doi.org/10.1007/978-981-15-4902-1_3

For space robots, motion planning can be done in joint space or Cartesian space, provided that the planned trajectory function must be continuous and smooth to ensure the smoothness of the robot's motion. Motion planning in joint space is to plan the angular displacement, angular velocity and angular acceleration of one single joint, while motion planning in Cartesian space is to plan the position/attitude, velocity/angular velocity, and acceleration/angular acceleration of the robot end effector. The corresponding joint position, velocity, and acceleration can be obtained through inverse kinematics calculation.

3.1.2 Selection of Motion Trajectory

In order to ensure the smoothness of a space robot motion, both the joint trajectory and the end-effector trajectory will be interpolated in joint space and Cartesian space during motion planning. This section will introduce two interpolation methods commonly used for space robots.

3.1.2.1 Cubic Polynomial Interpolation

Taking joint space planning as an example, the value $\theta_{i\ \mathrm{int}}$ at the moment $t = t_0$ is the starting joint angle, and the value $\theta_{i\ \mathrm{end}}$ at the end time t_f is the ending joint angle. Obviously, there are many smoothing functions that can be used as joint interpolation functions, and in most cases, cubic polynomials are used for interpolation.

In order to achieve a smooth motion of the joint, the trajectory function $\theta(t)$ should satisfy at least four constraints. Among them, two constraints are the joint angles corresponding to the starting point and the ending point:

$$\begin{cases} \theta(t_0) = \theta_0 \\ \theta(t_f) = \theta_f \end{cases}. \tag{3.1}$$

In order to meet the requirement for joint movement speed continuity, two other constraints on the joint angular velocities at the starting point and the ending point should also be satisfied. In the current situation, they are given by

$$\begin{cases} \dot{\theta}(t_0) = 0 \\ \dot{\theta}(t_f) = 0 \end{cases}. \tag{3.2}$$

The above four boundary conditions define a unique cubic polynomial:

$$\theta(t) = a_0 + a_1 t + a_2 t^2 + a_3 t^3, \tag{3.3}$$

where a_0, a_1, a_2, and a_3 are four coefficients to be determined based on the above-mentioned four constraints. The angular velocity and angular acceleration of the joint on the trajectory are

$$\begin{cases} \dot{\theta}(t) = a_1 + 2a_2 t + 3a_3 t^2 \\ \ddot{\theta}(t) = 2a_2 + 6a_3 t \end{cases}. \tag{3.4}$$

Substituting the four constraints listed in Eqs. (3.1) and (3.2) into Eqs. (3.3) and (3.4), four linear equations are obtained for solving the four coefficients:

$$\begin{cases} \theta_0 = a_0 \\ \theta_f = a_0 + a_1 t_f + a_2 t_f^2 + a_3 t_f^3 \\ 0 = a_1 \\ 0 = a_1 + 2a_2 t_f + 3a_3 t_f^2 \end{cases}. \tag{3.5}$$

By solving the above equations, the four coefficients can be obtained as

$$\begin{cases} a_0 = \theta_0 \\ a_1 = 0 \\ a_2 = \dfrac{3}{t_f^2}(\theta_f - \theta_0) \\ a_3 = -\dfrac{2}{t_f^3}(\theta_f - \theta_0) \end{cases}. \tag{3.6}$$

3.1.2.2 Displacement Curve with Continuously Changing Speed (Trapezoidal Interpolation)

There are many kinds of paths in Cartesian space, such as linear motion path, circular motion path, and constrained curve motion path. The difference among these paths lies mainly at the end of the trajectory. The overall algorithms of all paths are the same. Here, the trapezoidal interpolation method is described by linear motion.

When the linear motion of a robot is planned in Cartesian space, the trapezoidal interpolation method can be used to ensure the continuous velocity change of the robot end effector. Sometimes, the parabolic excess based on the trapezoidal interpolation is introduced to ensure the acceleration continuity, but it will increases the cost of system calculation and puts forward higher requirements on the calculation ability of the controller. In practice, the commonly used trapezoidal interpolation can satisfy the system requirements. Therefore, this section only introduces the trapezoidal interpolation method.

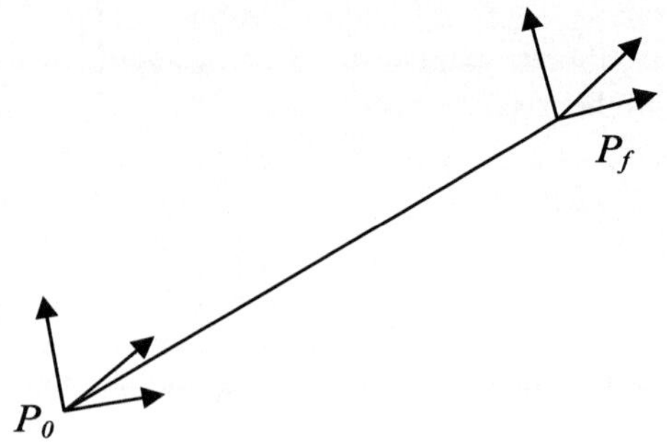

Fig. 3.1 Motion trajectory of robot end effector

Calculate the length of the straight line from the starting point P_0 to the ending point P_f as (Fig. 3.1)

$$\text{dist} = \sqrt{\sum_{i=x,y,z} \left(P_{f,i} - P_{0,i}\right)^2}. \tag{3.7}$$

By using the trapezoidal method (continuous speed), the motion path can be planned. Let t_0 and t_f be the starting time and ending time of the process; let t_s and $t_f - t_s$ be the transition time of the acceleration curve segment and that of the deceleration curve segment; let v_m be the maximum speed; let d_f be the displacement at the end point. Since the curve is symmetric, the acceleration time is equal to the deceleration time. The motion path of the robot is shown in Fig. 3.2.

If $t_a = t_f - t_s$, then $v_m = d_f / t_a$.

From the above known variables, other variables and curve functions can be derived. The function expression for each segment of the displacement curve d_t is

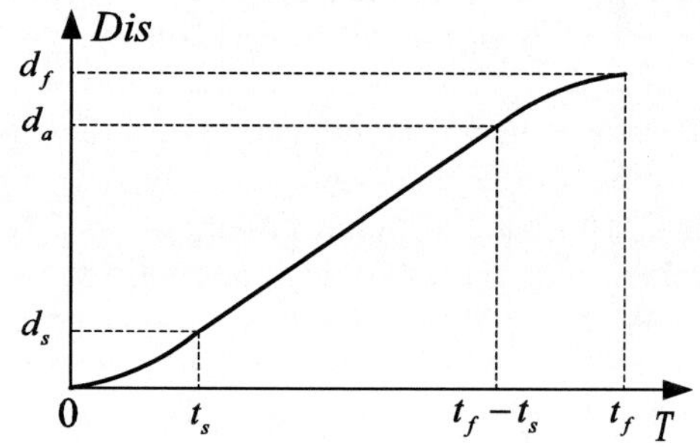

Fig. 3.2 Position curve

$$d_t = \begin{cases} (v_m / (2t_s)) \cdot t^2 & if\ (t < t_s) \\ \left[v_m \cdot t_s / 2 + v_m \cdot \left(t_a^2 + t_f^2\right) / (2t_s) - v_m \cdot t_f \cdot t_a / t_s - d_f\right] \cdot (t - t_s) / (t_s - t_a) & \\ + v_m \cdot t_s / 2 & if\ (t_s \le t < t_a) \\ - v_m \cdot t^2 / (2t_s) + v_m \cdot t_f \cdot t / t_s + d_f - v_m \cdot t_f^2 / (2t_s) & if\ (t_a \le t \le t_f) \end{cases} \quad (3.8)$$

3.1.3 Motion Planning in Joint Space

Motion planning in joint space generally refers to the process of designing a set of trajectories for spatially moving robot joints from their initial angles to the desired angles. Since the inverse kinematics is not involved, the calculation of this motion planning is relatively simple and free from singularity problem. In this section, the planned motion path of a space robot in joint space is generated by the cubic polynomial fitting trajectory. However, in practical applications, trapezoidal interpolation is more commonly used for trajectory generation.

The single joint position fitted by cubic polynomial is used to calculate the planned angle and angular velocity for each joint according to the initial joint angle, target joint angle, and motion time.

According to the known starting joint angle and ending joint angle, the single joint position planning algorithm can be designed as follows:

a. Acquire the starting joint angle θ_0, target joint angle θ_f, motion time t_f, and motion planning period Δt;
b. Calculate the total number of steps $N = \frac{t_f}{\Delta t}$;
c. Determine the next step (step k) and the corresponding time $t_k = k \times \Delta t, k \le N$;
d. Calculate the joint angle $\theta(t_k)$ at the step k according to Eq. (3.6) and Eq. (3.3);
e. Calculate the joint angular velocity based on the joint angle difference between the step k and its previous step $k - 1$:

$$\dot{\theta}(t_k) = \frac{\theta(t_k) - \theta(t_{k-1})}{\Delta t};$$

f. Check the relationship between the current number of steps k and the total number of steps N. If $k = N$, the planning is completed; otherwise, increase the number of steps by 1 and return to step c.

The specific algorithm flowchart is shown in Fig. 3.3.

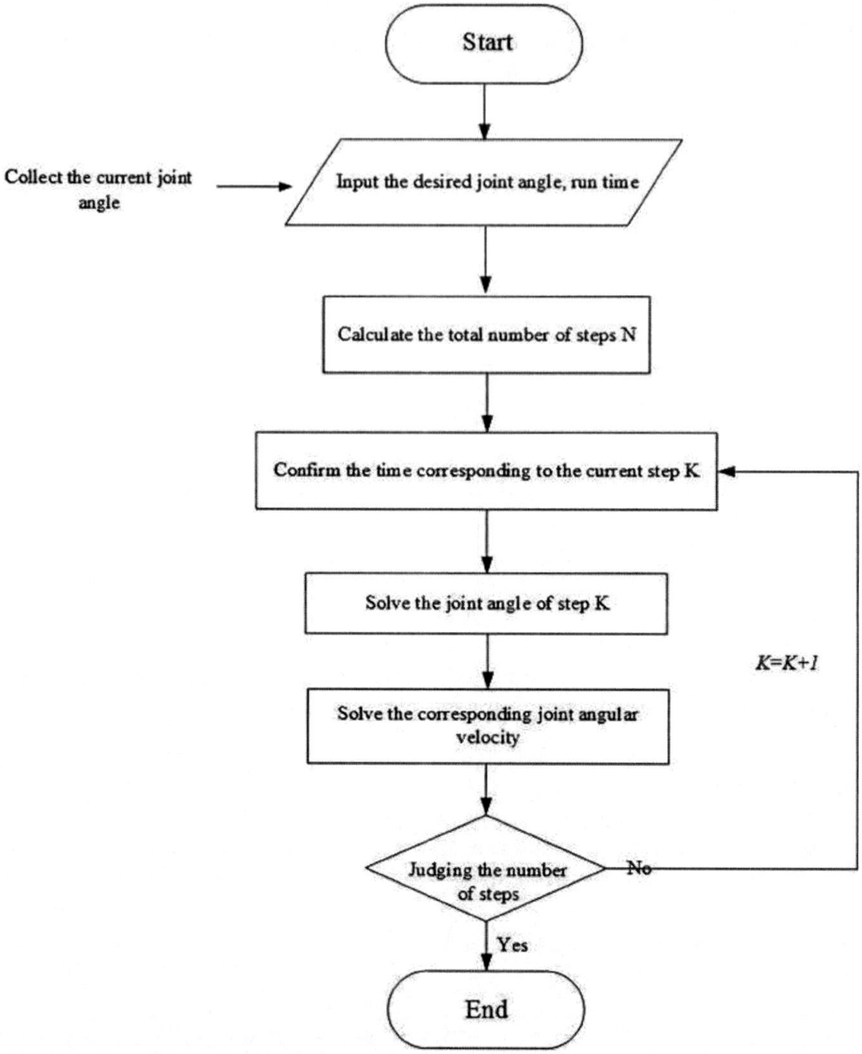

Fig. 3.3 Flowchart in single joint position mode

3.1.4 Motion Planning in Cartesian Space

Motion planning in Cartesian space mainly aims to generate a trajectory for the end effector of a space robot to reach the desired destination position according to the given initial pose and attitude in the Cartesian coordinate system. Motion planning methods in Cartesian space can be divided into PTP motion planning and trajectory-tracking motion planning according to the motion trajectory. The PTP motion planning only cares about the pose (position/attitude), speed/angular velocity, and other motion characteristics of the robot end at a specific point. The trajectory-tracking motion planning is concerned about the pose, speed/angular velocity, and other motion characteristics of the robot end in the entire path. The methods of trajectory-tracking motion planning can be further divided into linear motion planning, circular motion planning, and constrained curve motion planning. The difference between these planning methods lies mainly at the end of the trajectory. The overall algorithm flows of all these planning methods are the same. Here, linear motion planning is taken as an example for description.

3.1.4.1 Input Conditions

The initial pose of the robot base in the base coordinate system is defined as $PE_{\text{int}}([P_{\text{int}}, E_{\text{int}}])$, with which the desired pose $PE_{\text{des}}([P_{\text{des}}, E_{\text{des}}])$, the robot end's desired linear velocity v_m and the angular velocity ω_m, the robot end's desired linear acceleration a and the angular acceleration α, motion planning period Δt, and the robot end's current pose PE_{POR} are determined.

3.1.4.2 Planning Process

The position and attitude of the end effector of the robot in Cartesian space are planned, respectively, and the total planning time t_f is determined according to the maximum time principle. The linear velocity and angular velocity at the Kth step are calculated according to the motion planning period. The angular velocity of the robot joint is obtained by using the Jacobian matrix J, and the corresponding joint angle is obtained by integration.

a. Calculate the time of linear velocity planning

The end position deviation is obtained from the initial robot-end position P_{int} and the desired robot-end position P_{des}.

$$d = \sqrt{\sum_{i=x,y,z} \left(P_{\text{des},i} - P_{\text{int},i}\right)^2}. \tag{3.9}$$

Then the acceleration time can be obtained as

$$t_{s1} = v_m / a. \tag{3.10}$$

The total time of linear planning is

$$t_{f1} = d / v_m + t_{s1}. \tag{3.11}$$

b. Calculate the time of angular velocity planning

First, the orientation deviations from the beginning to the final attitude of the end effector are obtained and converted into the axis–angular relationship. Then, the trapezoidal method is used to plan the axial speed at the end of the manipulator so that the angular velocity of the robot end effectors in each control cycle can be obtained.

Plan the angular velocity of the robot end, define the initial attitude of the robot end and the desired pose as $E_{\text{int}} = (\alpha_{e0}\ \beta_{e0}\ \gamma_{e0})$ and $E_{\text{des}} = \left(\alpha_{ef}\ \beta_{ef}\ \gamma_{ef}\right)$ according to the ZYX Euler angle, then

$$\begin{cases} A_{e0} = R_z(\alpha_{e0})R_y(\beta_{e0})R_x(\gamma_{e0}) = [n_0, o_0, a_0] \\ A_{ef} = R_z\left(\alpha_{ef}\right)R_y\left(\beta_{ef}\right)R_x\left(\gamma_{ef}\right) = \left[n_f, o_f, a_f\right] \end{cases}. \tag{3.12}$$

And the attitude deviation of the robot end is

$$e_0 = \frac{1}{2}\left(n_0 \times n_f + o_0 \times o_f + a_0 \times a_f\right). \tag{3.13}$$

If $r = \frac{e_0}{|e_0|}$, $\phi_0 = \arcsin(|e_0|)$, then

$$e_0 = r\sin(\phi_0). \tag{3.14}$$

The linear acceleration time under angular velocity constraints is

$$t_{s2} = \omega_m / \alpha. \tag{3.15}$$

The total line planning time under angular velocity constraints is

$$t_{f2} = \phi_0 / \omega_m + t_{s2}. \tag{3.16}$$

c. Calculate the desired speed of the robot end

If $t_{f1} \geq t_{f2}$, then $t_f = t_{f1}$, $t_s = t_{s1}$. In this case, the position curve should be planned according to the end linear velocity, and the end angle should be interpolated proportionally. The total number of steps in motion planning is given as $N = \frac{t_f}{\Delta t}$.

The initial linear speed of the end effector is set as $v_e = \left[0\ 0\ 0\right]$, and the desired angular velocity is chosen as $w_e = \left[0\ 0\ 0\right]$. The time corresponding to the next step (step K) is $t_k = k \times \Delta t (k \leq N)$. The desired position $d_t(t_k)$ at the step k can be obtained from Eq. (3.8). The current position of the robot end effector is $P_{\text{now}} = PE_{\text{POR}}(1:3)$, so the current straight line distance is

$$d_n = \sqrt{\sum_{i=x,y,z} (P_{\text{now},i} - P_{\text{int},i})^2}. \tag{3.17}$$

Then the linear velocity of the end effector at the step k can be expressed as

$$v_e(k) = (P_{\text{des}} - P_{\text{int}}) \cdot \frac{d_t(t_k) - d_n}{d} \cdot \frac{1}{\Delta t}. \tag{3.18}$$

The angular velocity of the end effector at the step k is obtained by the interpolation method and can be expressed as

$$\omega_e(k) = r\phi_0 \cdot \frac{d_t(t_k) - d_n}{d} \cdot \frac{1}{\Delta t}. \tag{3.19}$$

Similarly, if $t_{f2} \geq t_{f1}$, then $t_f = t_{f2}$, $t_s = t_{s2}$. In this case, the attitude planning should be carried out according to the attitude curve, and the position planning should be processed accordingly.

d. Solve the joint angular velocity

According to the planned robot-end motion speed, the angular velocities of the joints are obtained by the inverse kinematics method. The expected joint angular velocities of the robot are

$$\dot{q}_{md} = J^{+} \begin{bmatrix} v_e \\ \omega_e \end{bmatrix}. \tag{3.20}$$

where J^{+} is the Moore-Penrose generalized inverse of the Jacobian matrix.

Then the desired joint angles are

$$q_{md} = q_m + \dot{q}_{md} \Delta t. \tag{3.21}$$

e. Determine the relationship between the current number of steps K and the total number of steps N

If $K = N$, then complete the planning; otherwise, increase the number of steps by 1 and return to step c.

The specific flowchart is shown in Fig. 3.4.

3.1.5 Redundant Design for Avoidance of Joint Limits and Singularities

Space robots with more than six DOFs are called redundant space robots. The outstanding advantage of redundant space robots is the ability to use redundant DOFs to perform additional functions required in complex space environments, such as avoiding the joint limits, singularities, and obstacles and optimizing the joint torque.

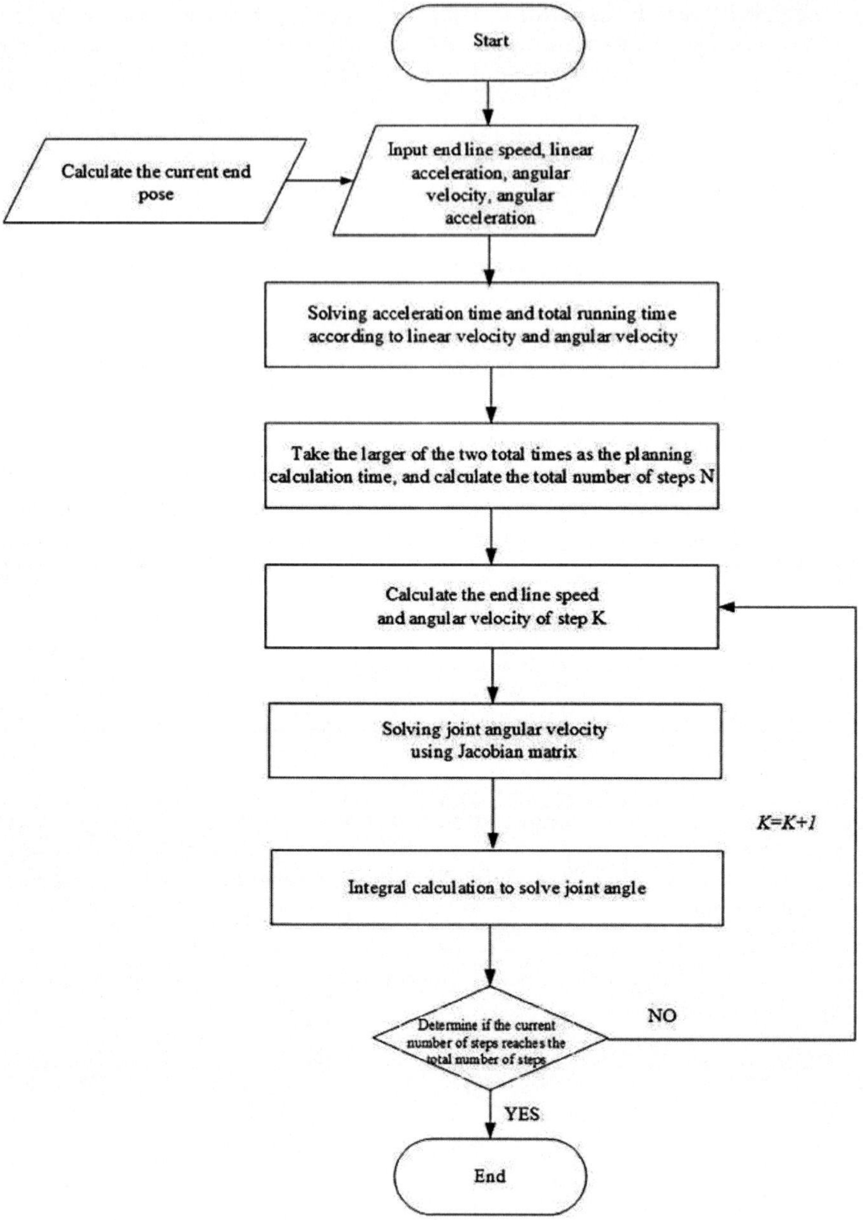

Fig. 3.4 Flowchart of end-effector trajectory planning in Cartesian space

A 7-DOF space robot is used as an example to introduce the method of avoiding the joint limits and singularities.

3.1.5.1 Avoiding the Joint Position Limits

For redundant robots or nonredundant robots in singular positions, the Jacobian matrix is not full-ranked and can only be evaluated by its generalized inverse. Because there is an infinite number of speed-level inverse kinematics, only the minimum norm solution is obtained generally. However, since each joint of the robot actually has a motion limit position, each joint has a limited range of motions. Therefore, in some cases, the minimum norm solution does not satisfy the requirement, that is, the solution obtained according to the robot motion control formula cannot be realized. When the end of the robot has not reached the desired spatial position, one or more joints would exceed the limit that the robot can actually reach. In order to complete the predetermined task, it is necessary to optimize this situation using the redundant joint of the robot and thus the resulting unique self-moving ability. In the optimized task plan, the joints of the robot can always be within the allowed range while maintaining the end effector's position and posture.

According to the inverse kinematics equation of the space robot, the relationship between the velocity of the robot end and the joint velocity can be described by the following kinematic equation:

$$\dot{\mathbf{x}} = \mathbf{J}(\mathbf{q})\dot{\mathbf{q}}, \tag{3.22}$$

where $\dot{\mathbf{x}} \in \mathbb{R}^m$, $\dot{\mathbf{q}} \in \mathbb{R}^n$, and $\mathbf{J} \in \mathbb{R}^{m\times n}$ are the end-effector speed, joint speed, and Jacobian matrix of the robot. For redundant robots, $m < n$, Therefore, the above inverse kinematics equation has a myriad of solutions, that is, for the defined $\dot{\mathbf{x}}$, there are a myriad of joint angular velocity vectors in an array satisfying the above equation.

The joint speed can be calculated by

$$\dot{\mathbf{q}} = J^{+}\dot{\mathbf{x}} + (I - J^{+}J)\alpha, \tag{3.23}$$

where J^{+} is the Moore–Penrose generalized inverse of the Jacobian matrix, and $J^{+}\dot{x}$ is the minimum norm solution of the equation. That is to say, the solution $\dot{q}$ of the equation has the least Euler norm. $(I - J^{+}J)\alpha \in N(J)$ is the homogeneous solution of the equation, and $N(J)$ is the zero space of the Jacobian matrix J. $\alpha \in R^n$ is an arbitrary vector. The homogeneous solution corresponds to the self-motion of the robot operation arm and does not cause any motion of the robot end.

The joint position limits can usually be avoided by the redundant motion using the local optimization of robot performance indices because the global optimization method needs to know the complete trajectory information of the robot in advance, and it is a complex algorithm with poor real-time performance. If continuous trajectory correction is needed according to the feedback information of the sensors, the global optimization is not very suitable. In contrast, although the local optimization method may not be able to achieve optimal joint trajectory due to the factuality of its calculation, it is still the most suitable optimization method for online control.

Among various local optimization methods, the Gradient Projection Method (GPM) and the Weighted Least-Norm (WLN) method are most commonly used.

a. Gradient projection

Let $\Phi(q)$ be the performance function related to the task target, and replace the free vector α in Eq. (3.23) with $k\nabla\Phi(q)$:

$$\dot{\mathbf{q}} = J^{+}\dot{\mathbf{x}} + k(I - J^{+}J)\nabla\Phi(q). \tag{3.24}$$

The coefficient in the above equation is a constant real number, and $\nabla\Phi(q)$ is the gradient vector of $\Phi(q)$ in the following form:

$$\nabla\Phi(q) = \left[\frac{\partial\Phi}{\partial q_1}, \frac{\partial\Phi}{\partial q_2}, \cdots, \frac{\partial\Phi}{\partial q_n}\right]^{\mathrm{T}}. \tag{3.25}$$

The performance index $\Phi(q)$ for avoiding the joint position limits is

$$\Phi(q) = \sum_{i=1}^{n} \frac{1}{4} \frac{(q_{\max}[i] - q_{\min}[i])^2}{(q_{\max}[i] - q[i])(q[i] - q_{\min}[i])}. \tag{3.26}$$

In the above equation, when the joint angle approaches the limit position, $\Phi(q)$ tends to infinity and its weight can be automatically given. Each item in the above equation corresponds to one joint. For each joint, if the joint angle is in the middle of the joint motion range, the value of the corresponding item will be 1. If it is close to the limit position of the joint motion range, the item corresponding to it tends to be tremendous.

b. WLN method

To limit unfavorable joint self-motion, the weighted norm defining the joint velocity vector should be

$$|\dot{q}|_W = \sqrt{\dot{q}^T W \dot{q}}, \tag{3.27}$$

where W is the weighting matrix, which is a positive symmetric matrix. In most cases, it is a diagonal array for the sake of simplicity.

By introducing the transformations:

$$J_W = JW^{-1/2}\ and\ \dot{q}_W = W^{1/2}\dot{q}, \tag{3.28}$$

Equation (3.22) can be rewritten as

$$\dot{\mathbf{x}} = \mathbf{J}\dot{\mathbf{q}} = \mathbf{J}W^{-1/2}W^{1/2}\dot{\mathbf{q}} = J_W\dot{q}_W. \tag{3.29}$$

The minimum norm solution of this equation is

$$\dot{q}_W = J_W^{+}\dot{\mathbf{x}},\ |\dot{q}|_W = \sqrt{\dot{q}_W^T\dot{q}_W}. \tag{3.30}$$

According to the definition of pseudoinverse, there is $J^{+} = J^T(JJ^T)^{-1}$. So

$$\begin{aligned}\dot{q} &= (W^{1/2})^{-1}\dot{q}_W \\ &= (W^{1/2})^{-1}J_W^{+}\dot{\mathbf{x}} \\ &= (W^{1/2})^{-1}J_W^T(J_W J_W^T)^{-1}\dot{\mathbf{x}} \\ &= W^{-1/2}(JW^{-1/2})^T(JW^{-1/2}(JW^{-1/2})^T)^{-1}\dot{\mathbf{x}} \\ &= W^{-1/2}W^{-1/2}J^T(JW^{-1/2}W^{-1/2}J^T)^{-1}\dot{\mathbf{x}} \\ &= W^{-1}J^T(JW^{-1}J^T)^{-1}\dot{\mathbf{x}}\end{aligned} \tag{3.31}$$

This is the weighted minimum norm solution of the equation, where the Jacobian matrix is full rank.

The weight matrix is usually treated as a diagonal matrix in the following form:

$$W = \begin{bmatrix} w_1 & 0 & 0 \cdots & 0 \\ 0 & w_2 & 0 \cdots & 0 \\ \vdots & \vdots & \vdots \ddots & \vdots \\ 0 & 0 & 0 \cdots & w_n \end{bmatrix}, \tag{3.32}$$

where w_i is the element in the diagonal matrix W and is defined as

$$w_i = 1 + \left|\frac{\partial \Phi(q)}{\partial q_i}\right|, \tag{3.33}$$

where $\frac{\partial \Phi(q)}{\partial q_i} = \frac{(q_{\max}[i] - q_{\min}[i])^2(2q[i] - q_{\max}[i] - q_{\min}[i])}{4(q_{\max}[i] - q[i])^2(q[i] - q_{\min}[i])^2}$.

As can be seen from above, when the joint angle i is in the middle of the joint range, the value of $\partial\Phi(q)/\partial q_i$ is zero. When the joint angle is at any limit position, the value of $\partial\Phi(q)/\partial q_i$ tends to infinity. Therefore, according to the definition of w_i, when the joint angle i is in the middle of the joint range, the value of w_i is 1. When the joint angle value is at any limit position, the value of w_i tends to infinity. Therefore, if a joint approaches the limit position, its weight will become larger and its speed will be correspondingly smaller. When the joint is very close to the limit position, its weight will tend to infinity and it will stop moving to ensure not exceeding the limit position.

According to the WLN method described above, the processing method remains unchanged whether the joint moves toward the limit position or away from the limit position. If the joint is moving away from the extreme position, even if the value of $\left|\partial \Phi(q) / \partial q_i\right|$ is very large, it is not necessary to eliminate this movement. In this case, allowing the joint to move freely will make the redundant nature of the robot useful for other purposes, such as obstacle avoidance. Considering this factor, it can be redefined as follows:

$$w_i = \begin{cases} 1 + \left|\frac{\partial \Phi(q)}{\partial q_i}\right|, & \text{if } \Delta\left|\frac{\partial \Phi(q)}{\partial q_i}\right| \geq 0 \\ 1, & \text{if } \Delta\left|\frac{\partial \Phi(q)}{\partial q_i}\right| < 0 \end{cases}. \tag{3.34}$$

As can be seen from the above equation, w_i is not a continuous function of the joint angle. When $\Delta\left|\partial \Phi(q) / \partial q_i\right|$ changes its sign, w_i may become discontinuous. Note that the value of $\Delta\left|\partial \Phi(q) / \partial q_i\right|$ increases as the joint moves toward the extreme position. When the joint speed is zero, the value of $\Delta\left|\partial \Phi(q) / \partial q_i\right|$ will also be zero. When the joint is moving away from the limit position, $\Delta\left|\partial \Phi(q) / \partial q_i\right|$ will be negative. We consider the following two possible situations:

(1) When the joint is in the middle of its motion range, $w_i = 0$ whether $\Delta \geq 0$ or $\Delta\left|\partial \Phi(q) / \partial q_i\right| < 0$, so there is no discontinuity in w_i;
(2) If the joint leaves from the middle of the range, when $\Delta\left|\partial \Phi(q) / \partial q_i\right|$ changes the sign, the value of w_i will change from 1 into a value greater than 1, or from a positive value greater than 1 into 1. Since the corresponding joint velocity at these points is zero, this change does not affect its continuity.

Therefore, defining w_i as a discontinuous function does not affect the continuity of joint speed.

If continuous joint acceleration is required in a given task, a function can be determined as the smooth transition between $\Delta\left|\partial \Phi(q) / \partial q_i\right| \geq 0$ and $\Delta\left|\partial \Phi(q) / \partial q_i\right| < 0$. Since the weight matrix is defined as a diagonal matrix, its inverse matrix is also a diagonal matrix but in the form of the reciprocal of its corresponding term. Therefore, the WLN method is less computationally intensive than the gradient projection method in which the homogeneous solution is also calculated.

3.1.5.2 Avoiding the Singularities

The singular point is the edge or singularity of the robot motion space, which corresponds to the case where the Jacobian matrix is not full rank. If a space manipulator is singular (that is, the end-motion space is reduced in dimension), the arm will lose its ability to move in some directions so that its dexterity will be reduced. In addition, if the singularity is not properly handled, it may cause the processor error or the joint trembling, thus resulting in the hardware and software failures. Therefore, in order to avoid singularity, the space manipulator must have better dexterity. The dexterity of the space manipulator can be reflected by the operability, which is defined as $w = \sqrt{\det(JJ^{\mathrm{T}})}$, where J is the Jacobian matrix of the space manipulator system.

In order to improve the dexterity of the robot, the gradient projection method is used for robot motion planning. In the optimization planning, the free vector α in Eq. (3.23) is replaced by $k_c \nabla w$. The general optimization algorithm can be expressed as

$$\dot{q} = J^{+}\dot{\mathbf{x}} + k_c \cdot (I - J^{+}J)\nabla w, \tag{3.35}$$

where $\dot{q}$ and $\dot{x}$ are the joint velocity and end velocity, respectively, $J^{+}\dot{x}$ is the minimum norm solution that maintains the motion trajectory of the robot end, and $k_c \cdot (I - J^{+}J)\nabla w$ is the homogeneous solution that does not change the robot-end motion, where k_c is the optimization term coefficient and ∇w is the w gradient.

Let $\mathbf{q} = (q_1, \ldots, q_n)$ be the joint variable. Then w is a function of $\mathbf{q}$, and the gradient of w to $\mathbf{q}$ can be expressed as

$$\nabla w(\mathbf{q}) = \frac{\partial w}{\partial \mathbf{q}} = \left(\frac{\partial w}{\partial q_1}, \ldots, \frac{\partial w}{\partial q_j}, \ldots, \frac{\partial w}{\partial q_n}\right)^{\mathrm{T}}. \tag{3.36}$$

Since 6×7 matrix J is a non-square matrix, given $Jp = JJ^{\mathrm{T}}$, the bias derivative of w to q_j can be expressed as

$$\frac{\partial w}{\partial q_j} = \frac{\partial\sqrt{\det(JJ^{\mathrm{T}})}}{\partial q_j} = \frac{\partial(\det(JJ^{T}))/\partial q_j}{2\sqrt{\det(JJ^{\mathrm{T}})}} = \frac{\partial(\det(Jp))/\partial q_j}{2w}. \tag{3.37}$$

Let $Jp_i (i = 1 \sim n)$ be the column i of the matrix Jp. According to the matrix determinant properties,

$$\frac{\partial \det[Jp]}{\partial q_j} = \sum_{j=1}^{n} \det[\{Jp_1, Jp_2, \ldots, \frac{\partial Jp_i}{\partial q_j}, \ldots, Jp_n\}], \tag{3.38}$$

where $\frac{\partial Jp_i}{\partial q_j} = \left(\frac{\partial Jp}{\partial q_j}\right)_{[[i]]}$, in which the subscript i represents the column i of $\frac{\partial Jp}{\partial q_j}$, and the $\frac{\partial Jp}{\partial q_j}$ can be expressed as

$$\frac{\partial Jp}{\partial q_j} = \frac{\partial(JJ^{\mathrm{T}})}{\partial q_j} = \frac{\partial J}{\partial q_j}J^{\mathrm{T}} + J\frac{\partial J^{\mathrm{T}}}{\partial q_j} = \frac{\partial J}{\partial q_j}J^{\mathrm{T}} + J\left(\frac{\partial J}{\partial q_j}\right)^{\mathrm{T}} \quad (j = 1 \sim n). \tag{3.39}$$

Through the above optimization of the operability of the space manipulator, the operable value can be greatly improved, that is, it is away from the singular position of the space manipulator, thereby realizing the singularity movement of the space manipulator.

3.2 Motion Planning of Space Robot

Motion planning of mobile robot mainly refers to planning the robot motion path and controlling the robot to track the path. Path planning is to find an optimal motion path from the starting point to the target point in the working environment with obstacles. Trajectory tracking is to control the robot to track the planned path.

Motion planning of a mobile robot mainly involves the following two aspects: (1) Applying an algorithm to find an optimal or near-optimal collision-free path from the starting point to the target point; (2) Controlling the robot to track the planned path.

According to the understanding of environmental information, path planning can be divided into two planning methods: the global path planning based on a priori complete information of the environment, and the local path planning based on the real-time environmental information collected by sensors. Among them, the global path planning considers medium resolution and a distance of tens of meters, and focuses on the path search. In the local path planning, the obstacle avoidance is emphasized under the consideration of high resolution and a distance of several meters [1].

At present, the mobile robots applicable for aerospace engineering are the wheeled mobile robots. This section mainly focuses on motion planning of wheeled mobile robots. For motion planning of leg robots, readers can refer to the related monographs.

3.2.1 Global Path Planning

Global path planning is also called static path planning, that is, offline path planning is performed because environmental information is completely known and the obstacles are stationary. Most global path planning methods are based on the idea of geometric model search. Firstly, the geometric model of the roaming environment space is constructed, and then some kind of graph search algorithm is used to obtain the optimal path. The main methods are visibility graph method, grid method, Voronoi diagram method, etc. [2].

3.2.1.1 Visibility Graph

In the visibility graph, the robot is regarded as a point, the target point and vertices of the polygon obstacle are combined and connected, and the lines are not intersected with the obstacles. The resulting graph is shown in Fig. 3.5, where the black areas represent the obstacles. Since the intersections of any two lines are visible, all the paths from the starting point along the lines to the target point are the collision-free paths of the moving object. The problem of finding the optimal path is translated into the problem on finding the shortest distance from the starting point to the target

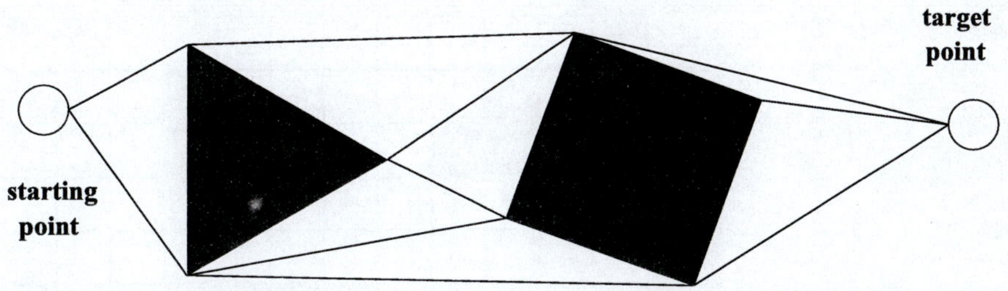

Fig. 3.5 Visibility graph

point through these visible lines. The optimization algorithm can remove unnecessary connections to shorten the view time and the search time. The view method can find the shortest path.

The visibility graph is widely used in robot path planning because this method is simple to implement, especially in a continuous or discrete sparse environment, where obstacles can be easily described by polygons so that the shortest distance and no-touch path can be quickly obtained. However, visibility graph also has certain limitations. First, the polygon size and the number of nodes defining the obstacle increase with the number of obstacles, resulting in a sharp increase in the feasible paths, which seriously affects the search efficiency. Second, the robot is assumed to be a point, regardless of the size of the robot itself. Therefore, there is a risk that the robot passing through the obstacle point is too close to the obstacle, or even contact with the obstacle.

3.2.1.2 Grid Method

In general, the grid method is based on the characteristic that the travel path of a mobile robot is built on the two-dimensional plane. The working plane of the mobile robot is divided into a plurality of rectangular grids of the same size, and the barrier-free grid and the obstacle grid are distinguished by different set values. By connecting feasible grids, the path of the mobile robot can be formed.

The simplified grid method is shown in Fig. 3.6. Assume that the grids only have binary information. Fill the grids where there are obstacles. The value of the filled grid is marked as 1. If there is no obstacle, the grid is not filled. The value of the unfilled grid is marked as 0. By increasing the grid density, any irregular objects can be rasterized. Although the method of marking the obstacles is less efficient, it is easy for the computer to calculate. As a very effective environment representation method, the grid method is more and more commonly used. The probe car Nomad developed by CMU University uses the grid map method to construct the environment map. A suitable map consists of several grids. Each grid contains two data: one is the suitability data indicating the adaptability of the car body to that place; the other is the credibility data indicating the confidence of the suitability. The Nomad generates

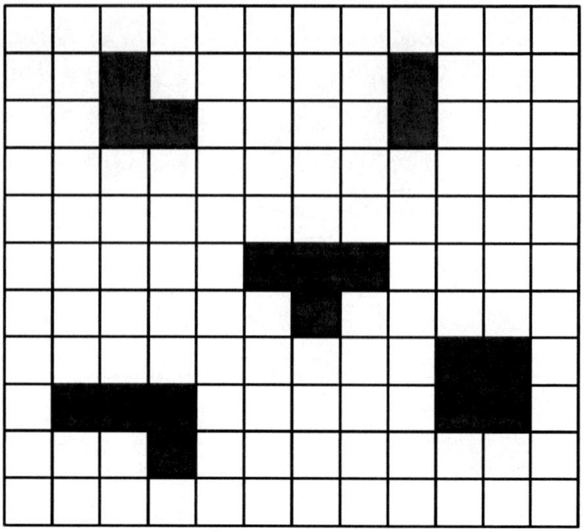

Fig. 3.6 Grid method

an appropriate map based on its measurement information from a stereo camera and a laser system. The appropriate map used in the obstacle avoidance module is the result obtained by weighting the calculation.

The grid method is a relatively mature environment modeling method, where the obstacle grids and non-obstacle grids have different recognition values. The environmental information is recorded in the unit rasters. The environment is quantized into a grid with a certain resolution, which can better handle obstacle boundaries and avoid complicated operations. However, this method also has certain defects. For example, the improper selection of grid granularity will affect the accuracy of the solution. When the search space is large, the storage space required by the algorithm will be relatively large. The searched path is a polyline, which will increase the time to reach the destination.

In order to solve the above shortcomings of the grid method, the concept of multi-resolution grid has been proposed. For example, in the frame quadtree method, a coarse grid is used to represent a large area of free space or obstacles that do not affect the planning, while a refined grid is used to describe the boundary area. In the variable resolution grid method, high resolution is used in the vicinity of the mobile robot and low resolution used in the distance. This greatly reduces the amount of calculation and maintains the high precision of the planning path of the mobile robot near the end, thus becoming an ideal online planning method with limited computing resources.

After the modeling of robot operation environment is completed based on the grid method, the heuristic A* algorithm can be used to complete the path search. The A* algorithm is to calculate the motion cost between nodes given in the raster graph and to search for the optimal path according to the valuation function. The A* algorithm is by far the fastest algorithm for calculating the shortest path. But it is just a "better" algorithm, that is, it can only find a better solution. Owing to its high efficiency, it has been widely used in real-time systems, artificial intelligence, and other areas.

The A* algorithm is the combination of heuristics and formal methods. It uses a Heuristic Function to estimate the distance (with weight) from the current point P of the mobile robot to its end point, thereby determining the search direction. When this path fails, it will try other paths.

The valuation function of A* algorithm is expressed as follows:

$$F(n) = Y(n) + P(n), \tag{3.40}$$

where $F(n)$ is the nth node's valuation function, $Y(n)$ is the actual cost incurred from the initial node to the nth node in the state space, and $P(n)$ is a heuristic function used to estimate the cost of the best path from the nth node to the target node.

In this search method, first, the cost of all nodes, the start point and target point is, respectively, is defined as $Y(n) = \infty$, n_S and n_E to establish a to-be-searched set and a search set. The to-be-searched set is used to store the nodes to be expanded, while the search set is used to store the expanded nodes. After defining $Y(n_S) = 0$ and putting it in the search set, each node placed in the to-be-searched set is sorted by value $F(n)$. The node with the smallest $F(n)$ is placed at the end of the queue. After that, the nth node at the end of the search set is taken out. If the cost of its neighbor node $Y\left(n^+\right)$ is greater than the cost of the current node $Y(n)$ plus the cost of the edges between the two nodes, then $Y\left(n^+\right)$ shall be updated, the nth node shall be put in the search set, and the above process shall be continued until the tail node is the target node n_E or the target node is not found. If the path is found, the remaining nodes in the search set shall be the points on the selected path, and shall be taken out and sequentially connected to obtain the optimal path to be searched.

3.2.1.3 Voronoi Diagram Method

The Voronoi diagram is a transformation of the visibility graph. The idea of the Voronoi diagram is exactly the opposite of the visibility graph. The ultimate goal of the visibility graph is to find the shortest distance between the starting point and the target point. The mobile robot sometimes moves close to the obstacle. Although the path is short, there is a risk of contact with the obstacle. The Voronoi diagram ensures that the mobile robot is as far away as possible from the obstacle, thus increasing the robot's moving distance. However, the robot can move safely according to this method. Even if a position error occurs, the mobile robot will not encounter an obstacle. Therefore, this method is extremely suitable for a robot moving in space.

As shown in Fig. 3.7, the black areas represent the obstacles, and the peripheral curve represents the feasible path. The Voronoi diagram is a full road map method based on the view method. All the boundaries of the obstacle are considered equidistant Voronoi edges, and then the points on each side are connected to each other. The intersecting points are called Voronoi intersections, and the obstacles cannot be crossed.

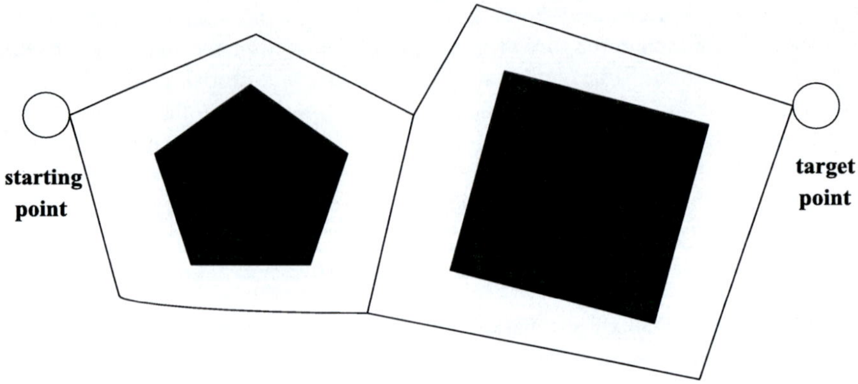

Fig. 3.7 Voronoi diagram method

3.2.2 Local Path Planning

Local path planning, also known as dynamics path planning, is the online planning method in which the environment is partially or completely unknown, or there are both unknown static obstacles and moving obstacles. Local planning is based on the limited range of environmental information fed back by various sensors (visual, laser, ultrasonic, sonar, tactile, and other sensors) installed on the mobile robot to locally plan the path within the environment. Due to the lack of global environmental information, local path planning cannot obtain the global optimal path, and can fall into the local minimum trap easily. Because of its low computational complexity, it is particularly suitable for the situations where the environment model changes frequently [3].

The local path planning methods differ from each other in many aspects, and they can be classified in different ways. By algorithm hierarchy, they can be classified into artificial potential field method, fuzzy algorithm, genetic algorithm, neural network algorithm, and other intelligent algorithms. According to the task hierarchy, they can be classified into behavior decomposition algorithm and terrain evaluation algorithm.

3.2.2.1 Artificial Potential Field Method

The artificial potential field method (Fig. 3.8) virtualizes the robot operation environment into an artificial potential field. The movement of the mobile robot is affected by the repulsive force generated by the obstacle and the gravitational force generated by the target point. The magnitude of the force field varies with the distance between the robot and the obstacle. Finally, the mobile robot relies on the resulting force in all directions to guide its advancement.

This algorithm is relatively simple and efficient in calculation, and easy to control in real time. Its disadvantage is that it is easy to stagnate or deadlock at the local

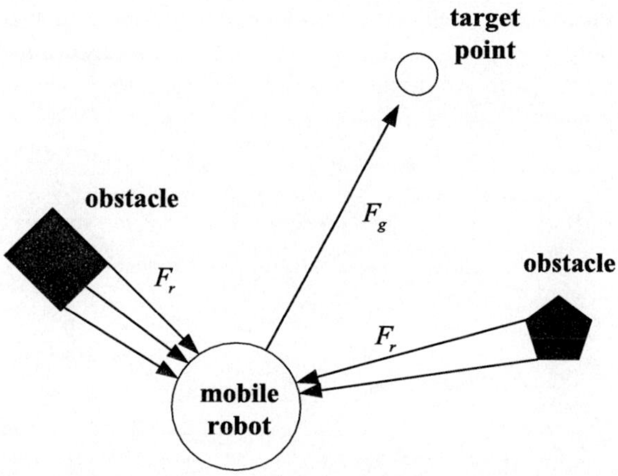

Fig. 3.8 Artificial potential field method

minimum value and cannot pass two adjacent obstacles. In addition, it is easy to oscillate near an obstacle or in a narrow passage. When the target point and the obstacle are adjacent, the target point cannot be reached.

3.2.2.2 Fuzzy/Genetic Algorithm

Although the traditional artificial potential field method is simple, it has the problems of local minimum and low robot intelligence level. With the rapid development of intelligent control algorithms, a new idea has been developed for the intelligent control of mobile robots. Intelligent methods such as fuzzy logic, neural network, and genetic algorithm can simulate human experience, approximate nonlinear self-organizing self-learning functions, and provide a certain fault tolerance. Mobile robots will be more flexible and intelligent in a dynamics environment by these methods of path planning.

a. Fuzzy algorithm

The idea of fuzzy algorithm is to install a large number of sensors for obtaining the information on the environment and robot self-pose. Then, according to the information on the sensor's self-return posture and the relative positional relationship with the obstacle, the feasible motion path of the robot can be planned online through the fuzzy rule reasoning.

Fuzzy algorithm is mainly based on the observation of the working process of the vehicle driver. The driver's collision avoidance action is achieved not by accurately calculating the environmental information, but by carrying out the local path planning according to the fuzzy environmental information. The advantage of this method is that it overcomes the local-minimum problem that the potential field method tends to be trapped in, and has a strong edge when dealing with the planning problem in the unknown environment. It is very effective to solve complex problems with quantitative methods, or to deal with the situations for which the outside world can only provide qualitative information. The fuzzy control algorithm can be represented by the program flowchart shown in Fig. 3.9.

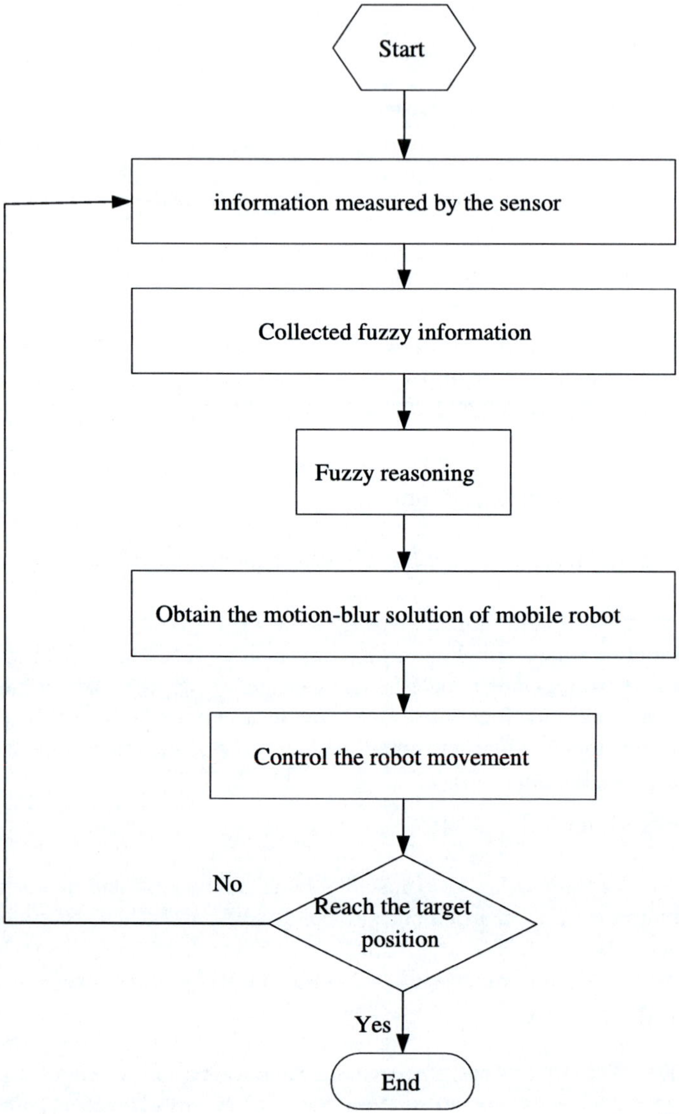

Fig. 3.9 Fuzzy logic control flow

b. Genetic algorithm

Genetic algorithm is the most commonly used algorithm in robot path planning. Its basic idea is to imitate the "survival of the fittest" process in the natural evolution of the biological world. Among several feasible solutions, the optimal solution is obtained through continuous evolution. Many different paths are grouped into a population (a set of possible solutions), each of which is regarded as an individual (each possible solution) and expressed as a series of path points by coding. Then multiple paths are randomly selected for hybridization and mutation and they evolve into the optimal next generation. After several generations of evolution, the path of fitness to a certain standard is finally selected. The flowchart of robot path planning based on genetic algorithm is shown in Fig. 3.10.

3.2.2.3 Behavior-Based Decomposition Algorithm

Traditional function-based path planning method is a deliberate top-down control system structure that controls the robot motion through the "perception-modeling-planning-action" approach, in which the action is the result of a series of stages. The specific control block diagram is shown in Fig. 3.11. The advantage of this method is that the idea is clear and easy to understand, and the implementation is simple. However, its problem lies in the difficulty in establishing an accurate environment map and in adapting to the dynamics path planning task in an unknown environment.

Behavior based path planning, also known as behavior decomposition method, is a common method to solve the local path planning for mobile robots. This method uses the bottom-up architecture similar to animal evolution to decompose the robot task into basic behavioral units. Each unit has its own sensor and actuator, which are coupled together. These basic units work together. Mobile robots use different sub-behaviors to deal with different situations in the encountered environment. By properly defining the sub-behaviors and setting their starting and ending conditions, robots can behave in different ways when encountering different situations in the environment. A better coping strategy can help complete the path planning tasks as quickly as possible and reduce the planning complexity. The advantage of this method is that each behavior function is simple, and that good operation results can be obtained through simple sensors and fast response processes.

3.2.2.4 Terrain Evaluation Algorithm

The path planning algorithm based on terrain evaluation generally determines the most suitable path by the statistical analysis of the local terrain environment around the mobile robot. This algorithm has strong applicability to space robots. A typical representative of this path planning algorithm is the Morphin algorithm.

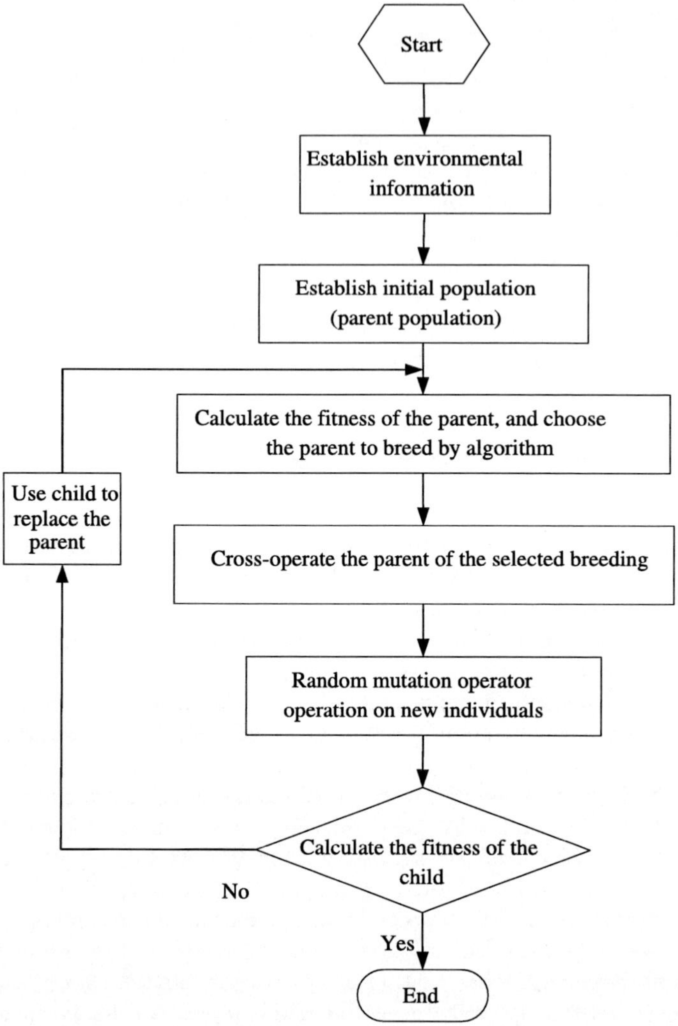

Fig. 3.10 Genetic algorithm control flow

The Morphin algorithm is a grid-based analysis method of local terrain traversability. According to the information measured by robot sensors, this algorithm searches for multiple barrier-free forward paths. Then, according to the current robot state (including position, attitude, and others) and the suitability and certainty of forward paths, it calculates the traversability of each path and finally finds the path with the highest traversability. This algorithm has been tested and verified on the ground mobile robots such as RATLER and Hyperion.

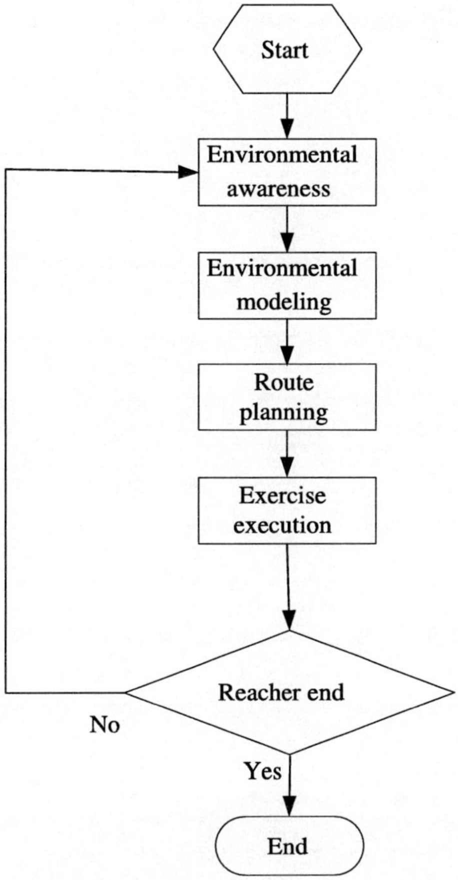

Fig. 3.11 Block diagram based on functional control system

Morphin algorithm can deal with the uncertainty in environmental modeling (such as the error in the 3D topographic map information measured by a navigation camera and the measurement error of laser range finder) and robot positioning error, so it is suitable for densely distributed obstacles. In addition, its search time is short, its real-time response ability is strong, and the global planning algorithm can be combined with it to jointly determine the underlying motion behavior of the mobile robot. The disadvantage of this algorithm is that the path of the robot is less smooth and more devious because the smoothness of the path and the optimality of the solution are restricted in order to ensure the security of the sought path.

Due to the accuracy and safety of planning, the terrain-based assessment method has been applied in many fields, especially in space exploration. In December 1996, the United States launched the "Mars Pathfinder" detector and conducted a field trip on Mars using the successfully carried "Sojourner" rover. The path planning method used by Sojourner was based on the combination of terrain evaluation and

D* algorithm. In this way, Sojourner walked freely and cautiously on the surface of Mars, independently identified the obstacles ahead, realized real-time re-planning, and made decision on the later action by truly "thinking carefully before acting".

References

1. B. Dai, X. Xiao, Z. Cai , Current status and future development of mobile robot path planning technology. Control Eng. China **12**(3), 198–202 (2005)
2. R. Siegwart, I.R. Nourbakhsh, D. Scaramuzza, *Introduce to Autonomous Mobile Robots*, vol. 2 (MIT Press, Cambridge, 2011)
3. B. Siciliano, L. Sciavicco, L. Villani et al., *Robotics: Modelling, Planning and Control* (Springer, Berlin, 2011)

Chapter 4
Motion Control of Space Robots

After the motion planning for space robots, the planned trajectory tracking will be implemented by joint motion. Therefore, the motion control of space robots ultimately falls into joint motion control. This chapter takes the single joint control as an example to illustrate the motion control method of space robots. The objective of motion control is to achieve the tracking of the desired joint states (angle, speed) with two main methods: three-loop servo motion control and dynamic model-based motion control.

The three-loop servo motion control method is to establish a three-loop control strategy of current, speed, and position based on the models of reducer, motor, etc. And then the output currents are calculated separately by the servo control algorithm according to the requirements of the robot for the desired state of each joint. In the dynamic model-based control method, the joint torque (current) is calculated by using the multi-body system dynamic model and a control algorithm based on nonlinear control theory. The advantages of the three-loop servo motion control method are that it has a simple and reliable structure and is easy to realize, and has been widely used in the field of robotics. In the dynamic model-based motion control method, the control torques of all joints are uniformly calculated according to the desired state and the actual state of robots, which is better than the three-loop servo motion control method in terms of bandwidth, stability and accuracy, and its robustness, adaptability, and intelligent control features can enhance the adaptability of the robot motion control.

4.1 Three-Loop Servo Motion Control

A complete servo control system includes a controller, a driver, drive components, transmission components, and sensors. As a core part of the whole servo control system, the controller generates control signals through the servo control algorithm; the driver acts as a power amplifier to convert the control signals into power drive signals; the drive components convert the electrical energy into mechanical energy

Y. Wang, *Space Robotics*, Space Science and Technologies,
https://doi.org/10.1007/978-981-15-4902-1_4

outputs; the transmission components output the torque and speed required by the servo control system in collaboration with the drive components according to the design requirements; and the sensors can detect the real-time position, force, and other information, control the feedback of the joints, and monitor the operating state of the system.

4.1.1 *Motor Drive and Joint Servo Control*

The single joint servo control system controls the current, speed, and position of a joint or a motor. The drive motors commonly used in a space robot servo control system include the Permanent Magnet Synchronous Motor (PMSM) and the Blushless DC Motor (BLDCM). This section will go deep into the driving methods of the two motors, and illustrate the joint servo control algorithm.

4.1.1.1 Space Vector Driving Method of PMSM

With the development of electronic power technology, PMSM has been rapidly popularized due to its smoother torque, and widely used in the field of space robots. The PMSM is developed from an electric excitation three-phase synchronous motor, which replaces the electric excitation system with a permanent magnet, thereby eliminating the excitation windings, the collector rings, and the brushes.

The structure of a two-pole surface-mounted PMSM is shown in Fig. 4.1, where AX, BY, and CZ are the three-phase windings of the motor. The phase voltages and currents are u_a, u_b, u_c, and i_a, i_b, i_c, respectively, and their magnetic field directions specify the motor stator coordinate (*a*, *b*, *c*). The $\boldsymbol{\alpha}$ axis is the space reference coordinate of the *abc* axis system, the three-phase windings are represented on the axis as coils. According to the principle of the motor theory, assuming that the positive direction of the induced Electro-Motive Force (EMF) in the phase winding is opposite to the positive direction of the current, the counterclockwise direction is taken as the positive direction of the rotation speed and the electromagnetic torque. The resultant stator winding voltage and current are u_s and i_s, respectively, and the angle between the generated magnetic field and α axis is θ_s. The rotor is a permanent magnet, and its flux linkage is expressed as ψ_f, its equivalent current is i_f. The angle between the direction of the rotor flux linkage and α axis is θ_r, and the angle in the direction of the resultant magnetic field of the stator is β. The direction of the rotor flux linkage is defined as the *D*-axis (Direct axis), and leading *D*-axis by 90° is represented as the *Q*-axis (Quadrature axis).

According to the motor theory, the electromagnetic torque of the PMSM can be expressed as

$$t_e = p_0 \psi_f i_s \sin\beta = p_0 \psi_f \times i_s, \tag{4.1}$$

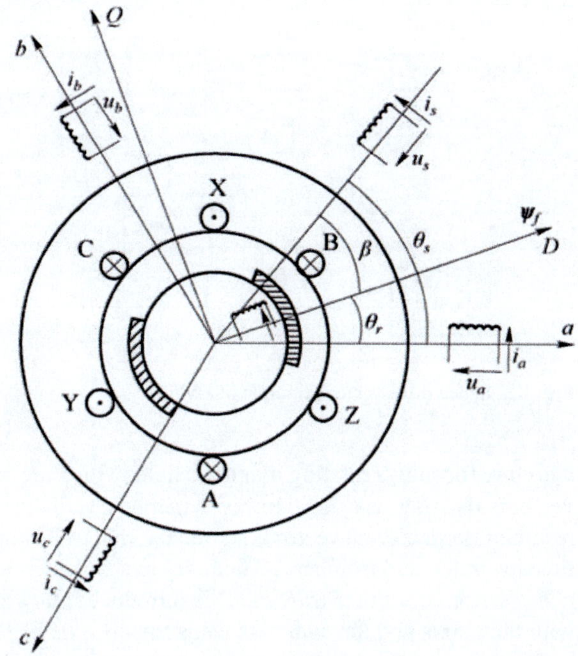

Fig. 4.1 Structure of a two-pole surface-mounted PMSM

where the flux linkage generated by the permanent magnet excitation magnetic field ψ_f is a constant value regardless of temperature, p_0 is the number of the pole pairs, i_s is the stator current vector, and β is the angle between ψ_f and i_s. It follows from Eq. (4.1) that the electromagnetic torque is determined by the component of the stator current on the Q-axis. If β is equal to 90°, then i_D is equal to 0, and ψ_f is orthogonal to i_s, and the stator current i_s is equal to the torque current i_Q. In this case, although the rotor rotates at the electrical angular velocity $\boldsymbol{\omega_r}$, ψ_f and i_s are always relatively stationary in the DQ-axis system.

(1) Space vector control of PMSM

In space vector control, also known as magnetic field orientation control, the stator current or voltage is controlled by detecting the position of the motor rotor, and the magnetic field generated by the stator winding is regulated to lead the rotor by 90°. According to Eq. (4.1), with space vector control the motor electromagnetic torque $\boldsymbol{T_e}$ is proportional to $\boldsymbol{i_q}$. Therefore, for a PMSM, the torque of the PMSM can be controlled by a control method similar to that of a DC motor to obtain a performance comparable to a DC motor.

In order to control the torque through the current like a DC motor, it is envisaged to establish a moving coordinate (D, Q) that rotates at the electrical angular frequency in the control of the PMSM. From the view of the stationary coordinate (a, b, c), the components of the resultant stator current vector are constantly changing with time, which causes the resultant vector to rotate in space at the electrical angular frequency,

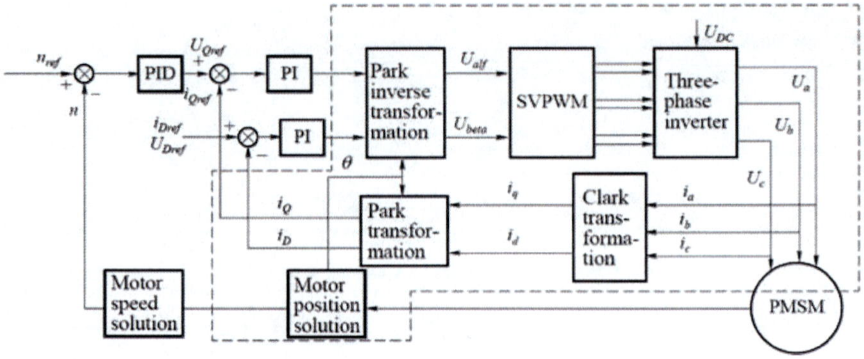

Fig. 4.2 Space drive vector principle of PMSM

therefore forming a rotating magnetic field, which means, the resultant stator current vector is also time-varying. However, in the view of the moving coordinate (D, Q), the resultant stator current vector is stationary, that is, it changes from a time variable to a time invariant, and from an AC quantity to a DC quantity. For the calculation, a two-phase stator coordinate (*alf*, *beta*) is introduced, in which the *alf* axis is consistent with the a-axis and the *beta* axis leads the *alf* axis by 90°.

By this way, the resultant stator current vector is transformed from a stationary coordinate to a rotating coordinate by transformation, and the value of the resultant stator current required for the torque control is calculated in this rotating coordinate, then this current value is inversely transformed into the stationary coordinate, and the virtual resultant current is converted into the actual winding current, thereby realizing the control of the motor torque.

Figure 4.2 shows the space vector drive principle of PMSM. The motor driving part is shown in the dotted line frame, which includes various modules, such as a motor position solution, a Park inverse transformation, a Space Vector Pulse Width Modulation (SVPWM), a Park transformation, and a Clark transformation.

The space vector drives of PMSM are shown inside the dotted frame in Fig. 4.3. The flow is as follows:

(1) Sampling and obtaining the current i_a, i_b of the current two-phase windings of the motor stator.

Generally, the current of the two-phase windings of the motor is collected by A/D. The signal interference existing in the phase current sampling can be processed by a filter, usually using an average filter or a first-order filter.

(2) Motor position acquisition.

The acquisition of the rotor position is essential for space vector control, which requires the help of a position sensor. In the field of space robots, because of the reliability requirements and mechanical characteristics, a resolver is usually used as

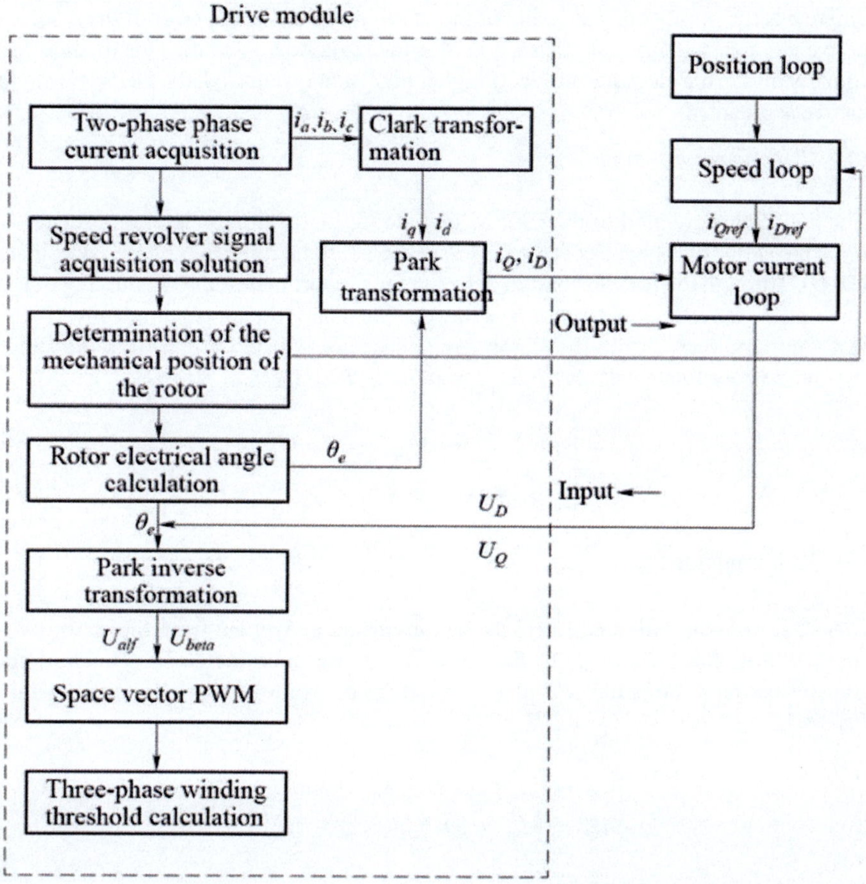

Fig. 4.3 PMSM drive flow

the position sensor, and the current angle of the motor θ_m is calculated by the circuit. This same angle can also be used as the input of motor speed calculation. The resolution of the resolver solution should be consistent with the resolver accuracy.

(3) Calculation of the motor electrical angle θ_e.

Since the motor electrical angle is used in the Park transformation and the inverse transformation, the mechanical angle should be converted into an electrical angle, which requires the calculation of the electrical angle of the motor rotor θ_e according to the mechanical position of the rotor. When a resolver is installed, the resolver zero is deviated from the motor rotor zero and corrected when calculating the electrical angle. The motor electrical angle is obtained according to Eq. (4.2):

$$\theta_e = (p \times (\theta_m - \theta_{\text{err}}))\%2\pi, \tag{4.2}$$

where p is the number of pole pairs of the motor rotor, θ_{err} is the error of the resolver zero compared with the rotor zero, and % is the remainder symbol. This method can roughly divide the electrical angle. If higher precision is required, the electrical angle can be segmented.

(4) Clark transformation.

The coordinates related to the PMSM include the stator three-phase coordinate (*a*, *b*, *c*), the stator two-phase coordinate (*alf*, *beta*), and the rotor two-phase coordinate (*D*, *Q*). Through the transformation between these coordinates, the decoupling of the three-phase current of the motor is achieved. The Clark transformation converts the stator current vector from the stator three-phase coordinate (*a*, *b*, *c*) into the stator two-phase coordinate (*alf*, *beta*), i.e., from i_a, i_b, i_c to i_{alf}, i_{beta}, as in Eq. (4.3):

$$\begin{cases} i_{\text{alf}} = i_a \\ i_{\text{beta}} = (i_a + 2 \times i_b) \Big/ \sqrt{3} \end{cases}. \tag{4.3}$$

(5) Park transformation.

The Park transformation converts the stator current or voltage from the stator two-phase coordinate (*alf*, *beta*) to the rotor two-phase coordinate (*D*, *Q*). The Park transformation obtains the two-phase current of the rotor i_D, i_Q. It is calculated by Eq. (4.4):

$$\begin{cases} i_D = i_{\text{alf}} \times \cos\theta_e + i_{\text{beta}} \times \sin\theta_e \\ i_Q = -i_{\text{alf}} \times \sin\theta_e + i_{\text{beta}} \times \cos\theta_e \end{cases}. \tag{4.4}$$

(6) Park inverse transformation.

The Park inverse transformation converts the stator current or voltage from the rotor two-phase coordinate (*D*, *Q*) to the stator two-phase coordinate (*alf*, *beta*), that is, the transformation from U_D, U_Q to U_{alf}, U_{beta}, as in Eq. (4.5):

$$\begin{cases} U_{\text{alf}} = U_D \times \cos\theta_e - U_Q \times \sin\theta_e \\ U_{\text{beta}} = U_D \times \sin\theta_e + U_Q \times \cos\theta_e \end{cases}. \tag{4.5}$$

(7) SVPWM.

The basic principle of the space vector drive is to obtain the resultant vector of the three-phase winding voltage vector of the stator through the switching state of the three-phase bridge circuit, and obtain the spatially approximate rotating magnetic field by controlling the ON-OFF time of two adjacent basic space vectors. The ON-time of each phase of the three-phase bridge circuit, i.e., the duty cycle of the PWM

Table 4.1 Switching time of different sectors

Sector	1	2	3	4	5	6
T_1	Z	Y	$-Z$	$-X$	X	$-Y$
T_2	Y	$-X$	X	Z	$-Y$	$-Z$

wave, is calculated and converted into a vector switching time at which triangular wave modulation can be performed. The calculation is as follows:

① Determine the sector in which the voltage vector is located.

Let: $U_{r1} = U_{\text{beta}}$, $U_{r2} = U_{\text{alf}} - \frac{1}{\sqrt{3}} U_{\text{beta}}$, and $U_{r3} = -U_{\text{alf}} - \frac{1}{\sqrt{3}} U_{\text{beta}}$,
If $U_{r1} > 0$, then $A = 1$, otherwise $A = 0$;
If $U_{r2} > 0$, then $B = 1$, otherwise $B = 0$;
If $U_{r3} > 0$, then $C = 1$, otherwise $C = 0$.
Let $N = A + 2B + 4C$.

② Calculate X, Y, Z and T_1, T_2.

Let: $X = \sqrt{3} U_{\text{beta}} \times T_{\text{pwm}} / U_{\text{dc}}$, $Y = \left(\sqrt{3} U_{\text{beta}}/2 + 3U_{\text{alf}}/2\right) \times T_{\text{pwm}} / U_{\text{dc}}$, and $Z = \left(\sqrt{3} U_{\text{beta}}/2 - 3U_{\text{alf}}/2\right) \times T_{\text{pwm}} / U_{\text{dc}}$,
where U_{dc} is the drive voltage and T_{pwm} is the PWM control period.

Obtain the sector number N according to step ①, set the method of obtaining intermediate value T_1, T_2. See Table 4.1.

Determine if they are saturated after assigning the values to T_1, T_2:
If $T_1 + T_2 > T_{\text{pwm}}$, then $T_1 = T_1 \times T_{\text{pwm}} / (T_1 + T_2)$, $T_2 = T_2 \times T_{\text{pwm}} / (T_1 + T_2)$.

③ Calculate the vector switching points T_{cm1}, T_{cm2}, and T_{cm3}.

Define: $t_{\text{Aon}} = \frac{T_{\text{pwm}} - T_1 - T_2}{2}$, $t_{\text{Bon}} = t_{\text{Aon}} + T_1$, and $t_{\text{Con}} = t_{\text{Bon}} + T_2$.

According to different sectors, the values of vector switching points T_{cm1}, T_{cm2}, and T_{cm3} of the stator three-phase winding are assigned according to Table 4.2.

④ Calculate the three-phase winding threshold output PWM.

The ON/OFF state of the three-phase bridge circuit is controlled by the PWM duty cycle. After the vector switching point of each phase winding is calculated, it should be converted into the actual PWM duty cycle to drive the motor.

Table 4.2 Vector switching points for different sectors

Sector	1	2	3	4	5	6
T_{cm1}	t_{Bon}	t_{Aon}	t_{Aon}	t_{Con}	t_{Con}	t_{Bon}
T_{cm2}	t_{Aon}	t_{Con}	t_{Bon}	t_{Bon}	t_{Aon}	t_{Con}
T_{cm3}	t_{Con}	t_{Bon}	t_{Con}	t_{Aon}	t_{Bon}	t_{Aon}

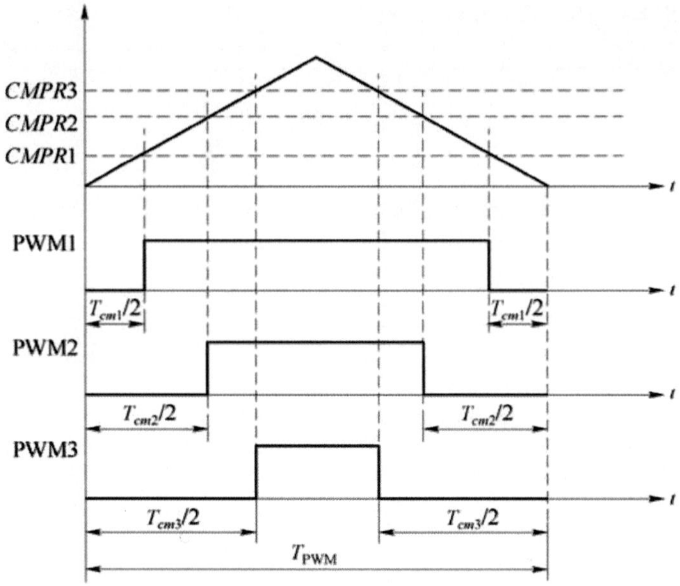

Fig. 4.4 PWM waves generated by triangle wave modulation

The PWM wave adopts a symmetrical triangular wave modulation method, as shown in Fig. 4.4. After the three-phase winding vector switching points T_{cm1}, T_{cm2}, and T_{cm3} are obtained, they can be converted into a threshold using a triangular wave modulation method. When the value on the counter matches the threshold, level switching is achieved.

When using the symmetrical triangular wave modulation method, the switching point thresholds can be obtained by

$$\mathrm{CMPR1} = \frac{\mathrm{FREQ} \times T_{\mathrm{cm1}}}{2}$$

$$\mathrm{CMPR2} = \frac{\mathrm{FREQ} \times T_{\mathrm{cm2}}}{2}$$

$$\mathrm{CMPR3} = \frac{\mathrm{FREQ} \times T_{\mathrm{cm3}}}{2},$$

where FREQ is the timing frequency of the modulation wave.

4.1.1.2 Drive Control of BLDCM

(1) Principle of BLDCM

BLDCM has the advantages of a simple and firm structure without brush, small size, lightweight, high efficiency, large starting torque, good overload capability, and high-speed operation performance. The BLDCM drive is mainly composed of a DC power supply, an inverter, a protection circuit, a motor body, a position sensor, current acquisition, and its peripheral circuits (controller), as shown in Fig. 4.5. The position sensor detects the rotor position signal of the motor, and then the controller performs logic processing according to the preset control command and the rotor position signal to generate a corresponding switch signal. The switch signal triggers the power switch tube of the inverter in a certain order, and turns on the three-phase windings of the stator according to the logical relationships, allowing the motor to produce a continuous torque output.

(2) Definition of the step

BLDCM generally uses Hall sensors for rotor position detection. There are 6 commutation phases in one electrical cycle, the corresponding phase is judged by the output signal of the Hall sensor. Step is the change of phase, and one step equals one phase change. The commonly used BLDCM phase is 60°.

(3) Determination of the rotor position

BLDCM uses a Hall sensor to determine the position of the rotor. Three Hall position sensors are mounted on the motor with an interval of 120°, and the polar arc width of the sensor permanent magnet is 180°. When the rotor rotates, the Hall components alternately output three rectangular wave signals with a width of 180° and a phase difference of 120°. The three-way Hall sensor signal can obtain six types of Hall pulse edge signals in one electrical cycle, which reflects the change of the rotor position, thus the rotor position is determined according to this Hall signal.

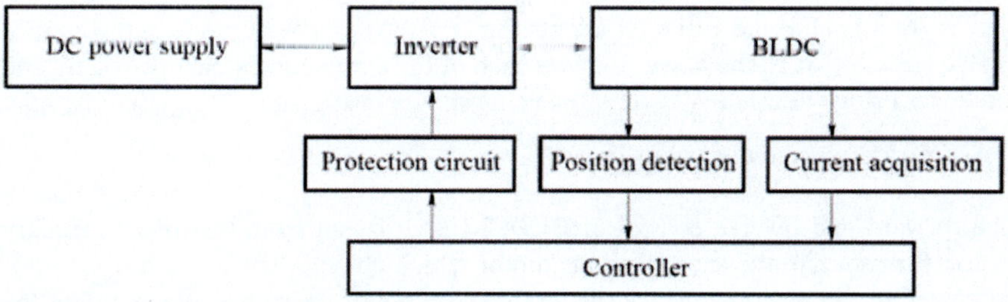

Fig. 4.5 Schematic of a BLDCM structure

(4) Calculation of motor angular velocity

The actual speed of the motor can be calculated by the time difference between the edges of two Hall signals (T method), or by measuring the number of pulses per unit time (M method). Here, we take the T method as an example. For a three-phase BLDCM with one pair pole, there will be 6 commutations per 360° angle rotation, and the time interval between two commutations is t, which can be obtained by measuring the counting difference between two commutations, then the angular velocity of the motor can be expressed as

$$\omega = \frac{2\pi}{6t} = \frac{\pi}{3t}.$$

Therefore, the calculation of the angular velocity can be simplified to the division, and only the time interval t between the two commutation signals is required to be measured to obtain the desired motor velocity.

(5) Three-phase six-state 120° conduction control

The BLDCM used by the space robot usually adopts a three-phase full-bridge control structure and a Y-type connection mode, and the conduction mode is the "two-two conduction mode", which means that two power tubes in the inverter are conducted at each instant, and it commutates once every 1/6 cycle (60° electrical angle); each power tube is conducted for an electrical angle of 120°, and each winding is conducted for an electrical angle of 240°, wherein each positive and reverse conduction are 120°. When the motor rotates clockwise, the conduction sequence of each power tube is T_1, T_4, T_1, T_6, T_3, T_6, T_2, T_3, T_2, T_5, T_4, and T_5; when the motor rotates counterclockwise, the conduction sequence of each power tube is T_4, T_5, T_2, T_5, T_2, T_3, T_3, T_6, T_1, T_6, T_1, and T_4. With this conduction control, the torque ripple is very small, changing only from 0.87 T_m to T_m. The above-described conduction control method is also referred to as the three-phase six-state 120° conduction control. The main circuit of the connecting winding three-phase bridge is shown in Fig. 4.6.

(6) PWM modulation control method

The PWM signal of the BLDCM adopts the H_PWM-L_PWM mode in the control bridge circuit, that is, the power tube on each of the conduction state controllers and the lower bridge arm is PWM modulated, as shown in Fig. 4.7.

(7) Forward/reverse control of BLDCM.

The forward and reverse control of BLDCM is different from that of an ordinary motor. It is not like the asynchronous motor which changes the phase of the motor by changing the phase sequence of the three phases of the motor, nor is it like the conventional DC motor that changes the polarity of the magnetic field by changing the conduction sequence of the rotor windings to change the motor rotation direction. During the operation of the BLDCM, its ON-OFF is always inseparable from the rotor

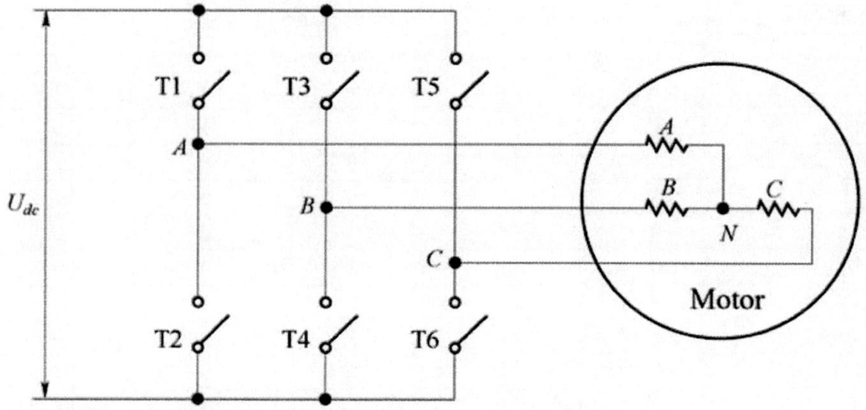

Fig. 4.6 Main circuit of the connecting winding three-phase bridge

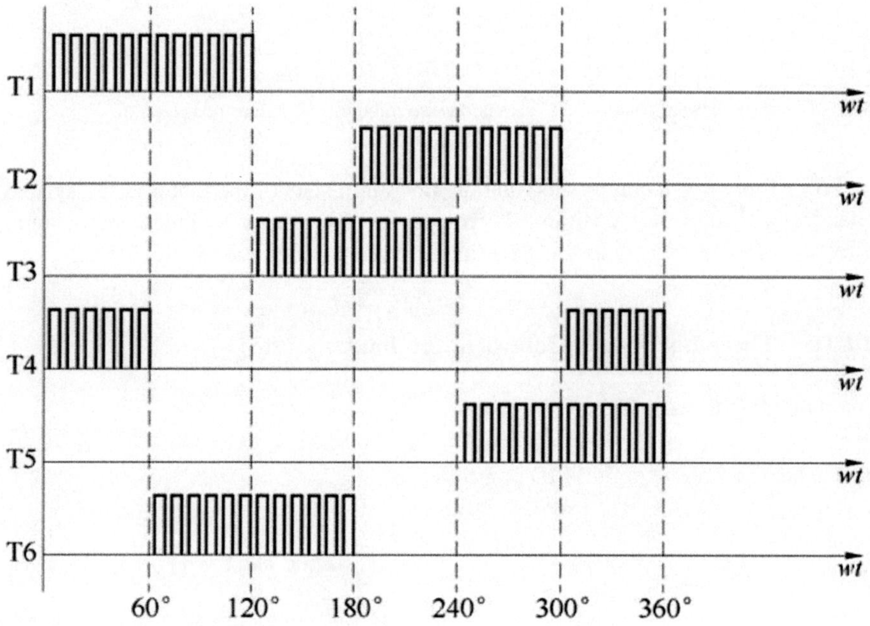

Fig. 4.7 Bipolar mode of PWM modulation

position signal of the motor. It is usually defined as that, looking from the output view of the motor, counterclockwise rotation is forward rotation and clockwise rotation is reverse rotation. The specific commutation relationships among motor rotation direction and rotor position signals and stator windings are shown in Table 4.3.

Table 4.3 Correspondence between Hall signal and commutation

Direction control	Hall signal			Logic drive signal					
	H_a	H_b	H_c	T_1	T_2	T_3	T_4	T_5	T_6
1	1	0	1	1	0	0	1	0	0
1	1	0	0	1	0	0	0	0	1
1	1	1	0	0	0	1	0	0	1
1	0	1	0	0	1	1	0	0	0
1	0	1	1	0	1	0	0	1	0
1	0	0	1	0	0	0	1	1	0
0	1	0	1	0	1	1	0	0	0
0	0	0	1	0	0	1	0	0	1
0	0	1	1	1	0	0	0	0	1
0	0	1	0	1	0	0	1	0	0
0	1	1	0	0	0	0	1	1	0
0	1	0	0	0	1	0	0	1	0
–	0	0	0	0	0	0	0	0	0
–	1	1	1	0	0	0	0	0	0

Note "1" means forward rotation; "0" means reverse rotation; "–" means no direction

From Table 4.3, it can be seen that the operation state of the motor in the system changes with the rotor position. Therefore, it is necessary to judge the rotation direction of the motor, so as to select the different commutation logic.

4.1.1.3 Three-Loop Servo Control of the Joints

(1) Digital PID controller

The expression of the digital PID controller is

$$u(k) = K_P e(k) + K_I \sum_{j=0}^{k} e(k) + K_D(e(k) - e(k-1)), \tag{4.6}$$

where $e(k)$ is the control system error at the kth step, it is the difference between the desired value and the actual feedback value; K_p is the proportional coefficient; K_I is the integral coefficient; K_D is the differential coefficient; and $u(k)$ is the output of the controller. Also, $K_I = K_P \frac{T}{T_I}$, $K_D = K_P \frac{T_D}{T}$, T is the system sampling period, T_I is the integral time constant, and T_D is the differential time constant.

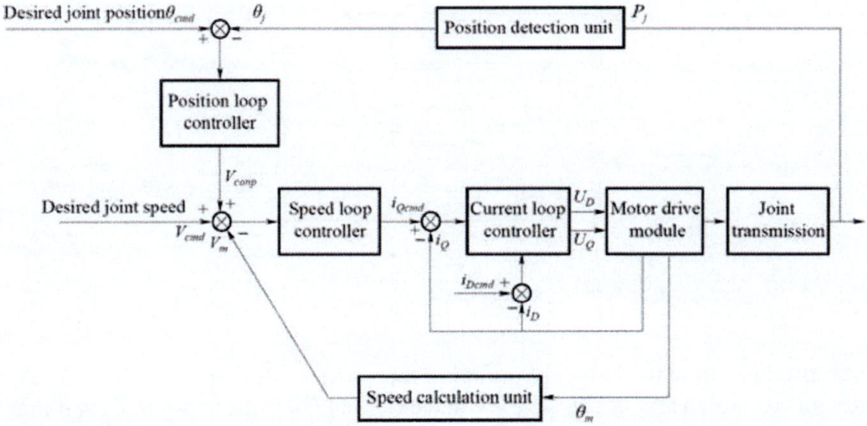

Fig. 4.8 Block diagram of the three-loop PID joint servo control

(2) Three-Loop control of the servo control system

The servo control algorithm uses a three-loop PID control method, as shown in Fig. 4.8. The three-loop PID includes a position loop, a speed loop, and a current loop.

(3) The period of the three-loop control of the servo control system

A higher current loop control period should be set according to the hardware capability. According to the characteristics of the currently used electronic components, 10, 20 kHz, and even 100 kHz control frequencies can be used to improve the stability of the system. The current acquisition period and the PWM period should not be larger than the current loop control period. The speed loop control period should be set according to the characteristics of the controller, and usually can be set to 1, 500 Hz, etc. Since the position loop is in the outermost loop, it can be selected in the range of 20–1 kHz.

The command of the servo controller generally comes from the upper control computer. Most space robots use CAN, 1553B, and other buses, and the period for sending the desired command is 1–20 Hz. In order to match the servo control system, the measures such as the equivalent, difference and filtering can be taken. Since the control of the PMSM is relatively complicated, the three-loop control method is described here by taking the PMSM as an example, and it can be used as a reference for the BLDCM.

(4) Features of the three-loop control of the servo control system

(1) Current loop.

The current loop controls the quadrature axis (*Q*-axis) current and the direct-axis (*D*-axis) current in the two-phase coordinate of the rotor. The input of the *Q*-axis

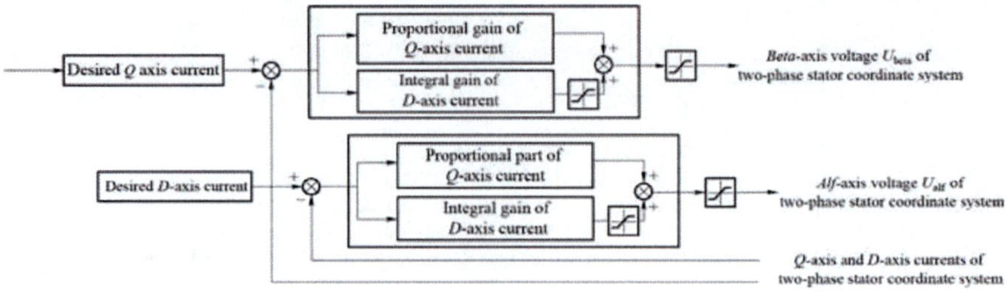

Fig. 4.9 Block diagram of current loop PI control

current comes from the output of the speed loop, and the desired value of the *D*-axis current is 0, which is converted to the voltage U_Q and U_D in the two-phase coordinate of the stator after passing the current loop controller. The current loop controller is in the innermost loop. In order to improve the response speed of the current, the PI controller is generally used. The control block diagram is shown in Fig. 4.9.

Because it involves the protection of the control circuit, the input and output of the current loop must be limited to avoid damage to the motor by overcurrent.

(2) Speed loop.

In the speed loop, the speed of the servo system is controlled. In order to avoid the influence of nonlinear factors such as backlash and friction in the transmission system, the speed control uses an inner closed loop, that is, the motor speed is used as the controlled object. The block diagram of the speed loop is shown in Fig. 4.10. The motor speed is controlled and the resulting output is used as the input of the current loop. For the differential link, the speed expectation can be selected as required, rather than using the speed difference as the input, so as to reduce the effects of noise. The motor speed can be calculated differentially according to the mechanical position of the motor, and the external noise can be reduced by a filter.

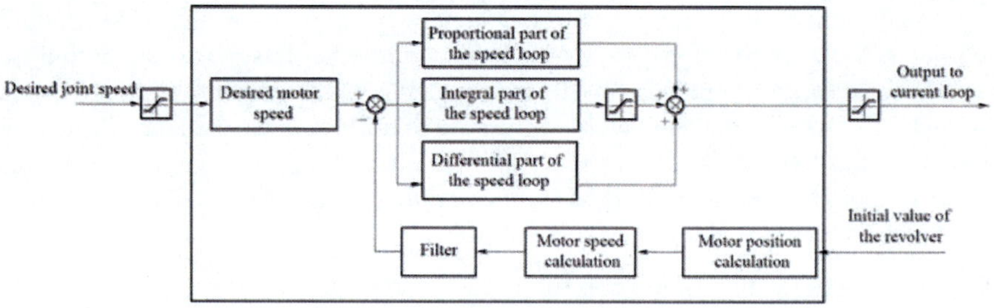

Fig. 4.10 Block diagram of the speed loop PID control

(3) Position loop.

The position loop controls the joint position. To ensure that the control accuracy meets the requirements, the position loop uses an outer closed-loop control, that is, it controls the angle of the output end of the transmission system. In order to improve the stability of the control system, a P or PD controller can be used to ensure steady-state error through the proportional gain of the speed loop.

4.1.2 Debugging and Testing of the Joint Servo Control System

Currently, the current loop and speed loop of the servo control system can automatically adjust the parameters and are widely used. Many commercial drivers have integrated the related functions. Space robots have special requirements for the servo system due to the particularity of their space environment (gravity/inertia, etc.). This section describes the debugging procedure for the joint servo control system.

The debugging/testing flow of the joint servo control system is shown in Fig. 4.11, including the following steps:

Firstly, perform single-motor debugging and confirm the initialization parameters of the motor through basic debugging.

Secondly, perform the motor open-loop debugging and current closed-loop debugging, and perform the current loop time-domain (TD) and frequency-domain (FD)

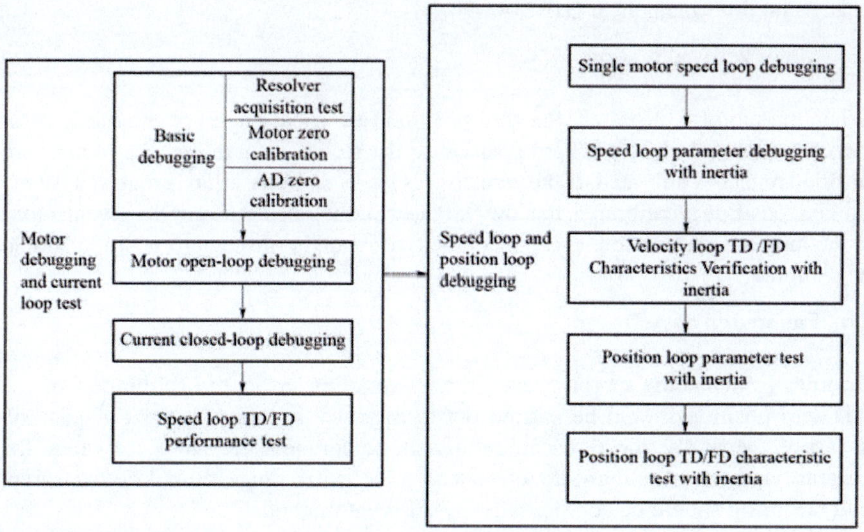

Fig. 4.11 Debugging/testing flow of the joint servo control system

characteristics tests to ensure that they meet the requirements of current loop debugging.

Thirdly, carry out single-motor speed closed-loop debugging and the motor zero-load debugging. For the space operation robot, its load inertia is often an important factor affecting its performance. Therefore, for such robots, the motor speed loop needs to be debugged with load inertia. Then performance testing of TD and FD characteristics is conducted.

Finally, debug the joint position loop parameters, and test the TD and FD characteristics with load inertia.

4.1.2.1 Motor Drive Debugging Method

Motor debugging is aimed to obtain the initial position error and current error of the motor. Compared with that of the BLDCM, the debugging procedure of the PMSM is more complicated, therefore, this section takes the PMSM configured with a resolver as the example to illustrate the debugging method, and the BLDCM can use it as a reference.

(1) Calibration of basic motor parameters

The state of the resolver, such as the number of resolver pole pairs and the coarse and fine channel zero (if it is a dual-channel) error, should be confirmed. Several coarse and fine channel data should be collected separately, then the number of fine channel cycles that pass in a coarse channel cycle range; the zero error of the resolver coarse channel and the fine channel should meet the specifications of the zero error of the coarse and fine channels in GJB2143-94.

(2) Motor zero calibration

When the motor is installed, the zero position (electrical angle) of the motor rotor deviates from the electrical zero position of the resolver, therefore, the motor zero position must be calibrated. If the zero error is large, compensation is required. When the zero position is calibrated, the low-voltage conduction method or the given D-axis value method can be used, and the compensation angle of the zero position should be recorded.

(3) Calibration of A/D zero

In order to avoid the motor phase current sampling error, the calibration of the AD zero position should be carried out to improve the phase current acquisition accuracy. Generally, the zero calibration can be compensated. In cases where the current requirement is high, the error within the entire range should be measured, and the fitting should be compensated by different methods.

(4) Motor open-loop test

The motor open-loop test is mainly designed to verify the space vector drive method and the correctness of the motor phase sequence and the motor drive logic. The motor open-loop test can be performed by supplying voltage to the *Q*-axis and *D*-axis.

(5) Current closed-loop debugging

The purpose of current closed-loop debugging is to make the current tracking performance stable, accurate, fast, and correspondently, give good system stability, small steady-state error, and quick response. Current closed-loop debugging can be performed by TD debugging and FD verification.

When adjusting the time-domain characteristics, the desired value of the *Q*-axis current can be given, and the desired value of the *D*-axis current is set to zero. If conditions permit, a torque can be applied to the motor. When debugging the TD characteristics, it should meet the requirements of overshoot, stabilization time, and steady-state error. Generally, the current loop has a fast-tracking performance, therefore, the stabilization time (or rise time) can be used as the debugging target in consideration of other technical indices.

In cases where there is an FD characteristics requirement, the FD performance should be tested. The technical indices for FD characteristics are bandwidth, amplitude margin, and phase margin. The FD can generally be measured by a sine wave signal of a different frequency of a given amplitude, or by a special tool for FD measurement.

When there is a situation in the TD or FD that does not meet the requirements, the parameters need to be compromised. The parameters of overshoot and time, bandwidth, and margin are themselves conflicting technical indices. The purpose of parameter debugging is to find the compromised parameters to meet all requirements.

4.1.2.2 Servo Control Debugging

Servo control debugging is mainly for speed servo and position servo. The servo control system mostly adopts double closed-loop control, in which the speed loop is the inner loop with the motor as the controlled object to ensure the speed response, and the position loop is the outer loop with the joint output as the controlled object to ensure the control precision. The sequence of the debugging is from inside to outside and from unloaded to loaded.

The speed loop is an important part of joint control performance. The speed loop is debugged, unloaded, and loaded in the TD. The load test equipment can be a torque motor or a dynamometer. The speed loop should meet the requirements of overshoot, stabilization time, and steady-state error. For the space robot motion system, the joint inertia characteristics may be a major characteristic, which have a great influence on the joint motion performance. For such systems, it is necessary to pay more attention to the debugging of the load inertia.

After the single-motor speed loop is debugged, the TD and FD characteristics should be verified and analyzed, and the unsatisfied requirements should be iterated so as to reach the desired target.

After the speed loop is verified to meet the requirements, the position loop parameter debugging and performance verification can be carried out. The most important requirement for the position loop is a minimum overshoot and a maximum steady-state error, because overshoot can cause interference collisions of the robot components, while the steady-state error is directly related to the accuracy of the control system. Similarly, the position loop can perform TD debugging in the sequence of no load, moment load, and inertial load, after that TD and FD performance verification and iteration are performed so as to meet the performance requirements.

4.2 Computed Torque Method-Based Motion Control

In order to improve the dynamic characteristics of the robot, model-based control such as computed torque method, sliding mode control and adaptive control are available, which can significantly improve the control accuracy, response speed and stability.

4.2.1 Control Method Based on Calculated Torque

The computed torque method can be used to track a continuous, time-varying desired trajectory under any initial condition. Unless otherwise stated, the determined desired trajectory is always considered to be a continuous time-varying trajectory $q_d(t)$ in joint space, without any loss of generality, and it can be quadratic differential, i.e., there is a bounded $\dot{q}_d(t)$ and $\ddot{q}_d(t)$.

The basic idea of the control design using a computed torque method is to first introduce a nonlinear compensation into the inner control loop, then make the robot a linear time-invariant system that is easier to control.

Specifically, first introduce the control law:

$$\tau = C(q, \dot{q})\dot{q} + G(q) + H(q)u. \tag{4.7}$$

Thus, the closed-loop equation is

$$H(q)\ddot{q} + C(q, \dot{q}) + G(q) = \tau = C(q, \dot{q}) + G(q) + H(q)u,$$

eliminating the nonlinear term into

$$H(q)\ddot{q} = H(q)u.$$

Because $H(q)$ is reversible, the above equation is equivalent to a decoupled linear constant system:

$$\ddot{q} = u.$$

Considering that when the desired trajectory $q_d(t)$ is given and both $\dot{q}_d(t)$ and $\ddot{q}_d(t)$ are known, a PD control with bias can be introduced into the linear constant system described above:

$$u = \ddot{q}_d + K_d(\dot{q}_d - \dot{q}) + K_p(q_d - q) = \ddot{q}_d + K_d\dot{e} + K_p e, \quad (4.8)$$

where K_d and K_p are both positive definite matrices (for convenience of calculation, it can be taken as a diagonal matrix), so the closed-loop system equation is

$$\ddot{e} + K_d\dot{e} + K_p e = 0.$$

Thus, because of the positive definiteness of K_d and K_p, it is known that $(e, \dot{e}) = (0, 0)$ is the equilibrium point of global asymptotic stability, that is, from any initial conditions, it always holds that $(q, \dot{q}) \to (q_d, \dot{q}_d)$, which achieves a global stable trajectory tracking. Substituting Eq. (4.8) into Eq. (4.7), the complete expression for the control law is

$$\tau = H(q)(\ddot{q}_d + K_d\dot{e} + K_p e) + C(q, \dot{q}) + G(q). \quad (4.9)$$

From Eq. (4.9), it can be seen that the control τ can be calculated in the robot inverse dynamics algorithm $\ddot{q} = \ddot{q}_d + K_d\dot{e} + K_p e$. Therefore, this control method is often referred to as the "computed torque method". It is the need to use the computed torque method for real-time control that has prompted many researchers to study many more effectively derived computed torque methods. The block diagram of the computed torque method control is shown in Fig. 4.12.

Computed torque method is a typical dynamics control scheme considering the robot dynamic model. It is the most basic method in the trajectory tracking control

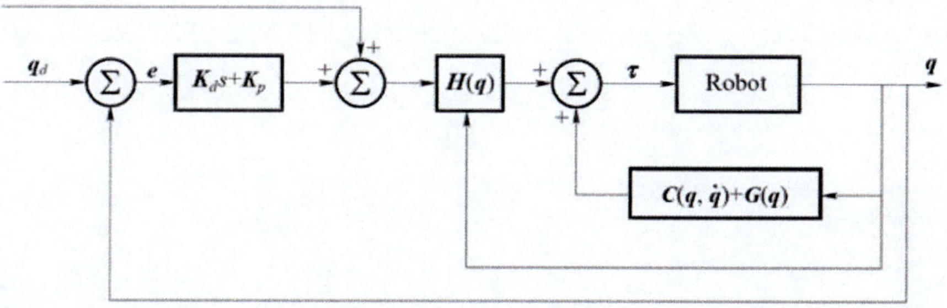

Fig. 4.12 Block diagram of the computed torque method control

of free-motion robots and plays an important role in the research of robot control problems. In addition, it is known from the previous analysis that the computed torque method uses a nonlinear compensation to achieve a global linearization and decoupling of the complex nonlinear coupled system of the robot, which has great inspiration and promotion to the development of the feedback linearization theory in the broader modern nonlinear control theory.

4.2.2 Sliding Mode Variable Structure Control

(1) Definition and mathematical expression of sliding mode dynamics

The sliding mode variable structure control (or sliding mode control for short) is essentially a special nonlinear control method, the most obvious feature of which, in comparison with conventional control approaches, is the discontinuity of control, i.e., a switching characteristic that changes the system "structure" over time. This control characteristic may cause a small-amplitude, high-frequency fluctuation along a specified state trajectory under certain characteristics, the so-called sliding mode or sliding mode motion. This sliding mode is designable and insensitive to the parameters and disturbances of the system. In this way, the system in sliding mode motion is to some extent robust.

The concepts and characteristics of sliding mode control are as follows:

Consider the general situation of the system: $\dot{x} = f(x)\ x \in R^n$, as shown in Fig. 4.13, the switching surface $s = 0$ divides the space state into upper and lower parts: $s > 0$ and $s < 0$. There are three cases of moving points on the switching surface:

(1) Normal point—when the system moving point moves to the vicinity of the switching surface $s = 0$, it passes through this point (point A).

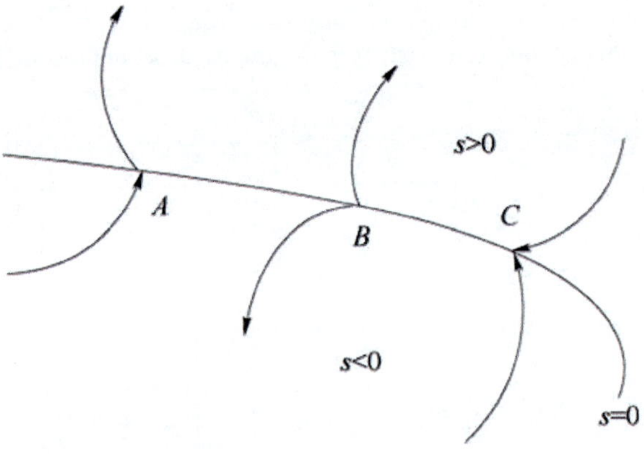

Fig. 4.13 Three point features on the switching surface

(2) Starting point—when the system moving point moves to the vicinity of the switching surface $s = 0$, it moves away from both sides of the switching surface (point B).
(3) Terminating point—when the system moving point moves to the vicinity of the switching surface $s = 0$, it approaches this point from both sides of the switching surface (point C).

If there is an area on the switching surface where all points terminate to the switching surface, once the moving point approaches this area, it is "attracted" into the area and moves. In this case, the area in which all moving points terminate on the switching surface $s = 0$ is called "Sliding Mode Area", or SMA. The motion of the system in the SMA is referred to as SM motion.

According to the requirement that the moving point on the SMA must be the terminating point, when the moving point moves to the vicinity of the switching surface, there must be

$$\lim_{s \to 0^+} \dot{s} < 0 \text{ and } \lim_{s \to 0^-} \dot{s} > 0. \tag{4.10}$$

It can also be written as

$$s\dot{s} < 0 \tag{4.11}$$

This inequality gives the system the necessary conditions for the Lyapunov function of (4.12).

$$v(x_1, x_2, \ldots, x_n) = \frac{1}{2}[s(x_1, x_2, \ldots, x_n)]^2. \tag{4.12}$$

Since Eq. (4.12) is positive definite in the adjacent area of the switching surface, and the derivative of s^2 from Eq. (4.12) is negative semi-definite, that is, v is a nonincreasing function in the vicinity of $s = 0$, so if the conditional expression (4.11) is satisfied, Eq. (4.12) is a conditional Lyapunov function of the system. The system itself is also stable to the condition $s = 0$.

Design of sliding mode surface and approach rate.

For linear systems:

$$\dot{x} = Ax + bu, \ x \in R^n, \ u \in R. \tag{4.13}$$

The sliding surface can be designed as

$$s(t) = ce(t) + \dot{e}(t), \tag{4.14}$$

where $e(t)$ and $\dot{e}(t)$ are the tracking error and its rate of change, respectively, c must be greater than zero.

Sliding mode motion includes two processes of approaching motion and sliding mode motion. The system moving from any initial state toward and until reaching the switching surface is called the approaching motion, that is, the motion process of approaching $s \to 0$. The commonly used exponential approach law is

$$\dot{s} = -\varepsilon \mathrm{sgn} s - ks \ \varepsilon > 0,\ k > 0, \tag{4.15}$$

where $\dot{s} = -ks$ is the exponential approach law term, and its solution is $s = s(0)e^{-kt}$. In the exponential approach, the approaching speed is gradually reduced from a larger value to zero, which not only shortens the approach time, but also reduces the velocity of the moving point when reaching the switching surface. For a simple exponential approach, it is an asymptotic process for the moving point to approach the switching surface, so it cannot be guaranteed to arrive in a limited time, and no dynamics mode will exist on the switching surface. Therefore, a constant velocity approach item $\dot{s} = -\varepsilon \mathrm{sgn} s$ is added, so that when s approaches zero, the approach speed is ε instead of zero to ensure that it arrives in a limited time.

Here, the control algorithm is derived by taking the computed torque method as an example. The model of the robot is

$$H(q)\ddot{q} + C(q, \dot{q})\dot{q} + F(\dot{q}) = \tau, \tag{4.16}$$

where $H(q) \in R^{n \times n}$ is a symmetric, bounded positive definite inertial matrix; $C(q, \dot{q}) \in R^n$ represents the centripetal force and the Coriolis force vector; and $f(\dot{q}) \in R^n$ represents the friction torque vector.

The robot dynamic equation satisfies the following properties:

Property 1: $H(q)$ is a symmetric positive definite matrix, $H(q)$ and $H^{-1}(q)$ are bounded uniformly.

Property 2: Matrix function $\dot{H}(q) - 2C(q, \dot{q})$ is an oblique symmetric matrix for any $q, \dot{q}$.

When the inertia parameters of the space robot are unknown, the control rate is taken by the computed torque method as

$$\tau = \hat{H}(q)\upsilon + \hat{C}(q, \dot{q})\dot{q} + \hat{F}(\dot{q}), \tag{4.17}$$

where $\hat{H}(q)$, $\hat{C}(q, \dot{q})$, and $\hat{F}(\dot{q})$ are the estimated values of H, c, and F which are calculated by the inertia parameter estimates, then the closed-loop system equation is

$$H(q)\ddot{q} + C(q, \dot{q})\dot{q} + F(\dot{q}) = \hat{H}(q)\upsilon + \hat{C}(q, \dot{q})\dot{q} + \hat{F}(\dot{q}). \tag{4.18}$$

That is

$$\hat{H}(q)\ddot{q} = \hat{H}(q)\upsilon - [\tilde{H}(q)\ddot{q} + \tilde{C}(q, \dot{q})\dot{q} + \tilde{F}(\dot{q})] = \hat{H}(q)\upsilon - Y(q, \dot{q}, \ddot{q})\tilde{p}, \tag{4.19}$$

where $\tilde{H} = H - \hat{H}, \tilde{C} = C - \hat{C}, \tilde{F} = F - \hat{F}, \tilde{p} = p - \hat{p}$.

If the estimated value of the inertia parameter $\tilde{p}$ makes $\hat{H}(q)$ reversible, the closed-loop system (4.20) can be written as

$$\ddot{q} = \upsilon - [\hat{H}(q)]^{-1} Y(q, \dot{q}, \ddot{q})\tilde{p} = \upsilon - \varphi(q, \dot{q}, \ddot{q}, \hat{p})\tilde{p}. \tag{4.20}$$

Defining

$$\varphi(q, \dot{q}, \ddot{q}, \hat{p})\tilde{p} = \tilde{d},$$

where $\tilde{d} = \left[\tilde{d}_1, \ldots, \tilde{d}_n\right]^{\mathrm{T}}, d = [d_1, \ldots, d_n]^{\mathrm{T}}$.

The sliding surface is set as

$$s = \dot{e} + \Lambda e,$$

where $e = q_d - q, \dot{e} = \dot{q}_d - \dot{q}, s = [s_1, \ldots, s_n]^{\mathrm{T}}$, Λ is a positive diagonal matrix, then

$$\dot{s} = \ddot{e} + \Lambda \dot{e} = (\ddot{q}_d - \ddot{q}) + \Lambda \dot{e} = \ddot{q}_d - \upsilon + \tilde{d} + \Lambda \dot{e}.$$

Taking

$$\upsilon = \ddot{q}_d + \Lambda \dot{e} + d, \tag{4.21}$$

where d is the vector to be designed, then

$$\dot{s} = \tilde{d} - d.$$

Taking $d = (\bar{d} + \eta)\mathrm{sgn}(s)$, $\left\|\tilde{d}\right\| \leq \bar{d}$, where $\eta > 0$, then:

$$\dot{s}s = (\tilde{d} - d)s = \tilde{d}s - \bar{d}\mathrm{sgn}(s) - \eta\mathrm{sgn}(s) \leq -\eta|s| \leq 0.$$

From Eqs. (4.17) and (4.21), the sliding mode control law can be expressed as

$$\tau = \hat{H}(q)\upsilon + \hat{C}(q, \dot{q})\dot{q} + \hat{F}(\dot{q}), \tag{4.22}$$

where $\upsilon = \ddot{q}_d + \Lambda \dot{e} + d, d = (\bar{d} + \eta)\mathrm{sgn}(s)$.

From the control law (4.22), it can be seen that the more accurate the parameter estimation $\hat{p}$, the smaller the $\|p\|$ and chattering caused by the sliding mode control.

Chapter 5
Force Control of Space Robot

Position control is capable of handling the tasks in which the robot is not interacting significantly with the environment. However, in many applications, the robot will inevitably contact and interact with the objects in its workspace and thus generate the forces of interaction. In this case, it is necessary not only to control the movement of the robot, but also to ensure that the contact force meets the requirements.

The robot force control is a method of modifying the contact force between the robot and the environment by controlling the joint output. The main purposes of force control are to protect the robot or the objects in contact, and to provide the required contact forces in specific tasks. Since the 1980s, a number of force control methods have been proposed, most of which fall into two categories, namely hybrid force/position control and impedance control.

The concept of hybrid force/position control (or hybrid control for short) was initially proposed by Raibert and Craig [1], and later developed into dynamics hybrid force/position control by Yoshikawa et al. [2]. According to the type of constraints, the task space of the robot is divided into two orthogonal subspaces. The position is controlled in one selected subspace, called position-controlled subspace. The contact force is controlled in the other orthogonal subspace, called force-controlled subspace. Then by projecting the dynamic equation into the two subspaces, the joint output can be divided into two parts so that motion control and force control can be designed separately.

The introduction of force-mixed control was very enlightening and aroused widespread interest and research at that time. However, with the deepening of research, the problem in force/position control was quickly exposed. It ignores the robot's work on the contact environment and believes that the robot actuator is confined to the constrained surface. In fact, under nonideal situations, the interaction between the end effector and the environment often switches between contact and noncontact states, thus resulting in kinematic discontinuities and instability problems.

One of the most essential problems in force control is the contradiction between the high stiffness required by the robot in free space motion and the high flexibility

Y. Wang, *Space Robotics*, Space Science and Technologies,
https://doi.org/10.1007/978-981-15-4902-1_5

required in the contact operation. Hybrid force/position control obviously does not handle this contradiction well.

In contrast, Hogan proposed impedance control in 1985, arguing that robotic force control should be designed, not to track a motion or force trajectory alone, but rather to regulate the mechanical impedance of the manipulator [3]. Impedance control provides a uniform method for collision avoidance in constrained and unconstrained motion, enabling stable transitions from noncontact phase to contact phase. Therefore, impedance control is more widely used in practical engineering.

Theoretically, with impedance control, the stiffness in different degrees of freedom (DOFs) of the task space can be arbitrarily configured. Since no constraint-based space division is made like that in hybrid control, the motion in high-accuracy-demanded dimensions is disturbed by interaction force, causing the loss of position control accuracy. According to Anderson and Spong, hybrid impedance control divides the force-controlled and position-controlled subspaces with a selection matrix, thereby ensuring the flexibility in force-controlled subspace and the precision in position-controlled subspace [4]. In a certain sense, hybrid impedance control is the combination of hybrid control and impedance control.

In space conditions, the robot is mounted on a free-floating or attitude-controlled spacecraft, that is, its mounting base has 6 or 3 under-actuated DOFs, so that the dynamic model of the entire system is quite different from that in the fixed-base situation. The control of a space robot is a problem of under-actuated system control. Since the 1980s, many modeling and control theories about space robots have been proposed. The virtual arm method equates the kinematics of free-floating space robots with a virtual robotic system in which the base is located at the mass center of the system. The generalized Jacobian matrix method derives the velocity transformation relationship between joint space and task space from the theory of momentum and angular momentum conservation. According to this theory, one can first solve the path planning problem in task space, and then take advantage of joint space control. However, the research work on the force control of space robots is still little. In order to simplify the algorithm, the force control of the space robot can be approximated by the fixed-base model in the following two cases:

(1). The inertia of the base spacecraft is much larger than that of the robot, and the target is located on the base spacecraft. In such a case, the operation of the robot has little influence on the position and attitude of the spacecraft, and the robot and the target are approximately on the same fixed base.
(2). Only the compliance, rather than accuracy, of the position or force control is required in the task. In some cases, the force control is carried out only to reduce the impact without a high requirement for precision, so the compliant function of force control on the ground is still valid on the space robot.

In other cases, the force control strategies of space robots should be based on the free-floating or free-flying model. Since the above two assumptions cover many situations in space tasks, this chapter will explore the force control strategies for both the fixed-base and free-floating situations.

5.1 Hybrid Force/Position Control

This section introduces the hybrid force/position control method based on the fixed-base model. Hybrid force/position control is a systematic robotic force control strategy developed decades ago, but still plays an important role in modern robot control. The basis of the implementation of force/position control is that the task space can be divided into two orthogonal subspaces, in which position control and force control can be accomplished separately.

As shown in Fig. 5.1, suppose the robot system is subject to m independent constraints, and the robot end effector is always confined to the constrained surface. Then the constraint equation is

$$\Phi(\boldsymbol{x}) = 0 \tag{5.1}$$

where $\boldsymbol{x}$ is a 6-D coordinate in Cartesian coordination, $\Phi(\boldsymbol{x})$ is m-D vector function. Differentiating (5.1) with respect to time yields

$$\frac{\partial\Phi(\boldsymbol{x})}{\partial\boldsymbol{x}}\dot{\boldsymbol{x}} = 0 \tag{5.2}$$

Equation (5.2) indicates that the projection of the end-effector velocity into the space spanned by column vectors of $\frac{\partial\Phi(\boldsymbol{x})}{\partial\boldsymbol{x}}$ is zero, thus we know $\frac{\partial\Phi(\boldsymbol{x})}{\partial\boldsymbol{x}}$ is the normal direction of the constrained surface, the space spanned by which is the force-controlled subspace $N(\boldsymbol{x})$ at $\boldsymbol{x}$, and the orthogonal subspace is position-control subspace $T(\boldsymbol{x})$. It follows that

$$\begin{aligned} N(\boldsymbol{x}) \cap T(\boldsymbol{x}) &= \{\mathbf{O}_{6\times 1}\} \\ N(\boldsymbol{x}) \oplus T(\boldsymbol{x}) &= \mathbb{R}^6 \end{aligned} \tag{5.3}$$

holds.

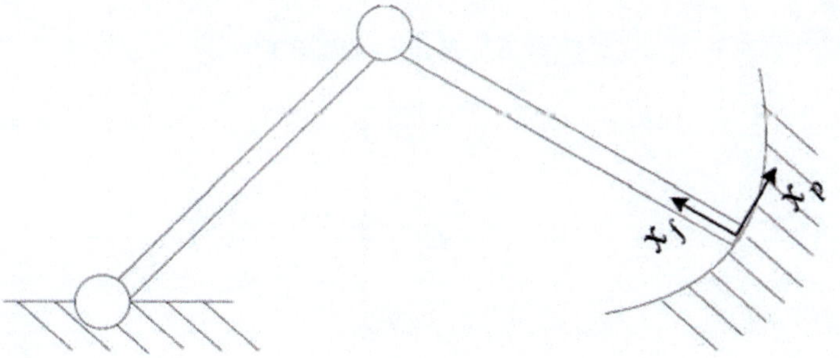

Fig. 5.1 Compliance frame

Take a set of standard orthogonal bases of $T(\boldsymbol{x})$ to form a matrix $\boldsymbol{E}_p$,another set of standard orthogonal bases of $N(\boldsymbol{x})$ to form a matrix $\boldsymbol{E}_f$, then the coordinate system composed of these two groups of bases is the so-called compliance frame. Let $\boldsymbol{E}_c = \left[\boldsymbol{E}_p, \boldsymbol{E}_f\right]$. The column vectors of $\boldsymbol{E}_c$ are the standard orthogonal bases of the compliance frame. The velocity of the end-effector in the task space $\dot{\boldsymbol{x}}$ expressed in the compliance frame is $\dot{\boldsymbol{x}}_c$; projection into the position-controlled subspace and the force-controlled subspace are, respectively, $\dot{\boldsymbol{x}}_p$ and $\dot{\boldsymbol{x}}_f$. The relationship of the coordinates in different frames is given by (5.4) as

$$\dot{\boldsymbol{x}}_c = \boldsymbol{E}_c^T \dot{\boldsymbol{x}} = \begin{bmatrix} \boldsymbol{E}_p^T \dot{\boldsymbol{x}} \\ \boldsymbol{E}_f^T \dot{\boldsymbol{x}} \end{bmatrix} = \begin{bmatrix} \dot{\boldsymbol{x}}_p \\ \dot{\boldsymbol{x}}_f \end{bmatrix} \tag{5.4}$$

The basic idea of hybrid force/position control is to express the constrained robot dynamic equation in the compliance frame, and to design the control laws of position and force respectively according to the orthogonality of the position constraints and force constraints. In addition, the joint torque is calculated through the transformation from task space to joint space.

From (5.4), it yields

$$\dot{\boldsymbol{x}} = \boldsymbol{E}_c^{-T} \dot{\boldsymbol{x}}_c = \boldsymbol{E}_c \dot{\boldsymbol{x}}_c \tag{5.5}$$

$$\ddot{\boldsymbol{x}} = \boldsymbol{E}_c \ddot{\boldsymbol{x}}_c + \dot{\boldsymbol{E}}_c \dot{\boldsymbol{x}}_c \tag{5.6}$$

Substituting (5.5), (5.6) into (2.94) and multiplying $\boldsymbol{E}_c^T$ on the left lead to

$$\underline{\boldsymbol{H}} \ddot{\boldsymbol{x}}_c + \underline{\boldsymbol{C}} \dot{\boldsymbol{x}}_c + \underline{\boldsymbol{G}} = \boldsymbol{u} - \boldsymbol{F}_c \tag{5.7}$$

where

$$\underline{\boldsymbol{H}} = \boldsymbol{E}_c^T \overline{\boldsymbol{H}} \boldsymbol{E}_c \tag{5.8}$$

$$\underline{\boldsymbol{C}} = \boldsymbol{E}_c^T \overline{\boldsymbol{H}} \dot{\boldsymbol{E}}_c + \boldsymbol{E}_c^T \overline{\boldsymbol{H}} \overline{\boldsymbol{C}} \tag{5.9}$$

$$\underline{\boldsymbol{G}} = \boldsymbol{E}_c^T \overline{\boldsymbol{G}} \tag{5.10}$$

$$\boldsymbol{u} = \boldsymbol{E}_c^T \boldsymbol{J}^{-T} \boldsymbol{\tau} \tag{5.11}$$

$$\boldsymbol{F}_c = \boldsymbol{E}_c^T \boldsymbol{F}_e. \tag{5.12}$$

Taking the natural constraints into account, it follows that

$$\dot{\boldsymbol{x}}_c = \begin{bmatrix} \dot{\boldsymbol{x}}_p \\ \dot{\boldsymbol{x}}_f \end{bmatrix} = \begin{bmatrix} \dot{\boldsymbol{x}}_p \\ \boldsymbol{O} \end{bmatrix} \tag{5.13}$$

$$\boldsymbol{F}_c = \begin{bmatrix} \boldsymbol{F}_p \\ \boldsymbol{F}_f \end{bmatrix} = \begin{bmatrix} \boldsymbol{O} \\ \boldsymbol{F}_f \end{bmatrix} \tag{5.14}$$

holds. Denoting $\underline{\boldsymbol{H}} = \begin{bmatrix} \underline{\boldsymbol{H}}_{11} & \underline{\boldsymbol{H}}_{12} \\ \underline{\boldsymbol{H}}_{21} & \underline{\boldsymbol{H}}_{22} \end{bmatrix}$, then (5.7) becomes

$$\begin{bmatrix} \underline{\boldsymbol{H}}_{11} & \underline{\boldsymbol{H}}_{12} \\ \underline{\boldsymbol{H}}_{21} & \underline{\boldsymbol{H}}_{22} \end{bmatrix} \begin{bmatrix} \ddot{\boldsymbol{x}}_p \\ \boldsymbol{O} \end{bmatrix} + \underline{\boldsymbol{C}}\dot{\boldsymbol{x}}_c = \boldsymbol{u} - \begin{bmatrix} \boldsymbol{O} \\ \boldsymbol{F}_f \end{bmatrix}. \tag{5.15}$$

Let $\boldsymbol{u} = \boldsymbol{u}_1 + \boldsymbol{u}_2$, then with the control law

$$\begin{cases} \boldsymbol{u}_1 = \begin{bmatrix} \underline{\boldsymbol{H}}_{11}\left[\ddot{\boldsymbol{x}}_p^d + \boldsymbol{K}_d\left(\dot{\boldsymbol{x}}_p^d - \dot{\boldsymbol{x}}_p\right) + \boldsymbol{K}_p\left(\boldsymbol{x}_p^d - \boldsymbol{x}_p\right)\right] \\ \underline{\boldsymbol{H}}_{21}\left[\ddot{\boldsymbol{x}}_p^d + \boldsymbol{K}_d\left(\dot{\boldsymbol{x}}_p^d - \dot{\boldsymbol{x}}_p\right) + \boldsymbol{K}_p\left(\boldsymbol{x}_p^d - \boldsymbol{x}_p\right)\right] \end{bmatrix} - \underline{\boldsymbol{C}}\dot{\boldsymbol{x}}_c - \underline{\boldsymbol{G}} \\ \boldsymbol{u}_2 = \begin{bmatrix} \boldsymbol{O} \\ \boldsymbol{F}_f^d \end{bmatrix} \end{cases} \tag{5.16}$$

position track in position-controlled subspace and force track in force-controlled subspace can be enforced, where $\boldsymbol{K}_p$, $\boldsymbol{K}_d$ are positive definite matrices, $\boldsymbol{x}_p^d$ is the desired value of $\boldsymbol{x}_p$, and $\boldsymbol{F}_f^d$ is the desired value of $\boldsymbol{F}_f$.

Then the control input can be expressed in joint space with $\boldsymbol{\tau} = \boldsymbol{E}_c\boldsymbol{J}^T\boldsymbol{u}$.

Figure 5.2 shows the structure of the above control algorithm. This hybrid force/position control algorithm does not need to measure the contact force, but to know the exact shape of the constraint that limits its application. There is another hybrid force/position algorithm that doesn't divide the subspace according to the

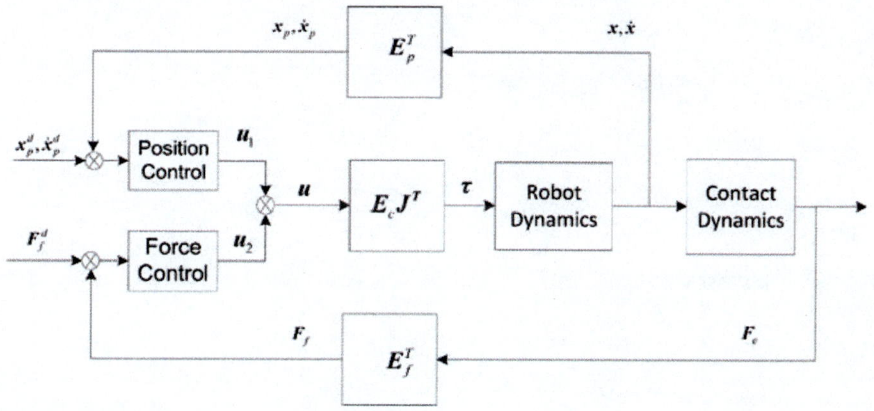

Fig. 5.2 Principle of force/position control

shape of the constraint but directly defines the dimensions in Cartesian space as position-controlled or force-controlled by using a diagonal selection matrix S. Therefore, it is not necessary to know the shape of the constraint, but to couple the outputs for position control and force control in the joint space. As the shape of the constraint changes, the accuracy of position control will be affected by the component of the contact force, which cannot be precisely controlled in the normal direction.

5.2 Impedance Control

The impedance control for robots refers to a method of configuring the dynamics response of the end effector to the environment by controlling the motion or force/torque of the joints. According to the task requirements, the impedance can be configured in many forms. The most commonly used impedance for robots is a second-order linear model. In other words, a robot is often configured as a multidimensional mass-spring-damping system, as shown in Fig. 5.3.

The impedance control can be divided into position-based impedance control (implicit impedance control) and force-based impedance control (explicit impedance control) according to the control output, as shown in Fig. 5.4. The position-based impedance control is achieved by tracking the positional response of the desired impedance system. A force feedback signal is introduced to the position controller to adjust the joint position or velocity, i.e., to modify the motion according to the force. The force-based impedance control is realized not by the position controller but by calculating the actuating force and the recovery force according to the motion of the desired system, that is, by modifying the force according to the motion. Generally, the force-based impedance control can also be divided into the impedance control using Jacobian transposition and the dynamics-based impedance control.

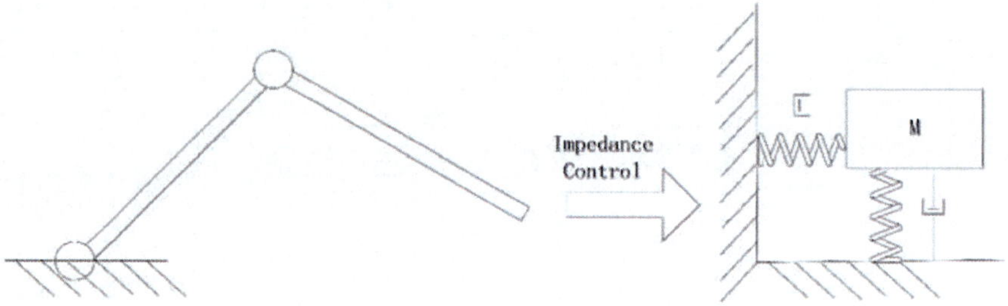

Fig. 5.3 Impedance control

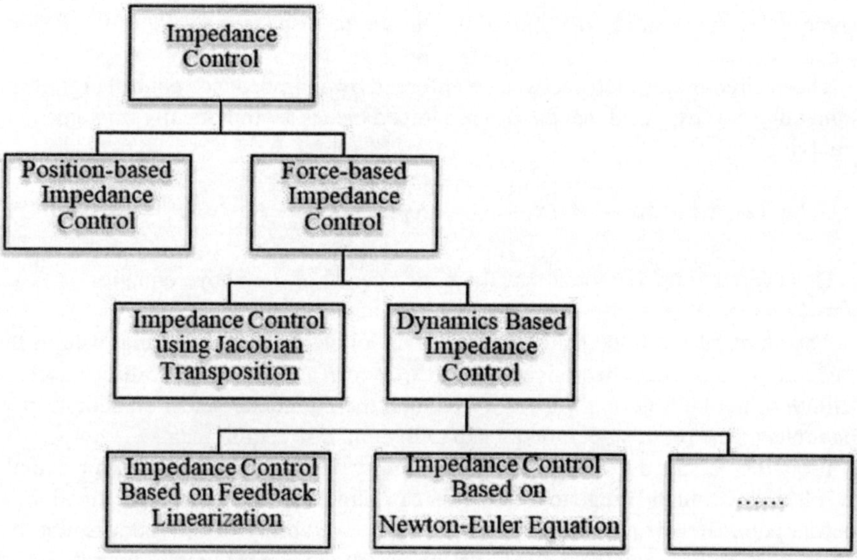

Fig. 5.4 Classification of impedance control

5.2.1 Selection of Desired Impedance

The impedance of a dynamics system is defined as the ratio of the Laplace transform of the external force to the Laplace transform of the velocity as

$$Z(s) = \frac{F(s)}{V(s)} \tag{5.17}$$

where $\boldsymbol{Z}$ is the impedance of the dynamics system, $\boldsymbol{F}$ the outer force exerted to the system, and $\boldsymbol{V}$ the velocity of the system.

We expect the robot to perform like a second-order linear system in the task space, with the desired impedance in the frequency domain of

$$\boldsymbol{Z}(s) = \boldsymbol{M}_d(s) \cdot s + \boldsymbol{B}_d(s) + \frac{\boldsymbol{K}_d(s)}{s} \tag{5.18}$$

where $\boldsymbol{M}_d$ is a positive definite desired inertial matrix, $\boldsymbol{B}_d$ is a positive or semi-positive definite desired damping matrix, and $\boldsymbol{K}_d$ is a positive or semi-positive definite desired stiffness matrix.

Considering the desired motion and the desired contact force bias, the expected impedance equation in the time domain is

$$\boldsymbol{M}_d(\ddot{\boldsymbol{x}} - \ddot{\boldsymbol{x}}_c) + \boldsymbol{B}_d(\dot{\boldsymbol{x}} - \dot{\boldsymbol{x}}_c) + \boldsymbol{K}_d(\boldsymbol{x} - \boldsymbol{x}_c) = \boldsymbol{F}_d - \boldsymbol{F}_e \tag{5.19}$$

where $\boldsymbol{x}_c$ is the desired position/attitude in the task space, and $\boldsymbol{F}_d$ is the desired contact force.

The desired impedance needs to be enforced by an impedance control algorithm. Sometimes it is expected that the desired impedance is asymptotically implemented, satisfying:

$$\lim_{t\to\infty}\left[\boldsymbol{M}_d(\ddot{\boldsymbol{x}}-\ddot{\boldsymbol{x}}_c)+\boldsymbol{B}_d(\dot{\boldsymbol{x}}-\dot{\boldsymbol{x}}_c)+\boldsymbol{K}_d(\boldsymbol{x}-\boldsymbol{x}_c)-\boldsymbol{F}_d+\boldsymbol{F}_e\right]=0. \quad (5.20)$$

However, it must be noted that the convergence of the above equation is faster than the speed of response of the desired impedance system.

The value of the impedance parameter undoubtedly plays a crucial role in the execution of the task. Theoretically, $\boldsymbol{M}_d$, $\boldsymbol{B}_d$, and $\boldsymbol{K}_d$ can take any value within the definition, but considering the reality factors, the parameter selection should meet the limits of the robot's operational capability and task requirements.

From the perspective of operational capability, the impedance parameters should be selected within the range that the robot can achieve. For example, the smaller the inertial parameter $\boldsymbol{M}_d$ is, the stronger the maneuverability of the robot is, and the faster the robot responds to the external force, but the larger the joint output is. Due to factors such as sensing error and communication delay, too small $\boldsymbol{M}_d$, too large $\boldsymbol{B}_d$ and $\boldsymbol{K}_d$ are likely to cause oscillation.

From the perspective of task requirements, the technical indicators should be optimized by selecting the optimal impedance, and different indicators may correspond to different optimal impedances. Hogan pointed out that if the end of the robot is expected to deliver the most power to the environment, the robot impedance should be equal to the environmental impedance. This is easy to understand from the analogy with an electric circuit. As shown in Fig. 5.5, the dynamics circuit composed of the robot and the environment is very similar to the electric circuit. The force corresponds to the voltage, the speed corresponds to the current, and the mechanical impedance corresponds to the circuit impedance. Then the maximum power output must occur when the two impedances are equal. However, generally we do not pursue the maximum power output by the robot, and we are often concerned with the accuracy of the control. Anderson and Spong proposed the duality theorem, claiming that the impedance of the robot and the environment can be divided into three types: inductive (inertial) ($\lim_{s\to 0} Z(s)=0$), resistive ($\lim_{s\to 0} Z(s)=\text{constant}$), and capacitive ($\lim_{s\to 0} Z(s)=\infty$), and in order to obtain the force or position control without

Fig. 5.5 Impedance control circuit

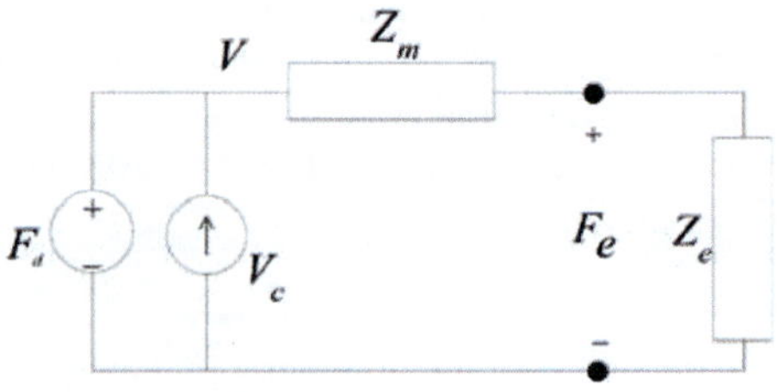

the steady-state error, with a capacitive environment it should take inductive robot impedance, and an inductive environment it should adopt capacitive robot impedance, which can also be derived from the analogy with the circuit. Hogan pointed out that to get the highest position and force accuracy, the impedance of the robot should be the reciprocal of the environmental impedance. However, in general, the impedance of the environment cannot be accurately obtained. Therefore, the selection of the impedance parameters should satisfy the above conditions as much as possible, and can be generally selected qualitatively. For example, for a stiff environment, it is expected to take a softer robot impedance; for a flexible environment, a more rigid robot impedance is suitable.

5.2.2 *Position-Based Impedance Control*

In the position-based impedance control, the force feedback signal is directly used in the motion control loop. When switched to the position adjustment function, the force signal is a stiffness control signal. When switched to the velocity adjustment function, the force signal is a damping control signal. When switched to both position and velocity adjustment functions, the force signal is an impedance control signal. The basic idea of position-based impedance control is to solve the Cartesian space velocity (Eq. (5.21)) according to the expected impedance in Eq. (5.19), then transform it into the joint velocity irrespective of acceleration, and finally enforce the reference velocity with the joint position controller. The implementation structure of position-based impedance control is shown in Fig. 5.6.

The control law with velocity as the control output is given by Eqs. (5.21) and (5.22):

$$\dot{\boldsymbol{x}}_r = \dot{\boldsymbol{x}}_c - \boldsymbol{B}_d^{-1}[\boldsymbol{K}_d(\boldsymbol{x} - \boldsymbol{x}_c) - (\boldsymbol{F}_c - \boldsymbol{F}_e)] \tag{5.21}$$

$$\dot{\boldsymbol{q}}_r = \boldsymbol{J}^{-1}\dot{\boldsymbol{x}}_r \tag{5.22}$$

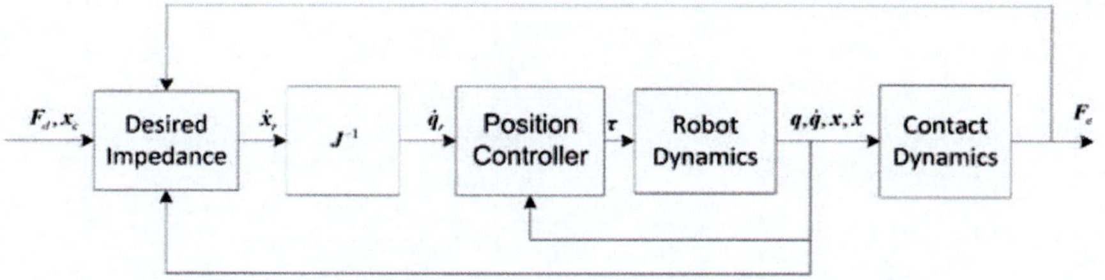

Fig. 5.6 Position-based impedance control

The above equation gives the reference velocity in Cartesian space. This impedance control law generally does not specify the desired inertial matrix. Since the force control loop is realized by the position loop, the system will show some inertia due to the position loop response. Also because of the limitation of position loop response speed, this impedance control can only be approximated, and global convergence generally cannot be obtained. Its advantage is an easy implementation by only slightly changing the position loop. So far, most of the industrial robots only support position control. It is difficult to achieve force-based control due to insufficient response speed and accuracy of joint force control and unmodeled factors (such as reducer).

5.2.3 Impedance Control Using Jacobian Transposition

Under the consideration of the relationship between joint space torque and Cartesian space static force irrespective of the inertia term, another impedance control can be implemented by solving the expected force of Cartesian space from Eq. (5.19), as shown in Eq. (5.23). The control block diagram is shown in Fig. 5.7.

The control law is expressed as Eq. (5.23) and Eq. (5.24):

$$\boldsymbol{F}_r = -\boldsymbol{B}_d(\dot{\boldsymbol{x}} - \dot{\boldsymbol{x}}_c) - \boldsymbol{K}_d(\boldsymbol{x} - \boldsymbol{x}_c) + \boldsymbol{F}_d \tag{5.23}$$

$$\boldsymbol{\tau} = \boldsymbol{J}^T \boldsymbol{F}_r \tag{5.24}$$

This algorithm is equivalent to the PD control with force bias in Cartesian space. It assumes that the driving force of the joint is finally completely converted into the contact force in Cartesian space, while ignoring the influence of inertial force, gravity, friction, and other factors. The impedance achieved by this algorithm is also approximate and can be improved by friction and gravity compensation or by full feedforward.

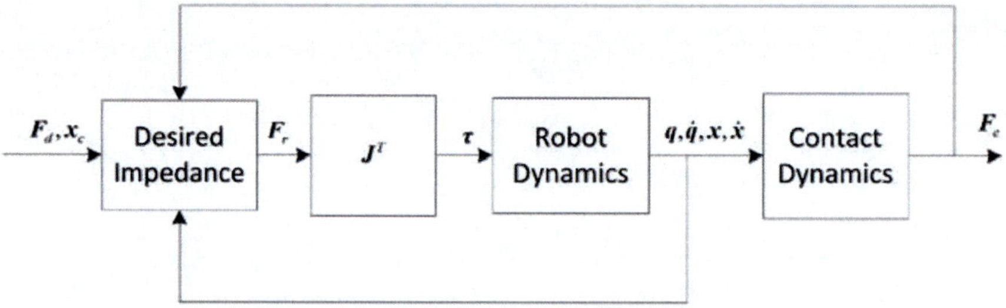

Fig. 5.7 Impedance control using Jacobian transposition

5.2.4 Dynamics-Based Impedance Control

The above two impedance control features are simple and easy to implement, but they are both established under a large degree of assumption. The convergence of the desired impedance is not obtained from the closed-loop equation. In essence, the impedance control does not utilize the dynamics of the system. The closed-loop equation is an impedance control that does not consider the dynamic equation. The dynamics-based impedance control applies the dynamic model to the control algorithm. Thus the controller can be designed with better control theory. The stability and convergence of the control algorithm can be strictly proved. The control block diagram is shown in Fig. 5.8, where the expected impedance feedback loop is often referred to as the outer loop and the controller feedback loop is referred to as the inner loop. There are many kinds of algorithms based on dynamic model. For free-floating robots, the control strategy may be quite different from that for fixed-base robots due to the under-actuated characteristics. This section describes several impedance control methods for fixed-base and floating robots.

5.2.4.1 An Impedance Control Algorithm for Fixed-Base Robot Based on Feedback Linearization

Feedback linearization is one of the most commonly used design methods for nonlinear systems. By introducing system state feedback or output feedback into the control law, the nonlinear part of the system can be eliminated to obtain a linear closed-loop system, which is then designed with linear system theory. For the robot system represented by Eq. (2.95), the control law is

$$\boldsymbol{\tau} = \boldsymbol{J}^T\left\{\overline{\boldsymbol{H}}\boldsymbol{u} + \overline{\boldsymbol{C}}\dot{\boldsymbol{x}} + \overline{\boldsymbol{G}} + \boldsymbol{F}_e\right\}. \tag{5.25}$$

Substituting it to (2.95), the closed-loop system becomes

$$\ddot{\boldsymbol{x}} = \boldsymbol{u} \tag{5.26}$$

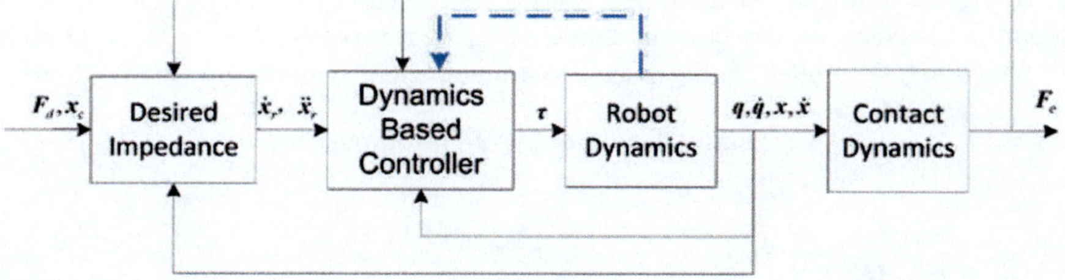

Fig. 5.8 Impedance control based on kinematics model

The nonlinear part is eliminated by feedback. And then by designing the intermediate variables $\boldsymbol{u}$, we can achieve different control purposes. Solving the acceleration from the desired impedance Eq. (5.19) yields

$$\ddot{\boldsymbol{x}} = \ddot{\boldsymbol{x}}_c - \boldsymbol{M}_d^{-1}[\boldsymbol{B}_d(\dot{\boldsymbol{x}} - \dot{\boldsymbol{x}}_c) - \boldsymbol{K}_d(\boldsymbol{x} - \boldsymbol{x}_c) + \boldsymbol{F}_d - \boldsymbol{F}_e] \tag{5.27}$$

Then take the linear control law as

$$\boldsymbol{u} = \ddot{\boldsymbol{x}}_c - \boldsymbol{M}_d^{-1}[\boldsymbol{B}_d(\dot{\boldsymbol{x}} - \dot{\boldsymbol{x}}_c) - \boldsymbol{K}_d(\boldsymbol{x} - \boldsymbol{x}_c) + \boldsymbol{F}_d - \boldsymbol{F}_e] \tag{5.28}$$

Impedance control can be realized. All the variables used in the control law are measurable or can be calculated using measurable variables.

If we let the desired inertial matrix equal to the passive inertial matrix of the robot, i.e., $\boldsymbol{M}_d = \overline{\boldsymbol{H}}$, the control law based on feedback linearization can be rewritten as

$$\boldsymbol{\tau} = \boldsymbol{J}^T\left[\overline{\boldsymbol{H}}\ddot{\boldsymbol{x}}_c - \boldsymbol{B}_d(\dot{\boldsymbol{x}} - \dot{\boldsymbol{x}}_c) - \boldsymbol{K}_d(\boldsymbol{x} - \boldsymbol{x}_c) + \overline{\boldsymbol{C}}\dot{\boldsymbol{x}} + \overline{\boldsymbol{G}} + \boldsymbol{F}_d\right] \tag{5.29}$$

It can be seen that the contact force in the equation is eliminated, and the force measurement can be avoided, and the closed-loop equation becomes

$$\overline{\boldsymbol{H}}(\ddot{\boldsymbol{x}} - \ddot{\boldsymbol{x}}_c) + \boldsymbol{B}_d(\dot{\boldsymbol{x}} - \dot{\boldsymbol{x}}_c) + \boldsymbol{K}_d(\boldsymbol{x} - \boldsymbol{x}_c) = \boldsymbol{F}_c - \boldsymbol{F}_e \tag{5.30}$$

In essence, this algorithm is equivalent to impedance-based control using Jacobi transposition with feedforward. Although it is not necessary to use a force sensor, it is easily interfered by friction and modeling inaccuracy, resulting in a large steady-state error in contact force.

5.2.4.2 An Impedance Control Algorithm for Free-Floating Robot Based on Feedback Linearization

The above impedance control method is suitable for a fixed robot on a star land or for an orbit robot under certain conditions, as it does not consider the influence of the moving base. However, the microgravity and nonbasic features of space sometimes have a great influence on the robot system. In this case, the design of the control algorithm is based on the dynamic model of the space robot.

The dynamic model of the space-constrained robot is given by (2.50). Denote $H = \begin{bmatrix} \boldsymbol{H}_b & \boldsymbol{H}_{bm} \\ \boldsymbol{H}_{bm}^T & \boldsymbol{H}_m \end{bmatrix}$. Solving $\ddot{\boldsymbol{x}}_b$ from Eq. (2.50), it follows that

$$\ddot{\boldsymbol{x}}_b = \begin{bmatrix} \boldsymbol{I}_6 & \boldsymbol{O}_{6\times n} \end{bmatrix} \begin{bmatrix} \ddot{\boldsymbol{x}}_b \\ \ddot{\boldsymbol{q}} \end{bmatrix}$$

$$= [\boldsymbol{I}_6\ \boldsymbol{O}_{6\times n}]\boldsymbol{H}^{-1}\left(\begin{bmatrix}\boldsymbol{O}_{6\times n}\\ \boldsymbol{I}_n\end{bmatrix}\boldsymbol{\tau}+\begin{bmatrix}\boldsymbol{F}_b\\ \boldsymbol{O}_{n\times 1}\end{bmatrix}+\begin{bmatrix}\boldsymbol{J}_b^T\\ \boldsymbol{J}_m^T\end{bmatrix}\boldsymbol{F}_e-\begin{bmatrix}\boldsymbol{c}_b\\ \boldsymbol{c}_m\end{bmatrix}\right) \tag{5.31}$$

holds.

Denote $\Phi = [\boldsymbol{I}_6\ \boldsymbol{O}_{6\times n}]\boldsymbol{M}^{-1}\begin{bmatrix}\boldsymbol{O}_{6\times n}\\ \boldsymbol{I}_n\end{bmatrix} \in \mathbb{R}^{6\times n}$, Φ is a reversible matrix when $n = 6$; when $n > 6$, it can be easily proved that there exist $\Phi^+ \in \mathbb{R}^{n\times 6}$ so that $\Phi\Phi^+ = \boldsymbol{I}_6$. It is advisable to use the feedback linearization method to take

$$\boldsymbol{\tau} = \Phi^+\left\{\boldsymbol{u} + [\boldsymbol{I}_6\ \boldsymbol{O}_{6\times n}]\boldsymbol{H}^{-1}\left(-\begin{bmatrix}\boldsymbol{F}_b\\ \boldsymbol{O}_{n\times 1}\end{bmatrix}-\begin{bmatrix}\boldsymbol{J}_b^T\\ \boldsymbol{J}_m^T\end{bmatrix}\boldsymbol{F}_e+\begin{bmatrix}\boldsymbol{c}_b\\ \boldsymbol{c}_m\end{bmatrix}\right)\right\}. \tag{5.32}$$

Thus the closed-loop system equation is

$$\ddot{\boldsymbol{x}}_b = \boldsymbol{u} \tag{5.33}$$

and considering the relationship expressed by (2.28), the indirect output $\boldsymbol{u}$ is designed as

$$\boldsymbol{u} = \boldsymbol{J}_b^{-1}\left[\ddot{\boldsymbol{x}}_c - \boldsymbol{B}_d(\dot{\boldsymbol{x}} - \dot{\boldsymbol{x}}_c) - \boldsymbol{K}_d(\boldsymbol{x} - \boldsymbol{x}_c) + \boldsymbol{F}_d - \boldsymbol{F}_e - \dot{\boldsymbol{J}}_b\dot{\boldsymbol{x}}_b - \boldsymbol{J}_m\ddot{\boldsymbol{q}} - \dot{\boldsymbol{J}}_m\dot{\boldsymbol{q}}\right] \tag{5.34}$$

Rewrite $\boldsymbol{x}_e$ as $\boldsymbol{x}$ for convenience, it follows that

$$\boldsymbol{M}_d(\ddot{\boldsymbol{x}} - \ddot{\boldsymbol{x}}_c) + \boldsymbol{B}_d(\dot{\boldsymbol{x}} - \dot{\boldsymbol{x}}_c) + \boldsymbol{K}_d(\boldsymbol{x} - \boldsymbol{x}_c) = \boldsymbol{F}_d - \boldsymbol{F}_e. \tag{5.35}$$

Thus the impedance control algorithm for the free-floating robot system is obtained.

The expected impedance achieved by this method is accurate. Considering the complete dynamic model of the free-floating robot, it is no need to solve the generalized Jacobian matrix. In fact, the dynamic model (2.50) contains the constraints imposed by the conservation of momentum and angular momentum that the generalized Jacobian matrix can be derived from.

5.2.4.3 An Impedance Control Algorithm for Free-Floating Robot Based on Newton–Euler Equation

The modeling and control algorithms based on Newton–Euler equation are particularly advantageous for multi-joint systems, complex topological systems, and modular variable-structure systems. This section introduces a space-robot impedance control algorithm based on Virtual Decomposition Control (VDC) [5], which is also a control algorithm based on Newton–Euler equation.

According to Eq. (5.19), the reference velocity $\dot{\boldsymbol{x}}_r$ and reference acceleration $\ddot{\boldsymbol{x}}_r$ in the task space are defined as

$$\ddot{\boldsymbol{x}}_r = \boldsymbol{M}_d^{-1}\left[\ddot{\boldsymbol{x}}_c - \boldsymbol{B}_d(\dot{\boldsymbol{x}} - \dot{\boldsymbol{x}}_c) - \boldsymbol{K}_d(\boldsymbol{x} - \boldsymbol{x}_c) + \boldsymbol{F}_d - \boldsymbol{F}_e\right] \tag{5.36}$$

$$\dot{\boldsymbol{x}}_r = \boldsymbol{M}_d^{-1}\left\{\dot{\boldsymbol{x}}_c - \boldsymbol{B}_d(\boldsymbol{x} - \boldsymbol{x}_c) - \int_0^t [\boldsymbol{K}_d(\boldsymbol{x} - \boldsymbol{x}_c) + \boldsymbol{F}_d - \boldsymbol{F}_e]dt\right\} \tag{5.37}$$

Transformed into joint space, the reference velocity and reference acceleration become

$$\dot{\boldsymbol{q}}_r = \boldsymbol{J}^{-1}\dot{\boldsymbol{x}}_r \tag{5.38}$$

$$\ddot{\boldsymbol{q}}_r = \boldsymbol{J}^{-1}\left(\ddot{\boldsymbol{x}}_r - \dot{\boldsymbol{J}}\boldsymbol{J}^{-1}\dot{\boldsymbol{x}}_r\right). \tag{5.39}$$

The Newton–Euler dynamic equation is used to design the control law, and the above reference velocity is substituted for the real velocity in the dynamic Eqs. (2.78)–(2.83) to obtain the controlled kinematic recursive equations ($i = 0, 1, 2, \ldots n$):

$$\boldsymbol{V}_{r\hat{1}} = \boldsymbol{V}_b \tag{5.40}$$

$$\boldsymbol{V}_i = {}^{i}\boldsymbol{T}_{\hat{i}}\boldsymbol{V}_{r\hat{i}} + (\sigma_i z_3 + \bar{\sigma}_i z_6)\dot{q}_{ri} \tag{5.41}$$

$$\boldsymbol{V}_{r\hat{i+1}} = {}^{\hat{i+1}}\boldsymbol{T}_i\boldsymbol{V}_{ri} \tag{5.42}$$

$$\mathring{\boldsymbol{V}}_{r\hat{1}} = \dot{\boldsymbol{V}}_b \tag{5.43}$$

$$\mathring{\boldsymbol{V}}_{ri} = {}^{i}\boldsymbol{T}_{\hat{i}}\mathring{\boldsymbol{V}}_{r\hat{i}} + {}^{i}\boldsymbol{T}_{\hat{i}}\boldsymbol{V}_{r\hat{i}} + (\sigma_i z_3 + \bar{\sigma}_i z_6)\ddot{q}_{ri} \tag{5.44}$$

$$\mathring{\boldsymbol{V}}_{r\hat{i+1}} = {}^{\hat{i+1}}\boldsymbol{T}_i\mathring{\boldsymbol{V}}_{ri}. \tag{5.45}$$

According to the dynamic equation of the rigid body subsystem (2.73), the reference resultant force of the rigid body is defined as

$$\boldsymbol{F}_{ri}^{*} = \hat{\boldsymbol{M}}_i \cdot \mathring{\boldsymbol{V}}_{ri} + \hat{\boldsymbol{C}}_i \cdot \boldsymbol{V}_{ri} + \boldsymbol{K}_{Di}(\boldsymbol{V}_{ri} - \boldsymbol{V}_i) = \boldsymbol{Y}_i\hat{\boldsymbol{\theta}}_i + \boldsymbol{K}_{Di}(\boldsymbol{V}_{ri} - \boldsymbol{V}_i) \tag{5.46}$$

where $\hat{\boldsymbol{M}}_i, \hat{\boldsymbol{C}}_i, \hat{\boldsymbol{\theta}}_i$ are the estimates of $\boldsymbol{M}_i, \boldsymbol{C}_i, \boldsymbol{\theta}_i$, and $\boldsymbol{K}_{Di} \in \mathbb{R}^{6\times 6}$ is a positive definite matrix.

According to (2.84)–(2.87), the governed force recursion equation is ($i = n, n-1, \ldots, 1$):

$$\boldsymbol{F}_{r\widehat{n+1}} = \boldsymbol{O} \tag{5.47}$$

$$\boldsymbol{F}_{ri} = \boldsymbol{F}^{*}_{ri} + ({}^{\widehat{i+1}}\boldsymbol{T}_{i})^{T}\boldsymbol{F}_{r\widehat{i+1}} \tag{5.48}$$

$$\boldsymbol{F}_{r\hat{i}} = ({}^{i}\boldsymbol{T}_{\hat{i}})^{T}\boldsymbol{F}_{ri}. \tag{5.49}$$

Equation (5.46) uses the estimates of the dynamics parameters, so an adaptive law is needed to update the estimated parameters. Take the adaptive law:

$$\dot{\hat{\boldsymbol{\theta}}}_{i} = \boldsymbol{K}_{Ai}\boldsymbol{Y}_{i}^{T}(\boldsymbol{V}_{ri} - \boldsymbol{V}_{i}) \tag{5.50}$$

where adaptation gain $\boldsymbol{K}_{Ai}$ is a positive definite constant matrix. $\boldsymbol{\theta}_{i}$ is constant and $\dot{\hat{\boldsymbol{\theta}}}_{i} = \dot{\tilde{\boldsymbol{\theta}}}_{i}$. The joint output is calculated with (5.51):

$$\boldsymbol{\tau}_{i} = (\sigma_{i}z_{3} + \bar{\sigma}_{i}z_{6})^{T i}\boldsymbol{F}_{ri}. \tag{5.51}$$

Equation (5.42)–(5.51) constitute an impedance control algorithm based on the Newton–Euler equation.

To prove stability, first define a concept of virtual power flow. Let $\{A\}$ be a frame attached to a rigid body, the virtual power flow is defined as the inner product of the linear/angular velocity vector error and the force/moment vector error, that is

$$\boldsymbol{p}_{A} = (\boldsymbol{V}_{rA} - \boldsymbol{V}_{A})^{T}(\boldsymbol{F}_{rA} - \boldsymbol{F}_{A}) \tag{5.52}$$

Let $\{B\}$ be another frame attached to the rigid body. Then according to the second chapter, it follows that $\boldsymbol{V}_{A} = {}^{A}\boldsymbol{T}_{B}\boldsymbol{V}_{B}$, $\boldsymbol{F}_{B} = ({}^{A}\boldsymbol{T}_{B})^{T}\boldsymbol{F}_{A}$, $\boldsymbol{V}_{rA} = {}^{A}\boldsymbol{T}_{B}\boldsymbol{V}_{rB}$, $\boldsymbol{F}_{rB} = ({}^{A}\boldsymbol{T}_{B})^{T}\boldsymbol{F}_{rA}$ hold. The virtual power expressed in frame $\{B\}$ is

$$\begin{aligned}\boldsymbol{p}_{B} &= (\boldsymbol{V}_{rB} - \boldsymbol{V}_{B})^{T}(\boldsymbol{F}_{rB} - \boldsymbol{F}_{B}) \\ &= (\boldsymbol{V}_{rA} - \boldsymbol{V}_{A})^{T}({}^{A}\boldsymbol{T}_{B})^{\ T}({}^{A}\boldsymbol{T}_{B})^{T}(\boldsymbol{F}_{rA} - \boldsymbol{F}_{A}). \\ &= \boldsymbol{p}_{A}\end{aligned} \tag{5.53}$$

It is seen that the virtual power flow of the force/moment is independent of which coordinate system the expression is.

Regarding the base as rigid B_{0}, then $\boldsymbol{V}_{0} = \boldsymbol{V}_{r0} = \boldsymbol{O}_{6\times 1}$. For a space robotic system, choose the Lyapunov-like function as

$$\mathrm{V} = \sum_{i=0}^{n}\mathrm{V}_{i} \tag{5.54}$$

where

$$\mathrm{V}_i = \frac{1}{2}(\boldsymbol{V}_{ri} - \boldsymbol{V}_i)^T \boldsymbol{M}_i(\boldsymbol{V}_{ri} - \boldsymbol{V}_i) + \frac{1}{2}\tilde{\boldsymbol{\theta}}_i^T \boldsymbol{K}_{Ai}^{-1}\tilde{\boldsymbol{\theta}}_i. \tag{5.55}$$

Obviously that V_i, V are all positive definite, differentiating V_i with respect to time yield

$$\dot{\mathrm{V}}_i = ({}^i\boldsymbol{V}_{ri} - {}^i\boldsymbol{V}_i)^T \boldsymbol{M}_i\left({}^i\overset{\circ}{\boldsymbol{V}}_{ri} - {}^i\overset{\circ}{\underset{i}{\boldsymbol{V}}}\right) + \tilde{\boldsymbol{\theta}}_i^T \boldsymbol{K}_{Ai}^{-1}\dot{\tilde{\boldsymbol{\theta}}}_i \tag{5.56}$$

Substituting (2.73) and (5.46) into (5.56) leads to

$$\begin{aligned}\dot{\mathrm{V}}_i = & -(\boldsymbol{V}_{ri} - \boldsymbol{V}_i)^T \boldsymbol{K}_{Di}(\boldsymbol{V}_{ri} - \boldsymbol{V}_i) + (\boldsymbol{V}_{ri} - \boldsymbol{V}_i)^T(\boldsymbol{F}_{ri}^* - \boldsymbol{F}_i^*) \\ & -(\boldsymbol{V}_{ri} - \boldsymbol{V}_i)^T \boldsymbol{Y}_i(\boldsymbol{V}_{ri}, \dot{\boldsymbol{V}}_i)\tilde{\boldsymbol{\theta}}_i - (\boldsymbol{V}_{ri} - \boldsymbol{V}_i)^T \boldsymbol{C}_i(\boldsymbol{V}_{ri} - \boldsymbol{V}_i) + \tilde{\boldsymbol{\theta}}_i^T \boldsymbol{K}_{Ai}^{-1}\dot{\tilde{\boldsymbol{\theta}}}_i\end{aligned} \tag{5.57}$$

In view of (2.86) and (5.48), it yields

$$\begin{aligned}&(\boldsymbol{V}_{ri} - \boldsymbol{V}_i)^T(\boldsymbol{F}_{ri}^* - \boldsymbol{F}_i^*) \\ &= (\boldsymbol{V}_{ri} - \boldsymbol{V}_i)^T(\boldsymbol{F}_{ri} - \boldsymbol{F}_i) - (\boldsymbol{V}_{ri} - \boldsymbol{V}_i)^T({}^{\hat{i+1}}\boldsymbol{T}_i)^T\left(\boldsymbol{F}_{r\hat{i+1}} - \boldsymbol{F}_{\hat{i+1}}\right) \\ &= (\boldsymbol{V}_{ri} - \boldsymbol{V}_i)^T(\boldsymbol{F}_{ri} - \boldsymbol{F}_i) - \left(\boldsymbol{V}_{r\hat{i+1}} - \boldsymbol{V}_{\hat{i+1}}\right)^T\left(\boldsymbol{F}_{r\hat{i+1}} - \boldsymbol{F}_{\hat{i+1}}\right) \\ &= p_i - p_{\hat{i+1}}.\end{aligned} \tag{5.58}$$

From (2.75), we know that C_i is a skew-symmetric matrix, so that $(\boldsymbol{V}_{ri} - \boldsymbol{V}_i)^T \boldsymbol{C}_i(\boldsymbol{V}_{ri} - \boldsymbol{V}_i) = \boldsymbol{O}$, then it follows that

$$\begin{aligned}\dot{\mathrm{V}}_i = & -(\boldsymbol{V}_{ri} - \boldsymbol{V}_i)^T \boldsymbol{K}_{Di}(\boldsymbol{V}_{ri} - \boldsymbol{V}_i) + p_i - p_{\hat{i+1}} \\ & -(\boldsymbol{V}_{ri} - \boldsymbol{V}_i)^T \boldsymbol{Y}_i(\boldsymbol{V}_{ri}, \dot{\boldsymbol{V}}_i)\tilde{\boldsymbol{\theta}}_i + \tilde{\boldsymbol{\theta}}_i^T \boldsymbol{K}_{Ai}^{-1}\dot{\tilde{\boldsymbol{\theta}}}_i\end{aligned} \tag{5.59}$$

holds. Substituting the adaptation law (5.50) into (5.59) yields

$$\begin{aligned}\dot{\mathrm{V}}_i = & -(\boldsymbol{V}_{ri} - \boldsymbol{V}_i)^T \boldsymbol{K}_{Di}(\boldsymbol{V}_{ri} - \boldsymbol{V}_i) + p_i - p_{\hat{i+1}} \\ & + \tilde{\boldsymbol{\theta}}_i^T\left(\boldsymbol{K}_{Ai}^{-1}\dot{\tilde{\boldsymbol{\theta}}}_i - \boldsymbol{Y}_i^T(\boldsymbol{V}_{ri}, \dot{\boldsymbol{V}}_i)(\boldsymbol{V}_{ri} - \boldsymbol{V}_i)\right) \\ = & -(\boldsymbol{V}_{ri} - \boldsymbol{V}_i)^T \boldsymbol{K}_{Di}(\boldsymbol{V}_{ri} - \boldsymbol{V}_i) + p_i - p_{\hat{i+1}}.\end{aligned} \tag{5.60}$$

Then the Lyapunov-like function becomes

$$\dot{\mathrm{V}} = \sum_{i=0}^{n} \dot{\mathrm{V}}_i = -\sum_{i=1}^{n} (\boldsymbol{V}_{ri} - \boldsymbol{V}_i)^T \boldsymbol{K}_{Di} (\boldsymbol{V}_{ri} - \boldsymbol{V}_i) + p_0 - p_{\hat{n+1}}. \tag{5.61}$$

Noticing that the input virtual power flow of the base $p_0 = (\boldsymbol{V}_{r0} - \boldsymbol{V}_0)^T (\boldsymbol{F}_{r0} - \boldsymbol{F}_0) = 0$, and the output virtual power of the last body $p_{\hat{n+1}} = \left(\boldsymbol{V}_{r\hat{n+1}} - \boldsymbol{V}_{\hat{n+1}}\right)^T \left(\boldsymbol{F}_{r\hat{n+1}} - \boldsymbol{F}_{\hat{n+1}}\right) = 0$, it yields

$$\dot{\mathrm{V}} = -\sum_{i=1}^{n} (\boldsymbol{V}_{ri} - \boldsymbol{V}_i)^T \boldsymbol{K}_{Di} (\boldsymbol{V}_{ri} - \boldsymbol{V}_i) \le 0. \tag{5.52}$$

Considering the expression

$$\int_0^\infty \sum_{i=1}^{n} (\boldsymbol{V}_{ri} - \boldsymbol{V}_i)^T (\boldsymbol{V}_{ri} - \boldsymbol{V}_i) dt \le \frac{1}{\min_{i=1,2,\ldots,n}(\lambda(\boldsymbol{K}_{Di}))}$$

$$\int_0^\infty \sum_{i=1}^{n} (\boldsymbol{V}_{ri} - \boldsymbol{V}_i)^T \boldsymbol{K}_{Di} (\boldsymbol{V}_{ri} - \boldsymbol{V}_i) dt = \frac{1}{\min_{i=1,2,\ldots,n}(\lambda(\boldsymbol{K}_{Di}))} [V(0) - V(\infty)]$$

$$\le \frac{1}{\min_{i=1,2,\ldots,n}(\lambda(\boldsymbol{K}_{Di}))} V(0) < +\infty \tag{5.63}$$

where $\min_{i=1,2,\ldots,n}(\lambda(\boldsymbol{K}_{Di}))$ are the smallest eigenvalues for all $\boldsymbol{K}_{Di}$. According to the definition of $\boldsymbol{L}_p$ stability, it is known that $(\boldsymbol{V}_{ri} - \boldsymbol{V}_i) \in \boldsymbol{L}_2 \cap \boldsymbol{L}_\infty \forall i = 0, 1, 2, \ldots n$, among which it yields

$$(\boldsymbol{V}_{rn} - \boldsymbol{V}_n) = \dot{\boldsymbol{x}}_r - \dot{\boldsymbol{x}} \in \boldsymbol{L}_2 \cap \boldsymbol{L}_\infty. \tag{5.64}$$

In the view of (5.36) and (5.37), using Barbalat's lemma it follows that

$$\lim_{t\to\infty} \left[\boldsymbol{M}_d(\ddot{\boldsymbol{x}} - \ddot{\boldsymbol{x}}_c) + \boldsymbol{B}_d(\dot{\boldsymbol{x}} - \dot{\boldsymbol{x}}_c) + \boldsymbol{K}_d(\boldsymbol{x} - \boldsymbol{x}_c) - \boldsymbol{F}_c + \boldsymbol{F}_e\right] = 0 \tag{5.65}$$

holds. That is, impedance control is achieved asymptotically.

5.2.5 Multi-robot Cooperative Impedance Control

In this section, we discuss the impedance control in the tasks where multiple robots move a target together. Suppose m robots with 6 DOFs on a fixed base hold a rigid body. Since the target has only 6 DOFs, there are 6(m–1) hyperstatic DOFs in the multi-robot closed-chain system. In order to avoid the mutual conflict between branches that may cause too large internal force, the cooperation between robots

Fig. 5.9 Multi-robot cooperation handling

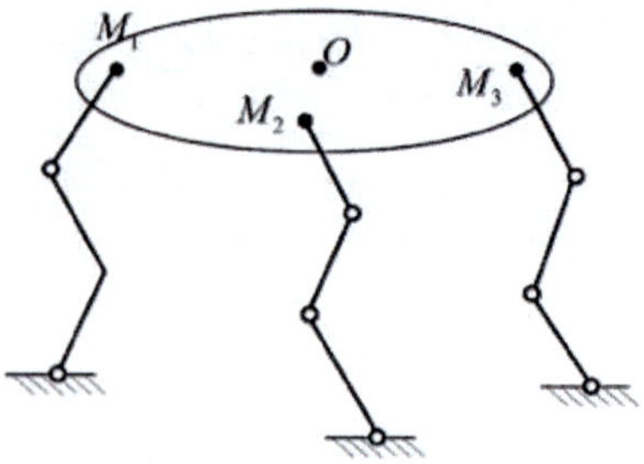

must be considered. Multi-robot impedance control is not equivalent to the simple superposition of each robot's impedance control. In order to arbitrarily configure the impedance parameter of the combined robot system, the impedance control law for the whole system must be uniformly designed according to the multi-arm system dynamic equation.

As shown in Fig. 5.9, a frame $\{O\}$ is established attached to the target as the reference coordinate system. Transformation of $\{O\}$ to the end of every robot forms $\{M_1\}$, $\{M_2\}$ (Fig. 5.9).

The velocity relationship from $\{O\}$ to each end effector can be obtained with the velocity transformation matrix, subject to

$$\boldsymbol{V}_{Mi} = {}^{Mi}\boldsymbol{T}_O \boldsymbol{V}_O \tag{5.66}$$

where $\boldsymbol{V}_{Mi}$ is the velocity of frame $\{M_i\}$ relative to the base coordinate system, and $\boldsymbol{V}_O$ is the velocity of the coordinate system $\{O\}$ relative to the base coordinate system. The velocity transformation matrix:

$${}^{Mi}\boldsymbol{T}_O = \begin{bmatrix} \boldsymbol{I}_3 & -(\boldsymbol{p}_{Mi}^{O})^{\times} \\ \boldsymbol{O}_3 & \boldsymbol{I}_3 \end{bmatrix} \tag{5.67}$$

The dynamic equation in task space of Robot i can be written as

$$\overline{\boldsymbol{H}}_i \ddot{\boldsymbol{X}}_i + \overline{\boldsymbol{C}}_i \dot{\boldsymbol{X}}_i + \overline{\boldsymbol{G}}_i = \boldsymbol{J}_i^{-T} \boldsymbol{\tau}_i - \boldsymbol{F}_i \tag{5.68}$$

where $\dot{\boldsymbol{X}}_i = \boldsymbol{V}_{Mi}$ is the velocity of the end effector; $\overline{\boldsymbol{H}}_i$,$\overline{\boldsymbol{C}}_i$, and $\overline{\boldsymbol{G}}_i$ are respectively the inertial matrix, the Coriolis centrifugal force matrix, and the gravity matrix of Robot i in the task space; $\boldsymbol{F}_i$ denotes the contact force at the end effector, $\boldsymbol{J}_i$ is the Jacobian matrix.

The impedance control algorithm is designed in frame $\{O\}$. The expected impedance equation is

$$\boldsymbol{M}_d\left(\ddot{\boldsymbol{X}}_o - \ddot{\boldsymbol{X}}_{Oc}\right) + \boldsymbol{B}_d\left(\dot{\boldsymbol{X}}_O - \dot{\boldsymbol{X}}_{Oc}\right) + \boldsymbol{K}_d\left(\boldsymbol{X}_O - \boldsymbol{X}_{Oc}\right) = -\boldsymbol{F}_O. \tag{5.69}$$

From the equivalent relationship between the coordinate systems, the total force of each point is equivalent to the O point as

$$\boldsymbol{F}_O = \sum_{I=1}^{m} {}^{Mi}\boldsymbol{T}_O^T \boldsymbol{F}_i \tag{5.70}$$

In order to balance the forces of the robots, the robot can evenly divide the contact force in the reference frame $\{O\}$, which also ensures that the robots do not interfere with each other, then it yields

$$\boldsymbol{F}_i = \frac{1}{m} {}^{O}\boldsymbol{T}_{Mi}^T \boldsymbol{F}_o \tag{5.71}$$

Taking (5.68) and (5.70) into consideration, the joint torques can be calculated by

$$\boldsymbol{\tau}_i = \boldsymbol{J}_i^T\left[\overline{\boldsymbol{M}}_i\left({}^{Mi}\boldsymbol{T}_O\ddot{\boldsymbol{X}}_O + {}^{Mi}\dot{\boldsymbol{T}}_O\dot{\boldsymbol{X}}_O\right) - \overline{\boldsymbol{C}}_i\dot{\boldsymbol{X}}_i - \overline{\boldsymbol{G}}_i + \boldsymbol{F}_i\right]. \tag{5.72}$$

The acceleration is solved by the expected impedance equation of (5.69) as

$$\ddot{\boldsymbol{X}}_o = \ddot{\boldsymbol{X}}_{Oc} + \boldsymbol{M}_d^{-1}\left(\boldsymbol{B}_d\dot{\boldsymbol{e}} + \boldsymbol{K}_d\boldsymbol{e} - \boldsymbol{F}_O\right) \tag{5.73}$$

where the position and posture deviation $\boldsymbol{e} = -(\boldsymbol{X}_O - \boldsymbol{X}_{Oc})$.

Using the principle of feedback linearization, the impedance control law is

$$\begin{aligned}\boldsymbol{\tau}_i = {} & \boldsymbol{J}_i^T\overline{\boldsymbol{M}}_i{}^{Mi}\boldsymbol{T}_O\left[\ddot{\boldsymbol{X}}_{Oc} + \boldsymbol{M}_d^{-1}\left(\boldsymbol{B}_d\dot{\boldsymbol{e}} + \boldsymbol{K}_d\boldsymbol{e} - m{}^{Mi}\boldsymbol{T}_O^T\boldsymbol{F}_i\right)\right] \\ & + \boldsymbol{J}_i^T\left(\overline{\boldsymbol{M}}_i{}^{Mi}\dot{\boldsymbol{T}}_O\dot{\boldsymbol{X}}_O - \overline{\boldsymbol{C}}_i\dot{\boldsymbol{X}}_i - \overline{\boldsymbol{G}}_i + \boldsymbol{F}_i\right).\end{aligned} \tag{5.74}$$

The feedback control law of each robot is obtained and structured as shown in Fig. 5.10. Through this control law, the multi-robot system becomes a six-dimensional mass-spring-damping system.

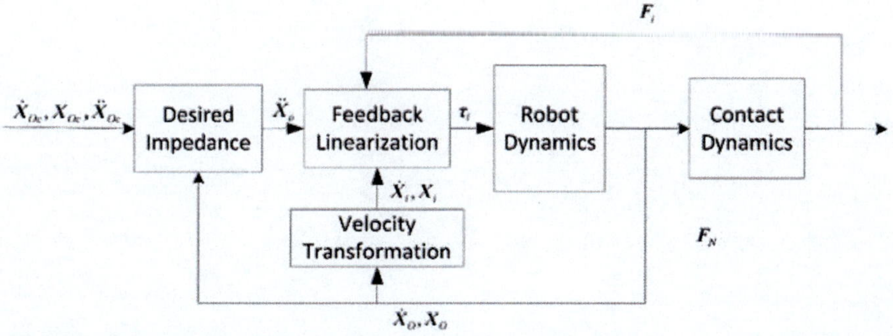

Fig. 5.10 Multi-robot impedance control

References

1. M.H. Raibert, J.J. Craig, Hybrid position/force control of manipulators. J. Dyn. Syst. Meas. Control **103**(2), 126–133 (1981)
2. T. Yoshikawa, T. Sugie, M. Tanaka, Dynamic hybrid position. Force control of robot manipulators–Controller design and experiment, 1987
3. N. Hogan., Impedance Control: an approach to manipulation, in *American Control Conference.* (IEEE, 2009), pp. 481–489
4. R.J. Anderson, M.W. Spong, Hybrid impedance control of robotic manipulators. IEEE Trans. Robot. Autom. **4**(5), 549–556 (1988)
5. W. Zhu, *Virtual Decomposition Control* (Springer, Berlin, Heidelberg, 2012)

Chapter 6
Space Robot System

6.1 System Design

Space robot system is a complex space system based on mechanics and control discipline, involving expertise and technologies in aerospace, materials, instruments, mechanics, optics, electronics, communications, computers, software, etc. [1]. The general processes of space robot system design include task requirement analysis, design feasibility study, preliminary system design, and detailed system design. The design of a space robot system includes overall design, configuration design, power supply and distribution design, information flow design, thermal design, interface design, ergonomic design, reliability design, safety design, maintainability design, and other design works. In addition, the system design also includes the development of technical processes and planning processes, and the design of system verification programs. System design is the top-level design of a space robot, which plays an important role in the process of system development.

6.1.1 Design Contents

Space robot system designs usually include the following tasks:

(1) Design elements and constraint requirements;
(2) Overall design;
(3) Configuration design;
(4) Power supply and distribution design;
(5) Information flow design;
(6) Thermal design;
(7) Interface design;
(8) Ergonomic design;
(9) Reliability design;
(10) Safety design;

Y. Wang, *Space Robotics*, Space Science and Technologies,
https://doi.org/10.1007/978-981-15-4902-1_6

(11) Testability design;
(12) Maintainability design;
(13) Supportability design;
(14) Selection of components, raw materials, and processes;
(15) Verification scheme design, etc.

6.1.2 Design Principles

(1) System optimization principle

Considering the weight, volume, power consumption, accuracy, and reliability, and other technical indices of the system, the system design is optimized to avoid performance degradation caused by the pursuit of local optimum.

(2) Inheritance principle

Fully inherit the mature design and process technology of space robots and even spacecraft, select mature products and components verified by on-orbit flight, and select materials and components with on-orbit application experience and long-term availability, reduce development costs and shorten the development cycle.

(3) Innovative principle

On the premise of ensuring the realization of system function and performance, new materials, new processes, new technologies, and new methods shall be appropriately adopted in the design to promote the development of space robots in lightweight, integration, and intelligence.

(4) Generalization principle

Generalized design shall be adopted as far as possible to reduce the variety of specifications and suppliers of raw materials, components, and parts, to reduce the management and control costs and risks, and the workload of development.

(5) Reliability principle

The space robot system is very complex and has a variety of failure modes. It is necessary to consider various reliability design measures from the perspective of improving system reliability to ensure the reliable operation of the product throughout its life cycle.

6.2 Design Elements and Constraint Requirements

In the design of a space robot system, resource constraints and space environment constraints shall be fully considered, such as the allowable dimensions, mass, power supply, and information by the spacecraft. In the design process, if it is very difficult or the cost is too high, it is necessary to fully communicate with the spacecraft program team to see if the design difficulty of the space robot can be reduced by optimizing the spacecraft system.

6.2.1 Functional Requirements

The main functional requirements of space robots include

(1) Motion function

After the space robot enters the predetermined working environment, it needs to have various motion functions: for the on-orbit operation robot, it usually requires the function of multi-DOF (usually more than two) movement; for the planetary exploration wheeled mobile robot, it is usually required to have multi-wheel steering and driving functions; for planetary exploration legged mobile robots, it is usually required to have multi-leg coordinated motion.

(2) End operation function

For on-orbit operation robots, they need to be configured with end effectors for specific purposes, which can perform the required operations for specific targets.

(3) Load handling function

Space robots are required to have the functions of carrying the load for position and attitude adjustment.

(4) Perception and measurement function

Space robots are required to have sensing and measuring functions for their working environment, and the shape, position/orientation, speed, temperature, and interaction force/torque of operating objects and other objects (astronauts, cooperative robots, etc.).

(5) Multimode control function

Space robots are required to have different modes to control their own motion according to the task requirements, especially position control, speed control, or force control.

(6) Decision-making and planning functions

Space robots are required to have analysis, judgment, decision, planning, and control abilities based on the perception and information of the tasks, such as task planning, motion planning, and troubleshooting.

(7) Multimode interaction function

Space robots are required to have multiple modes of operations, such as ground command operations, ground handling operations, and orbital astronaut operations.

(8) Hold-down and release functions

Space robots are required to be in the furled configuration during the launch stage, and be connected to the spacecraft through the hold-down mechanism to withstand the mechanical load during the launch; after entering the orbit, it is unlocked and released by the release mechanism, and the free motion functions are restored.

(9) Fault detection and handling functions

Space robots are required to have the ability to sense their own information, analyze the working state and produce analysis results in real time, and automatically take safety measures after detecting the fault to ensure the safety of itself and the operating object as well as the operating environment.

(10) Repair and upgrade functions

Space robots are required to have certain self-repair and on-orbit upgrade capabilities.

6.2.2 Performance Requirements

Different types of space robots have different performance requirements.

Performance indices of on-orbit robots usually include degrees of freedom, mass characteristics, envelope dimensions, power consumption, operation space, stiffness, positioning accuracy, speed/acceleration, end output force, and load capacity, etc.

The performance indices of space mobile robots usually include the number of wheels/legs, mass characteristics, envelope dimensions, power consumption, moving speed/acceleration, load capacity, obstacle-crossing capability, turning capability, and moving distance, etc.

6.2.2.1 Degree of Freedom (DOF)

This technical index is usually used for on-orbit operation robots. As the DOF requirements for space legged mobile robots are similar to that of the on-orbit operation robots, this technical index can be used as a reference. The number of DOFs reflects the dexterity of the robot's movements. For the on-orbit operation robot, the number of independent movements of the end effector relative to the base is called the freedom of motion of the robot, also known as Degree of Freedom.

6.2.2.2 Number of Wheels/Legs

This technical index is used for space mobile robots, and it specifies the number of wheels or legs on the robot. Generally, space wheeled robots are mostly of four wheels, six wheels, or eight wheels, while legged robots usually use two-, three-, four-, six-, or eight-leg forms.

6.2.2.3 Mass

This technical index specifies the maximum allowable mass of the space robot and, if necessary, the mass characteristics of the robot such as centroid, moment of inertia, or product of inertia.

6.2.2.4 Envelope Dimensions

This technical index specifies the maximum envelope dimension of the space robot when it is held down in launch state, and, if necessary, the envelope dimension requirements for other operation configurations.

6.2.2.5 Power Consumption

This technical index specifies the electric power that can be consumed by the space robot in different states, including the power consumption of the motor that actuates the robot, the thermal power consumption to keep the robot within the predetermined temperature range, the operation of electronic components or standby power consumption, etc. Commonly used power consumption indices include standby power consumption, average power consumption, steady-state power consumption, peak power consumption, etc.

6.2.2.6 Workspace

This technical index specifies the geometric range of all points that can be reached by the end of the robot arm or the center of the wrist, also known as the work area. Because the end effector has various sizes and shapes, in order to truly reflect the characteristic parameters of the robot, it usually refers to the workspace when the end effector is not installed. The workspace is classified into a reachable workspace and dexterous workspace. The former refers to the set of spaces formed by all the points that the robot can reach, also called the total workspace, and the latter refers to the set of spaces formed by all the points that the end of the robot can reach in any pose; it is a subset of the total workspace.

6.2.2.7 Stiffness

This technical index specifies the displacement of the characteristic part of the space robot under load. There are static stiffness and dynamics stiffness: The static stiffness is usually measured by the spatial displacement of the characteristic part of the space robot under static or quasi-static load, and the dynamics stiffness is generally measured by the natural frequency of the space robot in furled and expanded states. Space robots generally have both static and dynamics stiffness requirements.

6.2.2.8 Positioning Accuracy

This technical index is generally used for the robot on orbit and is usually measured by the position accuracy of the robot end, including absolute positioning accuracy and repeatable positioning accuracy. Absolute positioning accuracy refers to the maximum difference between the actual arrival position of the robot end and the theoretical target position; repetitive positioning accuracy refers to the maximum difference between the actual positions when the robot end moves repeatedly to the same theoretical target position.

6.2.2.9 Speed/Acceleration

This technical index specifies the motion characteristics of the robot, including the maximum stable speed and maximum acceleration that can be achieved.

6.2.2.10 End Output Force

This technical index specifies the maximum output force/torque at the end of the space robot in the specified configuration.

6.2.2.11 Load Capacity

This technical index specifies the maximum mass that a space robot can load in any pose in a specific environment or operating range. The load capacity is determined not only by the load mass, but also by the environmental conditions (ground gravity, orbital microgravity, and planetary surface gravity) and the motion parameters (speed, acceleration and direction of motion, etc.).

6.2.2.12 Obstacle-Crossing Capability

This technical index is generally used for mobile robots to specify the maximum obstacle size that can be passed in a particular environment.

6.2.2.13 Turning Capability

This technical index is generally used for mobile robots to specify mainly the maximum turning radius, maximum turning speed in a specific environment.

6.2.2.14 Climbing Ability

This technical index is typically used for mobile robots to specify the maximum slope angle on which robots can maintain normal movement.

6.2.2.15 Moving Distance

This technical index is generally used for mobile robots to specify the maximum mileage in a given environment.

6.2.2.16 Service Life

This technical index is used to specify the space robot's AIT (Assembly, Integration, and Test) time, ground storage time, on-orbit application time, and their sum.

6.2.2.17 Reliability

This technical index is used to specify the ability of a space robot to perform a predetermined function at a specified time and under specified conditions, which is usually assessed by probability.

6.2.3 Environmental Constraints

The environmental constraints of space robots mainly refer to the launch environment and work environment. The launch environment refers to the mechanical environment in the launch stage; the work environment is relatively complex; in addition to the mechanical load, the following space environmental factors need to be considered:

(1) Vacuum

The vacuum degree in the space environment can usually reach 10^{-2}–10^{-11} Pa. When choosing the materials used by space robots, it is necessary to consider the material outgassing, evaporation, sublimation, decomposition, dry friction, cold welding, and evaporation effect of liquid lubrication caused by the vacuum environment.

(2) Ultraviolet irradiation

The ultraviolet rays emitted by the sun have an impact on metals, ceramics, glass materials, and high polymers, among which the impact on the polymer is particularly significant; in severe cases, it can reduce the elasticity and strength of the material and deteriorate its mechanical properties. According to different mission orbits, the total UV dose should meet the requirements of the robot tasks.

(3) Atomic oxygen

It mainly refers to the residual gas environment in the form of atomic oxygen in the Earth's low orbit. The space robot in this orbit should meet the total atomic oxygen accumulation requirement during the mission period. The interaction between atomic oxygen and the spacecraft can cause erosion and aging of the spacecraft structural materials, damage the thermal control coating of the spacecraft, and seriously endanger the reliable operation of the spacecraft.

(4) Total ionization dose

It mainly refers to the effects of heavy ions and proton effects in space, which may cause the electronic system from not working normally and also to have physical and electrical effects on electronic devices. According to the mission orbit, it should meet the requirements of the robot task on the total ionization dose.

(5) Single-event effect

It mainly refers to the radiation effect that may cause an abnormal change of states when a single high-energy event passes through the sensitive region of the microelectronic device, including space single-event upset, single-event latch-up, single-event burnout, and single-event gate rupture effects. According to the mission orbit, it should meet the requirements of the robot task for single-event protection.

(6) Displacement effect

It mainly refers to the radiation effect due to which the high-energy radiation makes the atoms of semiconductor materials move from the original lattice position to another location. When selecting photoelectric devices such as CCDs and optocouplers for space robots, focus consideration should be attached to the anti-displacement effect requirements.

(7) Temperature alternating environment

It mainly refers to the temperature alternating load when the space robot alternately enters the illumination zone and the shadow zone, which will affect the performance of the structural and the lubrication material; in addition, it will cause thermal deformation of the structure and affect the motion performance of the space robot.

6.3 Overall Design

For the development of the new space robots, the overall design can be divided into four stages, including task requirement analysis, design feasibility study, preliminary system design, and detailed system design. The overall design work is gradually deepened and refined from the beginning to the end.

6.3.1 Task Requirement Analysis

After receiving the design and development tasks of space robot users, firstly, analyze the task requirement at the system level according to the user's requirements, design elements and constraints, including identifying the basic task requirements of the user, analyzing the rationality, correctness and integrity of the user requirements, and determining the basic technical approach required to fulfill the mission; then carry out a conceptual design of space robot systems based on the above analysis results, perform task profile analysis and preliminary planning, and initially propose the robot system configuration, functions, and interface requirements with other systems, identify the key technologies involved and clarify the technical approach; finally, form a number of conceptual solutions for the space robot system. Preliminary estimates of development costs and development cycles should also be conducted at this stage.

6.3.2 Design Feasibility Study

Based on the conceptual solutions formed in the task requirement analysis, carry out specific requirements and functional performance analysis of the space robot system; initially determine the system composition and configuration relationship, and raise the basic requirements for each subsystem and core equipment, including the mechanical system, control system, perceptual system, energy system, thermal control system, etc.; carry out initial system configuration and equipment layout design; make budget allocation according to overall parameters such as system mass and power; identify and determine key technologies, and organize key technology research; finally, formulate a feasible design scheme of the robot system.

6.3.3 Preliminary System Design

According to results of the feasibility study, determine and decompose the overall technical indices of the space robot system, complete the design of system configuration and layout; perform the preliminary design of the system components, such as the mechanical system, control system, perception system, energy system, thermal control, according to the system interrelationship; develop the system design specifications, identify requirements for reliability, safety, space environment adaptability, and testing; determine the technical requirements of each subsystem and equipment; complete key technical research, carry out interface coordination and design of the robot and related systems to form an overall scheme of the space robot.

6.3.4 Detailed System Design

The detailed system design is the optimization, deepening, and refinement of the preliminary system design, which is the specific implementation process of the overall scheme. The detailed system design supports the development and test verification of Structural and Thermal Model, Electrical Model, Qualification Model, and Flight Model of space robot. Detailed system design needs to determine the overall technical specifications and interface requirements, and form the mechanical, electrical, information, and thermal interface data sheets of each equipment. According to preliminary system design, complete the detailed system configuration, layout, and assembly design; complete detailed design of the components of the mechanical system, control system, perceptual system, energy system, and thermal control system; complete system space environmental analysis, failure mode and impact analysis, reliability and safety design according to design specifications; complete on-orbit task flow and flight procedure design; develop an experiment, test programs, and test matrix; prepare detailed design report of the system, subsystem and equipment, and establish a product baseline as the basis for subsequent product development.

6.4 Configuration Design

The configuration of a space robot refers to the connection relationship among various modules of the robot, including information such as types of modules, number of modules, and the connection relationship between the modules. It usually has at least one hold-down state configuration and one working state configuration, sometimes including long-term parking, overnight, health check, and other special requirement configurations.

The configuration design of a space robot system is an iterative optimization process that continuously deepens the overall design. In general, a space robot undertakes multiple tasks throughout its life cycle. In the initial stage of configuration design, it usually starts with the design of robot functions and performances. Based on this, the robot design constraints of specific requirements of each task, such as the workspace, motion trajectory, and obstacle-avoidance capability, are considered. In addition, the configuration design also needs to consider the mass characteristics, dimensions, power consumption, field of view, interface, thermal control, electromagnetic compatibility (EMC), assembly, testing, and other aspects of the various components of the space robot. The space robot system configuration is very important for the overall space mission. Reasonable system configuration design can achieve better performance at a lower cost and reduce the resource requirements of the spacecraft system. The configuration design of a space robot includes

(1) Determine the number of DOFs and configuration of the robot. In general, a robot is a multi-DOF system consisting of links and joint modules. Therefore, the robot configuration design should first determine the number of DOFs and configuration relationship of the links and joints. When configuring the DOFs, attention should be paid to avoid singularities and reduce its effects through proper motion planning during use.
(2) Determine the shape of the robot. This refers to the robot shape as a whole, including various modules of the robot. The working space of the robot should be expanded as much as possible through a reasonable layout of each module and make the robot aesthetic.
(3) Reasonably design the load paths. The load paths affect the force state of each module of the space robot, and reasonable load paths can help to reduce the weight and energy consumption of the system. Different from the ground robot, when designing the space robot, it is necessary to consider the load paths under the hold-down configuration during the launch stage.

There are two ways for robot configuration design: one is to modify and redesign adaptively according to the existing robot configuration; the other is to start from the topology design of the mechanism, and obtain a series of constraint conditions by analyzing the characteristics of the task objectives, list all optional configurations that satisfy the constraints, then analyze the characteristics of each optional configuration, and select the optimal configuration scheme according to certain principles.

Professor D.L. Pieper studied the inverse kinematics of six-DOF robots and found that the sufficient condition for the analytic solution of the robots is that the adjacent three joint axes intersect at one point, and then he proposed a design guideline that the robot arm structure and the wrist structure be considered separately—the principle of arm–wrist separation. Professor Lung-wen Tsai integrated the position configuration and attitude configuration of the robot arm, and proposed the optimal position configuration and the optimal attitude configuration forms [2].

By using the optimal position configuration and optimal attitude configuration, a six-DOF robot with optimal dexterity can be constructed. Robots with more than six DOFs (such as seven- or eight-DOF robots) usually have the best dexterity by adding several DOFs to the above-described six-DOF robot.

The system configuration design of complex robots often needs to integrate the above design ideas and consider the task requirements of different levels. Figure 6.1 shows a complex integrated mobile and operation robots with six sets of wheel–leg combined robot arms (legs), each of which includes an end effector and wheels. In performing tasks, the configuration can change according to different tasks, to achieve various applications such as hold-down launch configuration, wheeled mobile configuration, legged mobile configuration, and mobile operating configuration.

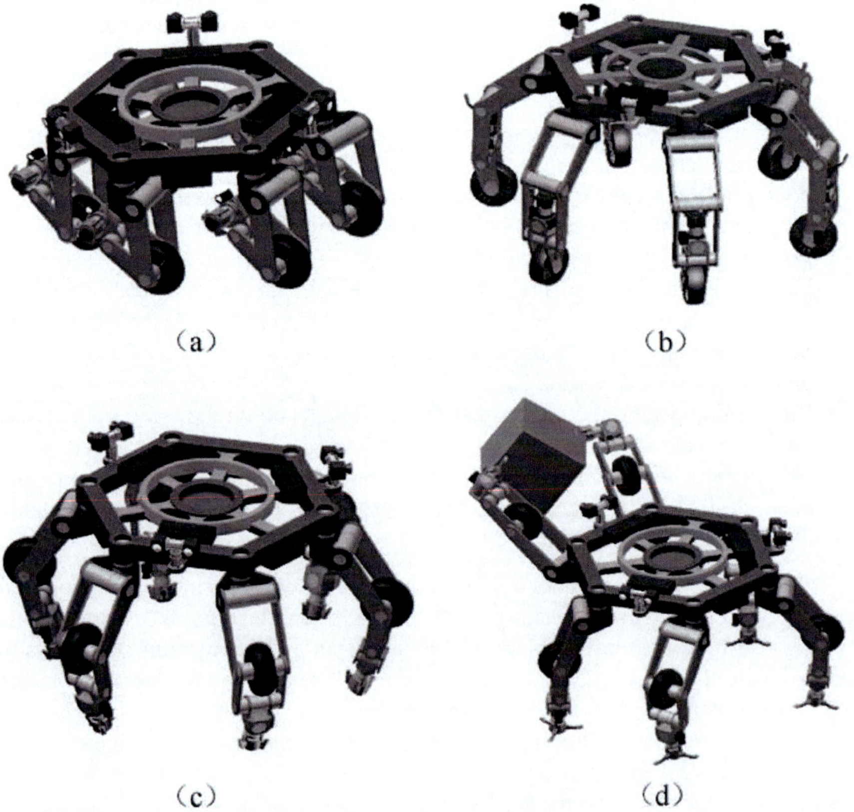

Fig. 6.1 Integrated mobile and operation robots **a** hold-down launch configuration; **b** wheeled mobile configuration; **c** legged mobile configuration; **d** mobile operation configuration

6.5 Power Supply and Distribution Design

The following should be taken into account in the power supply and distribution design of space robots:

(1) Meeting the energy supply requirements of the space robot during the mission.
(2) Internal power supply and distribution design of equipment should, as much as possible, be simple, using software functions to replace hardware functions, using recommended circuits in the interface design, ensuring the reliability and standardization of the design.
(3) Using verified standard components, standard unit circuits, standard electronic functional modules, and general-purpose components.
(4) The derating design of components used in power supply and distribution meets the derating criteria for electrical, electronic, and electromechanical components of the spacecraft.
(5) The initial power-ON state of each equipment of the robot should be known and reasonable, safe, consistent, and repeatable.
(6) The circuit of the hazardous system shall be of fail-safe design. It is not allowed that any one-time operation error or component failure may cause catastrophic consequences of the whole system.

6.6 Information Flow Design

Information flow design of space robots is generally considered as follows:

(1) When the system is a distributed architecture, the related equipment should be connected by the bus to reduce the number of cables, simplify software design, and improve reliability.
(2) Selecting appropriate bus type according to the requirements of communication rate, reliability, service life, cost, and working environment characteristics. The commonly used buses in the aerospace field include RS485, CAN, 1553B, IEEE 1394, etc.
(3) It should match with the hardware design to ensure that the performance of the hardware product meets the information flow design requirements.
(4) It should match the operation modes to meet the requirements of data transmission and fault monitoring under each operation mode.
(5) It should be designed according to the critical degree and change rate of data, focusing on the data that characterizes the critical state of the system.
(6) It should consider the functional performance of the system under fault conditions to ensure fault isolation.
(7) It should ensure the transmission reliability of key data, such as remote command, control data, and key status data. If necessary, measures such as redundant design, error detection, and correction coding design can be adopted.
(8) It should consider the impact of data delay and asynchronous on the system.

(9) It should improve data transmission efficiency and reduce data transmission irrelevant to the operation process.
(10) The results of command execution should be obtained clearly by telemetry as much as possible. The characteristic state of the important fault must be informed by telemetry and transmitted as soon as possible.
(11) The key instruction functions related to system safety shall be arranged with at least two independent instructions to ensure the execution of the instructions.

6.7 Thermal Design

Thermal design of a space robot system is to ensure the on-orbit temperature conditions of the components of the space robot. According to the thermal characteristics of space robots and the overall technical requirements of thermal control systems, a space robot thermal control design usually adopts a combination of passive thermal control technology and electric heating active thermal control technology. Specific design principles include

(1) Selecting low-power components and trying to use a robot structure that is conducive to heat transfer.
(2) Sufficient thermal analysis should be carried out for the entire robot, to ensure it meet the temperature derating requirements of the thermal control products, and the extreme conditions of each equipment of the robot should also be considered.
(3) If necessary, heat balance test to the robot should be conducted to verify the correctness of the thermal analysis and the rationality of the thermal design, and a basis for improving the thermal design should be provided.
(4) In order to ensure the on-orbit maintainability of the robot, it is necessary to design generalized thermal control for the same orbit replaceable parts.

6.8 Interface Design

Interface design is a very important technical state control target in space robot design. A space robot interface usually includes mechanical interface, power supply interface, information interface, thermal interface, etc. The spacecraft developmental model typically uses an interface data sheet (IDS) for configuration control.

6.8.1 Mechanical Interface

The mechanical interface includes the mechanical interface between the robot and the spacecraft, the interface to the equipment components, accessories, brackets, and the interface with ground installation, lifting and test equipment, etc. The requirements include the connection form, position, size, and surface accuracy.

6.8.2 Power Supply Interface

The power supply interface includes the power distribution interface between the robot and the spacecraft, including the types of power bus (primary power bus, command power bus, pyrotechnic power bus, etc.), interface voltage, voltage conversion constraints, etc.

6.8.3 Information Interface

Information interface is the information path between the robot and the spacecraft. Through this path, the interaction between control parameters and the measurement data of the robot can be realized. Information interfaces include bus interface (1553B bus, Ethernet bus, CAN bus, etc.), data type, telemetry, and instruction resources.

6.8.4 Thermal Interface

The thermal interface mainly refers to the thermal characteristics of the contacting surfaces between the robot and the spacecraft, the thermal characteristics of the outer surface, the operating temperature, the storage temperature, the control temperature, and the heating power. It is necessary to analyze and determine the operating temperature range requirements for equipment and components, and clarify the thermal control requirements according to the configuration of the space robot and the layout of the equipment, as well as other factors, such as materials, dimensions, power consumption, work cycle, and environmental conditions.

6.9 Ergonomic Design

The ergonomic design of space robots is generally considered as follows:

(1) For the equipment with operation and observation interfaces on the spacecraft, ergonomic design should be performed according to the shape, function, and flight mission of the product so as to meet the on-orbit operational requirements of the astronauts in the weightless state.
(2) Each equipment must meet the mechanical design requirements and eliminate mechanical hazards to prevent safety problems caused by burrs, sharp edges, sharp corners, and protrusions.
(3) Astronauts should take measures to prevent mis-operation.
(4) Space robot surfaces that cannot be stepped on or where operations may be carried out should be clearly marked.

(5) The ergonomic design of a space robot should be evaluated by the ergonomics of the astronaut system.

6.10 Reliability Design

The reliability design of space robots is generally considered as follows:

(1) The reliability design of each equipment should meet the specific environmental requirements of spacecraft, such as component derating, heat, anti-mechanical environment, anti-radiation environment, EMC, and antistatic and vacuum discharge.
(2) Redundant and backup measures should be taken for weak areas of reliability; the choices should be weighted by comprehensively considering the corresponding measures to optimize the system reliability design.
(3) The design should eliminate the single point failure mode as much as possible. If it cannot be eliminated, effective measures should be taken to improve the inherent reliability.
(4) All electronic products and non-electrical components with failure rate data are required to be assigned specific reliability index requirements and to carry out reliability estimation. Components with field failure rate data are estimated using existing data. Otherwise, for Chinese components, it is estimated by the method described in the Electronic Equipment Reliability Estimation Manual (GJB/Z299C-2006); for components that are imported from other countries, detailed estimations are made with reference to the stress method described in the Reliability Estimation Manual (MIL-HDBK-217F), and a certain degree of confidence should be considered during estimation [3].

6.11 Safety Design

The safety design of space robot products should enable the robot system or its equipment to prevent mis-operation and generation of dangerous sources, and the control and protection of hazards. The basic principles of spacecraft product design are safety first, minimum risk, sufficient margin, fault isolation, fault tolerance, fault alarm, and fail-safe design. The specific safety design requirements for the robot operation process are as follows:

(1) Software/hardware system safety check function.
(2) Independent control circuit for emergency termination function.
(3) The function of collision detection using image monitoring and measurement information of the camera.
(4) Overload protection for the joints and end effectors by software and independent control circuits.

(5) Brake and lock function for the joints and end effectors to prevent accidental release.
(6) Motion limit in software for the joints and end effectors.

6.12 Testability Design

The testability design of space robot products should ensure that the robot system or its equipment can determine its status timely and accurately (such as operational, inoperable, and degraded), and isolate the fault when it occurs. The following principles should be followed in system testability design:

(1) According to the characteristics of different types of robot products, comprehensively consider, coordinate, and weigh the performance, reliability, maintainability and safety design of the product, and propose corresponding testability requirements.
(2) According to the differences between the test conditions and methods in three different operating environments, including ground test, on-orbit unmanned flight, and on-orbit manned flight, propose corresponding qualitative and quantitative testability requirements.
(3) According to the testability requirements, by trade-off analysis and combination of the inherent testability design, built-in-test (BIT) design, automatic testing, and manual testing, and other available resources, jointly develop an optimal test plan and fault diagnosis solutions to support on-orbit maintenance.
(4) According to the test requirements, determine reasonable test content, test items, and test parameters to meet different levels of testing and maintenance requirements during product development and use.
(5) According to the determined test items and parameters, correct test points are selected, so as to measure and excite the internal functional circuits, thus reaching a higher level of inherent fault detection and isolation.

6.13 Maintainability Design

The maintainability design of a space robot is generally considered as follows:

(1) It should ensure that there are mis-operation prevention measures during ground and on-orbit maintenance. All types of connectors must be clearly identified without hidden risks of the operation.
(2) The testability design for maintenance should be performed, the test points should be properly configured; the automatic detection and fault isolation capabilities should be improved, and the demand for special test equipment should be reduced.
(3) The maintenance levels include modules (boards), component and assembly levels, and should be determined in combination with the testability design

of the equipment. The maintenance design should be done according to the requirements on different levels.

(4) Identifying the service life features and characteristic parameters of the product, and determining the maintenance categories in combination with the testability design: for products whose life expectancy does not meet design requirements, preventive maintenance should be taken; for products without wear-out, burstiness, and occasionality, repair maintenance should be adopted.
(5) On-orbit maintenance should not affect system safety.
(6) Robots and operating procedures that require on-orbit maintenance should meet ergonomic requirements, thereby improving the maintenance quality and efficiency, reducing maintenance workload, and lowering the requirements for astronaut maintenance skills.

6.14 Supportability Design

The supportability design of space robots is generally considered as follows:

(1) Reducing the complexity of on-orbit operations and cutting the mission time.
(2) Cutting the on-orbit abnormal recovery time by measures such as single-event protection design.
(3) Reducing the number of on-orbit maintenance personnel and the workload of fault handling.
(4) Reducing the demand and dependence on ground support equipment and facilities, such as the demand for monitoring and control resources.
(5) Reducing special maintenance equipment.

6.15 Selection of Components, Raw Materials, and Processes

6.15.1 Selection and Design of Raw Materials

(1) The metal and nonmetal raw material of each equipment of a space robot shall be flight-verified materials and shall meet the requirements for its corresponding space environment.
(2) It is preferred to use lightweight, high strength, and corrosion-resistant materials.
(3) The selection of raw materials should meet the requirements of the designed structure, including carrying capacity, stiffness, and manufacturing process.
(4) It should focus on identifying new materials, prohibited (restricted) materials, and frequently occurring problems, strictly control the selection, identification, and use of raw materials to meet the material requirements.

6.15.2 Selection and Design of Components

(1) The individual components of the space robot shall meet the spacecraft project requirements (such as quality grade, component policy, manufacturing and delivery schedule, budget, and quantity), design requirements (such as component type, casing, dimensions, material, performance, and parameters), production requirements (such as packaging and sealing, heat and storage conditions, and component mounting process), and operational requirements (such as power supply, mechanical environment, radiation, reliability, and service life).
(2) It is preferred to use domestically produced components and maximize the selection of components that have flight experience and reliability data or those successfully applied in spacecraft with the same designed service life.
(3) The selection of components should consider the requirements of performance, parameters, such as anti-mechanical environment design, derating design, thermal design, tolerance design, radiation protection design, and electrostatic protection design. The variation range of the parameters should be fully considered during component design and selection.
(4) Reduce the types, specifications, and suppliers of components as much as possible, and use standard components whenever possible.
(5) Select those components from qualified suppliers with reliable quality, good service, and on-time delivery.
(6) It is prohibited to use components that are known to be unstable or that may pose a safety hazard or reliability risk.

6.15.3 Process Selection and Design

Compared with the general process, the space process is characterized by high reliability, long life, and non-maintainability in application. The following principles should be followed in the processes of selection and control:

(1) The processes used in the production of equipment of space robots should be those listed in the spacecraft catalog. It is strictly prohibited to choose the forbidden process and the limited processes should be avoided.
(2) The following process selection sequence shall be followed: enterprise standards, national military standards, space industry standards, aviation standards, national standards, other industry standards, and specialty standards.
(3) Focus on identifying the problems and hidden risks of new processes, prohibited (limited) processes, common processes, and strictly control the selection, identification, and implementation of the processes to meet the specific requirements of products.

6.16 Verification Scheme Design

In the development of space robots, ground verification is one of the most difficult problems and also one of the most core technologies. Space robot verification is a systematic work, involving the verification ranging from materials to components, from parts to systems, and from missions to the verification of large space-ground loop operations. Although the ground verification (physical verification) of all physical flight models is the best verification method, in terms of system-level and task-level verification, it is difficult to achieve full physical verification in most cases due to ground gravity. Therefore, the ground verification of space robots generally adopts the combination of mathematical simulation verification, semi-physical simulation verification, and physical test verification to meet the test coverage requirements of space robots, and ensure the adequacy and comprehensiveness of product validation and task verification.

References

1. Y. Xiong, *Robotics* (Mechanical Industry Press, Beijing, 1993)
2. B. Siciliano, O. Khatib, *Handbook of Robotics* (Springer, New York, 2007)
3. J. Ellery, J. Kreisel, B. Sommer, The case for robotic on-orbit servicing of spacecraft: spacecraft reliability is a myth. Acta Astronaut. **63**, 632–648 (2008)

Chapter 7
Space Robot Mechanical System

7.1 Overview of Mechanical System

Mechanical system is the core of space robot, which is used to enable the motion functions of space robot, and its performance directly affects the application effect of space robot. Mechanical system design generally includes material selection and structural parts design, mechanism parts/components design, space lubrication design, and verification scheme design. The mechanical environmental conditions are the main constraints on the mechanical system design of space robot.

The mechanical system of space robot usually includes structure, joint, end effector, and/or mobile wheel, etc., see Table 7.1 and Fig. 7.1.

7.1.1 Structure

The structure typically provides the mounting interface for other components of space robot, and has the functions of connecting, supporting, and shape retaining.

7.1.2 Joint

The main function of the joint is to output a certain torque/force and rotation speed/displacement. In order to achieve this function, the joints are usually equipped with the drive and transmission components and the measuring components such as position, speed, and torque sensors.

Y. Wang, *Space Robotics*, Space Science and Technologies,
https://doi.org/10.1007/978-981-15-4902-1_7

Table 7.1 Main components of space robot mechanical system

No.	Parts	Space operation robot	Space mobile robot
1	Structure	Base, arm boom, connector, etc.	Rover body, suspension structure, wheel, etc.
2	Joint	Rotating joints, straight joints, etc.	Steering drive assembly, wheel drive assembly, etc.
3	End effector	Gripper, dexterous hand, sampler, etc.	Drilling, crushing devices (sometimes used for the wheel-legged mobile robot feet)
4	Mobile mechanism	/	Suspension mechanism, steering mechanism, etc.
5	Hold-down release mechanism	Hold-down device, release device	Hold-down device, release device
6	Drive controller	Joint and end effector drive controller	Wheel steering, drive and sampler controller
7	Cable assembly	Circuit and cable network	Circuits and cable network
8	Thermal control assembly	Thermal control coating, multilayer heat insulation components, thermistors, thermal control circuits, etc.	Thermal control coating, multilayer heat insulation components, thermistors, thermal control circuits, etc.

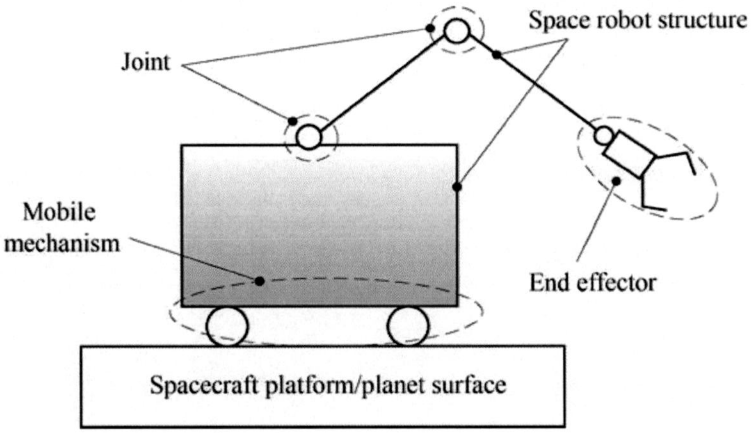

Fig. 7.1 Composition of space robot mechanical system

7.1.3 End Effector

The end effector is mainly used to perform various operational tasks, which is usually installed at the free end of a space operation robot. The end effector is usually equipped with various sensors for motion control.

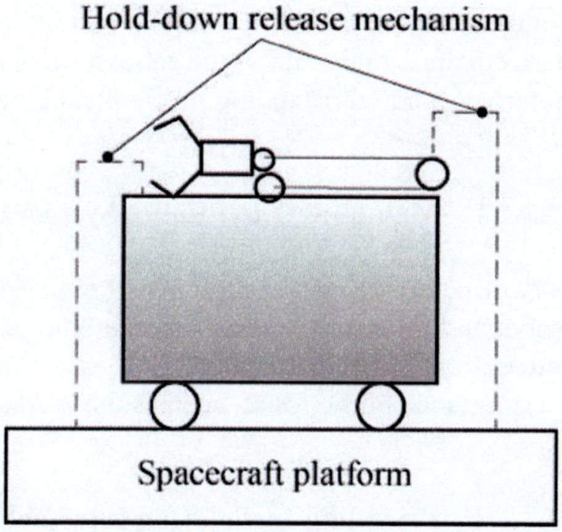

Fig. 7.2 Schematic diagram of the hold-down state of the space robot

7.1.4 Mobile Mechanism

The mobile mechanism is used to make the robot move freely. For example, through the Mobile Base System (MBS) of the International Space Station's Mobile Servicing System (ISS MSS), MSS can move on the truss structure of ISS. In addition, for the planetary exploration robot, the mobile mechanism is its main equipment, and the common forms are leg-type, wheel-type, and track-type mechanisms.

7.1.5 Hold-Down Release Mechanism

The main function of the hold-down release mechanism is to make the space robot furl in the spacecraft, so as to ensure that the space robot can withstand the mechanical load without damage; before the space robot starts to operate, the hold-down release mechanism unlocks and releases the constraints to the space robot, and resumes in its normal working conditions. Figure 7.2 shows the hold-down state of the space robot.

7.2 Structure of the Space Robot

7.2.1 Structural Function

Space robot structure is the general term of parts/components that connect all parts of the robot and form the system configuration, provide the installation basis for equipment, and maintain a certain rigidity and dimensional stability. Its main functions include maintaining the robot system configuration, providing the installation interface, and withstanding mechanical loads.

7.2.1.1 Maintaining the Robot System Configuration

The structure determines the installation relationship between the equipment of the robot and affects the main characteristic parameters of the robot, such as envelope dimensions, D-H parameters. The structure also has an important impact on the performance of the robot, such as the workspace and the natural frequency.

7.2.1.2 Providing Installation Interface

The structure provides installation and connection interface for other components of the robot, such as the joints, end effectors, cameras, controllers, and thermal control assemblies. It also ensures the accuracy of the equipment and payloads installed on it and provides protection for the internally installed equipment.

7.2.1.3 Withstanding Mechanical Loads

In the spacecraft launch stage (including the landing stage), space robots are usually fixed to the spacecraft by the hold-down release mechanism. In this stage, the space robot structure generally acts as the main load path to withstand the mechanical loads. In addition to ensuring that the structure itself is not damaged, it should also provide a good mechanical environment for the equipment and payloads mounted on it.

According to the shape, the robot structure generally includes bar-beam structure, shell structure, and box-plate structure. The bar-beam structure is usually a circular thin-walled boom, such as an arm boom (Fig. 7.3); the shell structure is usually a thin-walled structure of a revolving body or another curved shape, such as a joint shell; a box-plate structure is usually assembled by structural panels, such as a mobile robot vehicle structure.

7.2.2 Structural Materials

7.2.2.1 Metal Materials

The metal materials used in space robot structures are mainly low-density light metal materials. Commonly used materials include aluminum alloys, magnesium alloys, and titanium alloys. Transmission mechanisms, such as gears and bearings, usually

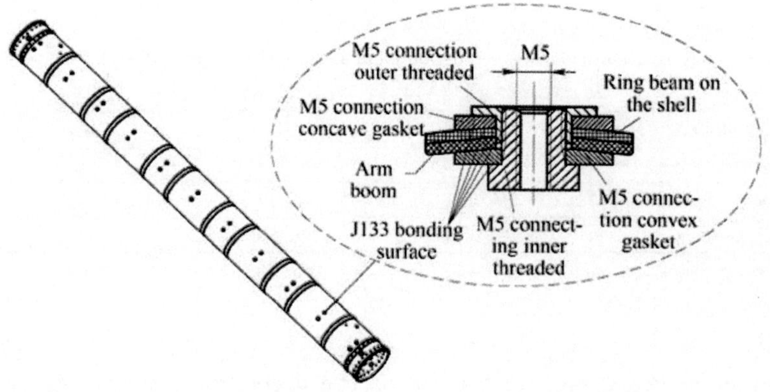

Fig. 7.3 Space robot structural parts (arm boom)

use high-strength stainless steel. The following mainly introduces aluminum alloys, magnesium alloys, and titanium alloys.

(1) Aluminum alloy

Aluminum alloy is the most widely used light metal material nowadays, its main characteristics: (a) low density, high specific strength, and specific modulus; (b) mature manufacturing technology and excellent machinability; (c) good thermal and electrical conductivity; (d) natural formation of oxidation protective film on the surface, with good anti-corrosion performance; (e) low-cost. Aluminum alloy is widely used in space robot structures, such as joints, end effector shells, connection flanges, and bases.

(2) Magnesium alloy

Magnesium alloy is the material with the lowest density in current light metal materials. Its main characteristics are: (a) high specific strength and specific modulus due to its low density; (b) good vibration damping capacity, able to withstand large impact loads; (c) good machining, casting, and forging characteristics; (d) good thermal and electrical conductivity. Magnesium alloys are mainly used on low-loaded structural parts on space robots, such as electronic equipment housings and small brackets.

(3) Titanium alloy

Compared with other metal materials such as aluminum, magnesium, and steel, titanium alloy has the following advantages: (a) very high specific strength due to its low density and high strength; (b) high- and low-temperature performance is satisfied, with the operating temperature range of −273 to +500 °C; (c) good anti-corrosion against atmosphere, seawater, acid, and alkali, with the best corrosion resistant material among current metal structural materials; (d) good anti-fatigue performance; (e) very low thermal conductivity, suitable for heat-insulating components; (f) low linear

Table 7.2 Performance of common light metal materials

Performance (Chinese brands)	Aluminum alloy	Magnesium alloy	Titanium alloy
	2A14	MB2	TC4
Density ρ/(g cm^{-3})	2.8	1.78	4.43
Tensile strength σ_b/(MPa)	451.3	235.4	900
Tensile modulus E/(GPa)	70.6	42.2	110
Elongation/(%)	14	5.0	10
Linear expansion coefficient/($\times 10^{-6}$ °C^{-1})	23.6	26	8.8

expansion coefficient, suitable for components with good dimensional stability. Titanium alloys are often used in structures that have stringent thermal deformation requirements, such as the locking mechanism of the end effector, the elastomer of the force/torque sensor.

The main performance indices of the above light metal materials are shown in Table 7.2.

7.2.2.2 Composite Materials

Composite material is a new kind of material composed of two or more materials with different properties. It not only retains the main characteristics of the original materials, but also obtains new properties that the original components do not have. The properties of the components can be complemented and correlated by design, so as to obtain new superior performance, which is the fundamental difference from the simple mixing of various materials.

The structural composite consists of matrix and reinforcement and their interface. The properties of the composite depend on the properties of the reinforcement and the matrix, the composition ratio, and the interface bonding. According to the classification of matrix materials, the composite materials commonly used in space robot structures are resin matrix and metal matrix composites.

(1) Resin matrix composite

Epoxy resin is generally used as the matrix of resin composite structural materials, which can support and protect the reinforcement and maintain the shape of the composite. The mechanical properties of the composite are mainly dependent on the reinforcement. Most of the current reinforcements are made of long-fiber materials, which directly affect the properties of the composite, such as density, strength, stiffness, coefficient of linear expansion, and the cost. The reinforcement for spacecraft structural composites is mainly carbon fiber. At present, the arm boom, rocker arm, and rover body are all made of carbon fiber reinforced composite.

In fiber reinforced composites, if all the fibers are arranged in the same direction, the composite is called unidirectional composite; if a composite is formed by many unidirectional composites laminated together with different fiber directions, it is called laminated composite. In fact, fibers can sometimes be braided into a form of cloth or cubic fabric and then compounded with the matrix material.

(2) Metal matrix composites

Metal Matrix Composite (MMC) is a kind of heterogeneous mixtures using metals or alloys as the matrix and metallic or nonmetallic wires, filaments, fibers, whiskers or granular components as the reinforcing phase, which commonly have continuous metal matrix. Since the matrix is a metal, MMC has the advantages similar to those of the metal, such as high strength, high elastic modulus, high toughness, low thermal shock sensitivity, low surface defect sensitivity, and good electrical and thermal conductivity. Generally, the reinforcing phase is a nonmetallic material with high strength and high modulus, such as carbon fiber, boron fiber, and ceramic material. The main purpose of adding the reinforcing phase is to compensate for the lack of certain properties of the matrix and to improve stiffness, wear resistance, high-temperature performance, and thermophysical properties. For the application of structural materials on space robots, more focus is given to the stiffness and lightweight design. Its operational environment is generally not higher than 450 °C, so the commonly used metal matrix composite is the Al-matrix composite.

7.2.2.3 Material Selection

The selection of structural materials is the first step in space robot design. Only when the materials are selected, the design and analysis of the structures can be carried out. Compared with ground robots, more attention should be given to the requirements for lightweight design and space environment adaptability in the selection of structural materials for space robots. In general, the following points should be considered [1].

(1) Select low-density materials

Space robots must be carried by the launch vehicle, but the thrust of the launch vehicle is limited. In order to reduce the launch cost, the weight of the space robot is strictly limited. In addition, since all the structural weight exists as the loads for the moving parts such as joints, too much mass often leads to a degradation in the dynamics performance of space robots, which is another major reason for structural weight reduction.

(2) Select materials with high modulus and/or strength

The most concerned mechanical properties of structural materials are elastic modulus and allowable strength. In practice, the selection should be made in conjunction with the low-density requirements mentioned above. For this reason, the two parameters, specific modulus and specific strength, are often used as the basis for structural material selection. The former is the ratio of the elastic modulus to the density of the material, and the latter is the ratio of the allowable strength to the density of the material.

(3) Select materials according to the specified thermal performance requirements

The material will respond to the ambient temperature, and exhibits different thermal properties, including thermal expansion, heat capacity, and thermal conduction. Their related parameters are linear expansion coefficient, specific heat capacity, and thermal conductivity. Because the space robot structures will stay in the space environment with a large temperature difference, in order to prevent the structure from being excessively stressed or deformed due to temperature change, it is desirable to use a material with smaller linear expansion coefficient, and adjacent structures are also expected to use materials with similar linear expansion coefficients. In addition, higher specific heat capacity and thermal conductivity are also required, because materials with higher specific heat capacity usually have larger heat capacity, and can reduce temperature variation of the structure. Materials with higher thermal conductivity can make the temperature distribution of the structure more uniform and prevent excessive stress or deformation.

(4) Select materials according to space environment requirements

Space robots usually have a long-life in orbit. For structural materials that are exposed to the space environment for a long time, the performance change should not exceed the designed allowable range, that is, the material should have good stability in the space environment. In addition, in the vacuum environment of the space orbit, it will release some substances with low relative molecular mass due to the sublimation of the material (especially of organic materials), thus forming a vacuum outgassing phenomenon, which may not only degrade the performance of the material, but more importantly may contaminate the surface of adjacent optical, thermal or electrical equipment. Therefore, the vacuum outgassing must be limited for the space robot structural material. Generally, it is defined that the Total Mass Loss (TML) of the material should not exceed 1%, and the content of Collected Volatile Condensable Material (CVCM) should not exceed 0.1%. Moreover, chrome plating or galvanizing on the surface of parts in space is prohibited, because chromium and zinc will sublimate in vacuum.

In addition to the above material properties considerations, other conditions related to the use of the material, such as the manufacturing technology of the material, cost control, supply, and processing conditions, should also be considered. In a word, the selection of structural materials for space robots is a comprehensive analysis process, and the appropriate materials should be finalized only after weighing various requirements and conditions.

7.2.3 Structure Design

7.2.3.1 Structure Design Process

The structure design of the space robot can be divided into two stages: system structure design and component structure design. The system structure design is usually combined with the system design of the space robot and is actually the preliminary design of the component structure. The complete process of structure design is shown in Fig. 7.4. Firstly, the system configuration is determined according to the relevant constraints of the system design, thus determining the preliminary state of each component design; then simulation analysis of the system structure is performed according to the payload conditions, so as to obtain the detailed requirements of each component; finally, the structure design of each component is optimized according to these requirements and determine the final state of each component structure design.

(1) Preliminary design of system structure

The preliminary design of the system structure is closely related to the system design of the space robot. The "system structure" here refers not only to all kinds of structural connectors in the space robot, but also the mechanism products such as joints and end effector in the power-off braking state, which are considered together as structures. The constraints of the preliminary design of the system structure mainly include spacecraft interface constraints, configuration layout constraints, and orbital

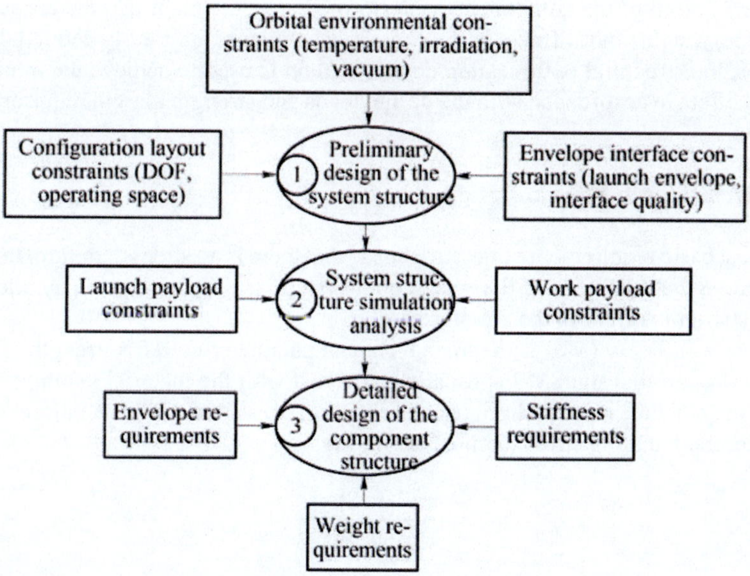

Fig. 7.4 Design process of space robot structure

environment constraints. Spacecraft interface constraints are requirements derived from the launch vehicle capability and the on-board layout, which are reflected mainly in the requirements on the weight, geometric envelope, and installation interface of the space robot. The configuration layout constraints are the requirements derived from the task demands. These tasks often require the space robot to have a certain workspace and dexterity, which will affect the selection of robot's DOF, type of joints and layout, as well as D-H parameters. Orbital environment constraints mainly include the temperature, vacuum, and irradiation conditions of the space robot's working orbit, which directly affects the selection of structural materials and connection modes. In a word, the important elements of the space robot are basically determined in the preliminary structure design, including structure materials, the overall layout, the contour envelope, and the connection modes.

(2) Structure analysis and simulation

According to the preliminary design state of the system structure, the structure analysis and simulation is performed based on the launch load and the working load. The former refers to the structural analysis of the space robot in hold-down and furled state; the latter refers to the dynamics analysis of the space robot in the on-orbit working state. By these two kinds of analysis and simulation, the main parameters such as geometric and physical parameters of each component structure can be further determined.

(3) Detailed design of the structure

Detailed design of the components is performed after the main parameters such as envelope, weight, and stiffness of the component structure are clearly defined. In this process, the structural optimization design method is used to achieve the minimum weight effect in accordance with the design loads and strength margin requirements.

7.2.3.2 Strength Margin Design

The most basic requirement of the structure is that there is no strength failure. In order to improve the reliability of the space robot system, it is necessary to pay attention to the strength margin in the structure design.

Margin of Safety (MS) is the main technical parameter used for strength calibration in structural design. MS is usually calculated with the material's failure stress, which may include material limit (fracture failure) stress, yield (plastic failure) stress, or structural buckling stress, defined as follows:

$$\mathrm{MS} = \frac{\text{Failure Stress}}{\text{Design Stress}} - 1. \tag{7.1}$$

The design load is defined as

Table 7.3 Minimum structural MS values [1]

MS of metal structure		MS of composite structure
Calculated by the material yield strength	0	0.25 (calculated based on the criterion of first layer failure)
Calculated by the material ultimate strength	0.15	
Calculated by component stability	0.25	0.3 (calculated based on the criterion of first layer failure)

$$\text{Design Load} = \text{Using Payload} \times \text{Design Safety Factor}. \quad (7.2)$$

For metal structures, the design stress can be calculated from the design load based on appropriate failure criteria. For composite structures, the design stress is generally calculated from the design load based on the first ply failure criterion of the laminate.

According to the structural strength design specification, the MS value must be greater than or equal to zero. The specific value is stipulated by the spacecraft design and construction specification. The recommended minimum MS values for different materials and failure modes are shown in Table 7.3.

7.2.3.3 Optimization of Structure Design

In the structure design of space robot, the lightweight of the structure should be paid more attention to, so the minimum weight is the main objective of structural design on the premise of satisfying the requirements of strength margin, stiffness, envelope, and technology. Therefore, the goal of the detailed design of the space robot structure is centered on the optimization for minimum weight.

The basic principle of optimization design is to establish an optimized model, use a variety of optimization methods, perform iterative calculation, get the limit value of the objective function, and work out the optimal design scheme.

The mathematical model of the optimization problem can be expressed as

$$\left.\begin{array}{l} \min F(X) = F(x_1, x_2, \ldots, x_n) \\ g_i(X) = g_i(x_1, x_2, \ldots, x_n), (i = 1, 2, \ldots, M) \\ X = (x_1, x_2, \ldots, x_n)^T \end{array}\right\}, \quad (7.3)$$

where $F(X)$ is the objective function for design variables; X is a design variable; $g_i(X)$ is a state variable. The design variable is an independent variable, and the optimization result is achieved by changing the values of the design variables. The state variable is a value to constrain the design, and is a function of the design variable, generally with upper and lower limits; the objective function is a value to be minimized and is a function of the design variable.

Structural design optimization can be divided into three categories: topology optimization, shape optimization, and dimension optimization. Topology optimization is to find the optimal distribution of materials in a given design space to achieve the lightweight structure. Shape optimization is to change the mechanical properties of the model to meet the relevant requirements (such as stress, displacement) by changing some shape parameters of the model (shape of geometric properties), to ultimately achieve the lightweight of the structure; dimension optimization is a detail design based on the shape of the model. It improves the local performance of the structure by changing the local parameters of the structure, such as the thickness of the partial casing, the cross section of the beam, so as to achieve the purpose of lightweight.

For the traditional machining processes, topology optimization is usually the preparation of shape optimization and dimension optimization, and it is in the preliminary structure design stage; while shape optimization and dimension optimization are the optimized design of the dimensions based on the basic structure determined by topology optimization. It is the detail design under the consideration of the process technology in the optimization process. Figure 7.5 schematically illustrates the relationship among the three, in which the rectangular region is optimized by topology under various constraints to form an irregular weight reduction groove, then a regular region is formed by shape optimization, and finally, the angular α is optimized to angle β by dimension optimization.

The disadvantages of traditional subtractive manufacturing processes are made up with the emergence of additive manufacturing (3D printing) technology, enabling the results of topology optimization to be directly manufactured by 3D printing technology. As shown in Fig. 7.6, a connection bracket is directly produced by 3D printing after topology optimization, and the weight of the structural member is reduced by 70% under the same payload conditions.

For composite materials, since the material properties can also be expressed by continuous design variables (such as the number of layers, fiber angle, layer thickness), they can also be optimized by the concept of parameterization.

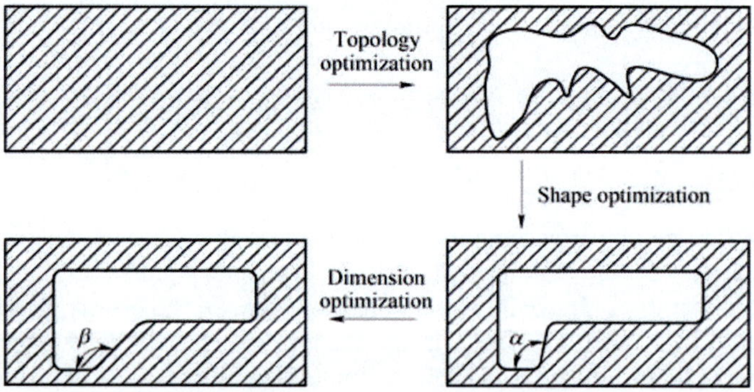

Fig. 7.5 Schematic diagram of optimization design

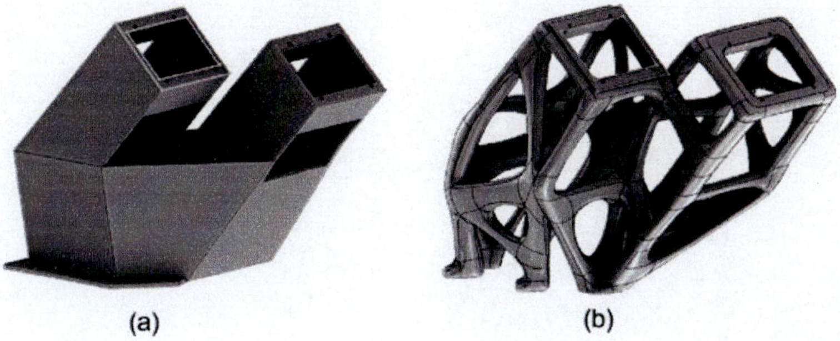

Fig. 7.6 Effect of the connection bracket before and after optimization a original design state; b optimized state

7.3 Joints

7.3.1 Joints Classification

Space robot joint is the drive assembly of the robot. It usually combines functions of mechanism, electric control, and thermal control to form integrated joints of various functions, hereinafter referred to as "joints".

Joints are the connecting parts of the rods on the robot and are the kinematic pairs that realize various movements of the robot. Joints are usually classified into rotary joints and rectilinear joints by their function. The modular design of joints is the development trend of space robotics, which can effectively shorten the development cycle of the robot. A single rotary joint or a rectilinear joint can be regarded as a single-DOF module. By free combinations of various rotary joint modules and rectilinear joint modules, a combination of multiple DOFs can be formed, such as "rotation + swing", "rectilinear motion + swing", "rectilinear motion + rotation" (as shown in Fig. 7.7), thus completing the multi-DOF robot.

Rectilinear joint modules are frequently used on industrial robots, and the moving distance can effectively increase the robots' workspace. Since space robots require small external envelope, they usually adopt compact rotary joints as long as they meet the requirements of the operating space. The following discussion mainly focuses on rotary joints.

7.3.2 Joint Composition

The joint of space robot has the characteristics of high integration, lightweight, and large loads. It can work in harsh space environments such as large temperature differences, strong radiation, and high vacuum. The joint of the space robot is generally composed of four parts: drive source, reducer, sensor, and drive controller. In order

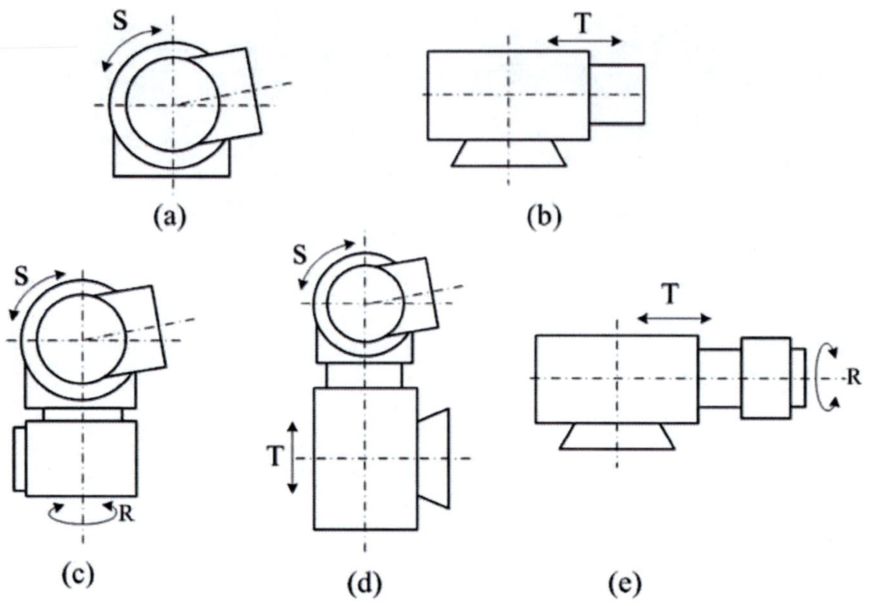

Fig. 7.7 Schematic of joint modules **a** rotary joint module **b** rectilinear joint module **c** rotation + swing module **d** rectilinear motion + swing module **e** rectilinear motion + rotation module

to adapt to the high- and low-temperature environment in space, thermal control measures (thermal control components) are usually taken, as shown in Fig. 7.8.

(1) Drive source: The commonly used drive sources on robot include hydraulic drive, pneumatic drive, and electric drive. Limited by the space vacuum environment, the main driving mode of space robots is electric motor drive.
(2) Reducer: Directly driven by a motor usually cannot meet the torque output requirements of the robot joints due to its high rotational speed and small torque, so a reducer is used to amplify the output torque and reduce the rotational speed.
(3) Sensor: Sensors are used to obtain the information of the joints state. Data measured by sensors include joint angles, speeds, currents, torques, etc.
(4) Drive controller: It is used to control the joints to output Required angles, speeds, torques.
(5) Thermal control assembly: It includes active and passive thermal control assembly. Active thermal control assembly generally consists of components such as heating sheets attached to the surface of the joint housing for heating the joint in a low-temperature environment; Passive thermal control assembly usually includes multilayer thermal control coating, heat pipes, heat sinks, etc., which are used to transfer the heat from the heat source to non-heat source, so as to ensure the proper temperature of the joint in high- or low-temperature environment.

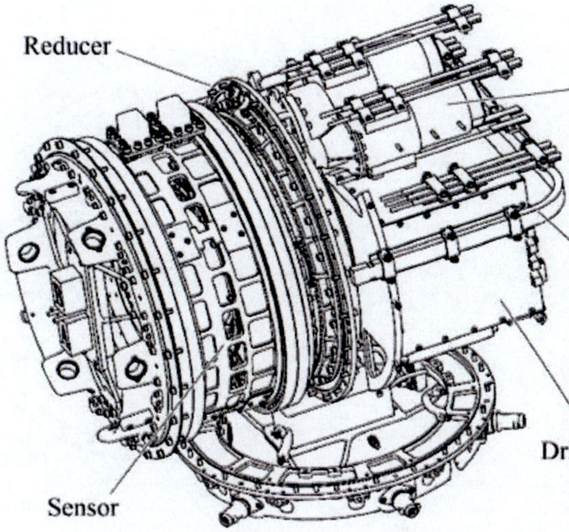

Fig. 7.8 General composition of space robot joints

7.3.2.1 Drive Source

Since space robots usually work in a gravity environment of less than 1 g, the output power is generally small. Commonly used motors are stepper motors, DC brush motors, and DC brushless motors.

(1) Stepper motor

Stepper motor is also called pulse motors, which is typically driven by a square wave voltage pulse, each pulse lasting a few milliseconds, when the motor winding receives one electrical pulse, the rotor will rotate one step angle. The angular displacement of the rotor and the magnitude of the rotational speed are proportional to the number of input electrical pulses and its frequency. Only by controlling the number and frequency of the input electrical pulses and the sequence of energization of the motor windings, the required angle of rotation, speed and direction can be obtained, and the motion of the motor can be easily controlled.

Stepper motor is mainly divided into two types: excitation stepper motor and reactive stepper motor. The difference is that the former has an excitation coil on the rotor, while the latter has no excitation coil on the rotor (permanent magnet). Since the control circuit of the reactive stepper motor is relatively simple, the motor is more frequently used in space applications, and its principle is shown in Fig. 7.9.

The stepper motor has the following advantages: high positioning accuracy, no cumulative error, high average rate stability for long-term operation; closed-loop or open-loop control can be adopted, if an open circuit control without digital input of position sensor is used, the control circuit can be simplified; a certain holding torque in a power-off state, which is very advantageous in space applications where power

Fig. 7.9 Composition and operational principle of the stepper motor

is limited. The disadvantages of the stepper motors are: not suitable for high-speed rotation; poor overload capability; large volume, mass, and power consumption; improper use may cause system vibration. When the stepper motor runs at a speed less than 10 steps/second (usually called low speed), it runs smoothly; when the number of steps is 10–200 steps/second, it is necessary to monitor the impact of speed on torque.

(2) DC brush motor

DC brush motor is a kind of DC motor widely used in the industry. It can be divided into permanent magnetic type and electromagnetic type according to the mode of excitation. Space robots generally use permanent magnet DC motor, in which a magnetic pole made of permanent magnet is mounted on the stator, an armature winding is mounted on the rotor, and the current is drawn out via the commutator by the brush. The advantages of DC brush motor are lightweight, small size, and low-cost; simple control mode by commutating the coil current with the brush; good starting performance and strong overload capability, capable of withstanding frequent starting, braking, and reversing. The disadvantage is that there is a contact between the brush and the commutator, so the brush is prone to wear in the space environment, which may affect its service life. In addition, sparks may occur at the brush, which may generate EM waves and affect the normal operation of other devices. Therefore, DC brush motors are not suitable for long-life working mechanisms, and shielding and filtering measures are required in the application.

(3) Brushless DC motor (BLDCM)

Unlike the DC brush motor, BLDCMs commutate the current in the winding by electronic means rather than mechanical means. Compared with the DC brush motor, the multiphase armature windings are mounted on the stator, and the rotor is made of a permanent magnet. The position sensor detects the position of the rotor magnetic

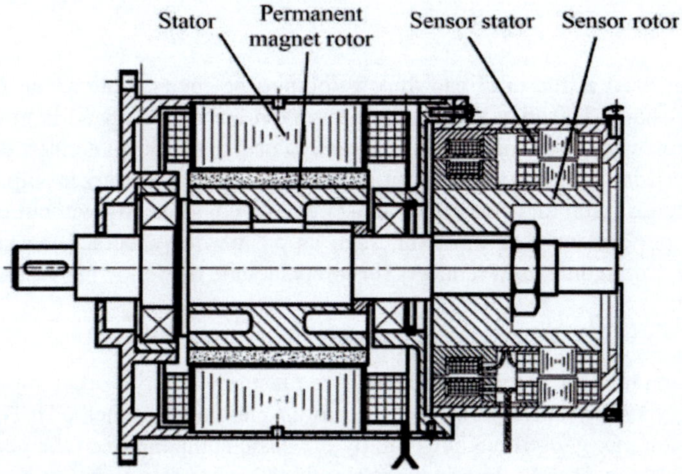

Fig. 7.10 Operational principle and composition of BLDCM

field relative to the stator winding and transmits the signal to the controller. The controller supplies power to the stator winding at an appropriate position follow an exact logic. The drive current can be divided into a square wave drive and a sine wave drive by the current form. The square wave drive only needs to detect the position of the commutation point, and can be configured with a Hall sensor; the sine wave drive needs to detect the absolute position of the rotor, and is generally configured with an absolute position sensor such as a resolver. Compared with the square wave drive mode, the sine wave drive mode has less torque fluctuation. The operational principle and composition of the BLDCM are shown in Fig. 7.10.

Compared with DC brush motors, BLDCM has two main advantages: no mechanical brushes, so there is no need to consider brush life problems due to wear and EMI(electromagnetic interference) problems caused by brush ignition; the windings are on the stator, so the heat is more easily dissipated, which reduces the temperature gradient across the bearing. The disadvantage is that a position sensor is added, which makes the drive circuit more complex.

The applicable speed of BLDCM is 3000 r/min, which can reach 1.5×10^4 r/min in space applications. In position control, the position accuracy and resolution are determined by both the drive electronic device and the position sensor. BLDCM has good instantaneous torque overload capability, and the instantaneous overload can be up to 5 times of the rated torque. Due to these excellent characteristics, BLDCM is often used to form the servo control system of the space robot joint.

7.3.2.2 Reducer

The motor used in the joint can directly output the rotary motion, but its output torque is usually less than the requirements, and the output speed is greater than the requirements. Therefore, it is necessary to use a reducer to convert the higher motor speed into a lower joint output speed, so as to obtain a larger output torque. This motion transfer and conversion must be realized efficiently without degrading the main performances of the joint, such as positioning accuracy, backlash, and reliability. Commonly used reducers for joints include the following types.

(1) Planetary reducer

As the name implies, planetary reducer uses planetary gear for deceleration. There are many types of planetary gearing and various classification methods. Two classifications commonly used in China are by the basic components of the gearing and by the meshing mode. According to the classification of basic components, the planetary reducers include 2*Z*-*X* and 3Z, as shown in Fig. 7.11, where Z represents the center wheel, and X represents the planet carrier.

For type 2*Z*-*X*, the basic components are two center wheels (*a*, *b*) and one planet carrier (*X*), the transmission ratio is calculated by

$$i_1 = 1 + \frac{z_b}{z_a}, \tag{7.4}$$

where Z_a is the number of teeth of the sun gear (*a*), Z_b is the number of teeth of the planetary gear (*b*).

For type 3Z, the basic components are only three center wheels (*a*, *b*, and *e*), the planet carrier (*X*) does not bear external torque, so it is not a basic component. The transmission ratio is calculated by

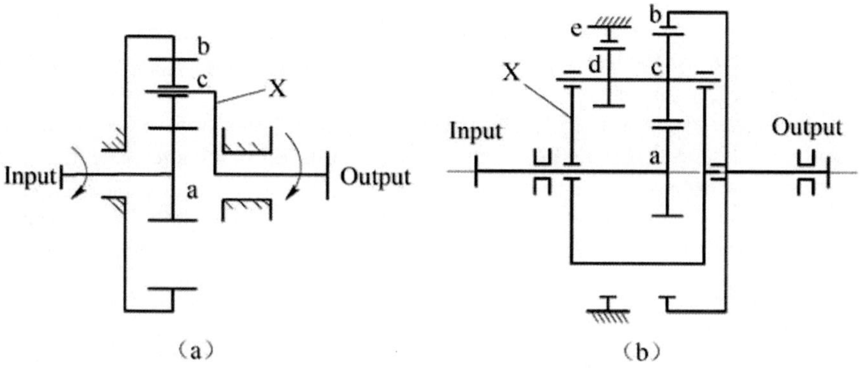

Fig. 7.11 Types of common planetary reducers for space robot joints **a** Type 2*Z*-*X*; **b** Type 3Z

$$i_2 = \frac{1 + \frac{z_c z_e}{z_a z_d}}{1 - \frac{z_c z_e}{z_b z_d}}, \tag{7.5}$$

where Z_a is the number of teeth of the sun gear (*a*); Z_b is the number of teeth of the output internal gear (*b*); Z_c is the number of teeth of the right double gear (*c*); Z_d is the number of teeth of the left double gear (*d*); Z_e is the number of teeth of the fixed internal gear (*e*).

Compared with the ordinary gear train, planetary reducer has the distinctive feature that its power can be shunt when transmitting the torque, and the input and the output shafts are coaxial, that is, the output and the input shafts are placed on the same axis. This feature gives the planetary reducer a compact structure, high reduction ratio, and large loading capacity. In addition, the symmetry of the planetary transmission structure ensures a smooth overall transmission and a high transmission efficiency, which can reach 0.97–0.99 under the reasonable structural arrangement.

(2) Harmonic reducer

The harmonic reducer consists of three basic components: a wave generator, a rigid gear, and a flexible gear. The rigid gear is a rigid structure with an internal ring gear (the number of teeth is z_1). The flexible gear is an elastic thin-walled gear with the same tooth profile as the rigid gear but with fewer teeth (z_2). The wave generator is composed of an elliptical cam and an elastic thin-walled bearing. When the cam rotates, it will force the bearing to make deforming motion along the elliptical shape within the elastic range. Since the long axis of the wave generator is slightly longer than the inner diameter of the flexible gear, when the wave generator is inserted into the flexible gear, the flexible gear is forced to change from the initial circular shape to an elliptical shape. As shown in Fig. 7.12, the wave generator forces the teeth near the two ends of the elliptical long axis of the flexible gear to fully mesh with the teeth of the rigid gear, while the teeth near the two ends of the elliptical short axis of the flexible gear are completely disengaged from the teeth of the rigid gear. With the continuous rotation of the wave generator, the positions of the long axis and the short axis on the flexible gear also change, so that the teeth of the flexible gear sequentially

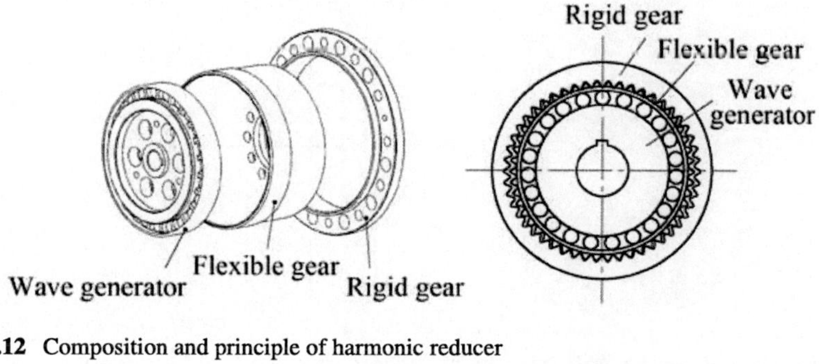

Fig. 7.12 Composition and principle of harmonic reducer

complete the cycle of "engagement → meshing out → disengagement → meshing → engagement" and the harmonic reducer realizes meshing transmission.

Because the waveform of the elastic deformation generated by the flexible gear during the transmission is similar to harmonic, it is called harmonic transmission. The number of protruding parts on the wave generator is called the wave number, which is represented by *n*. The difference in the number of teeth between the rigid gear and the flexible gear is usually equal to the wave number, i.e., $z_1 - z_2 = n$. According to the difference in wave numbers, harmonic transmission can be divided into double-wave transmission and triple-wave transmission. Most of the harmonic reducers used in space drive mechanisms adopt double-wave transmission.

The calculating equation of the harmonic transmission ratio is

$$i_3 = \frac{z_2}{z_1 - z_2}. \tag{7.6}$$

The main advantages of the harmonic reducer are large transmission ratio, wide single-stage transmission ratio range (from 50 to 300), preferably from 75 to 250; because more teeth are engaged at the same time, it has a smooth transmission, strong loading capacity, high transmission precision and efficiency, simple structure, and small volume. The main disadvantage is that the flexible gear will produce periodic elastic deformation, which is easy to produce fatigue failure.

(3) RV reducer

The Rot-Vector (RV) reducer is a new type of transmission developed on the basis of the cycloidal pinwheel transmission. The 3D model of the RV reducer and its mechanism diagram is shown in Fig. 7.13. The high-speed stage is a K-H type differential gear train composed of involute gears, which consists of a central wheel (1) and three evenly distributed planet wheels (2) in the radial direction and a tie bar (7); the low speed stage is a K-H-V type planetary gear train composed of a cycloidal

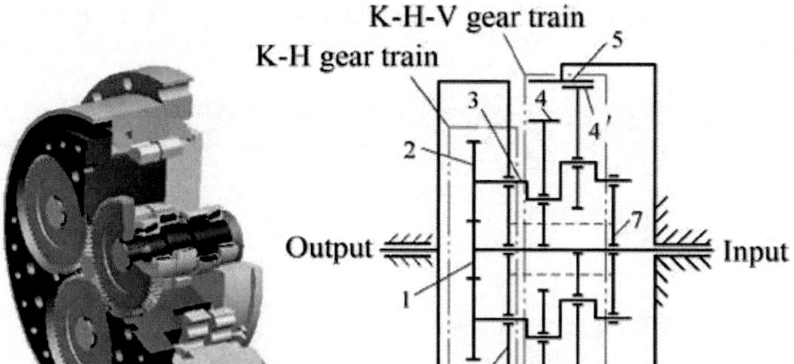

Fig. 7.13 Composition and principle of an RV reducer

gear, which consists of three evenly distributed crankshafts (3) in the radial direction, two symmetrically arranged cycloidal wheels (4, 4′), and several needle teeth (5) and an output disc (6). In the gear train, the tie bar (7) and the output disc (6) are firmly connected to the same member. The three crankshafts with the same offset direction are fixedly connected with the corresponding planetary gears at one end, and the other ends constitute rotating pairs with the corresponding pin holes on the output disc, and their middle portions constitute rotating pairs with the cycloidal wheels. When the gear train transmits motion, the planetary wheel drives the crankshaft to rotate, the rotation of the planetary gear is transmitted to the cycloidal wheel via the crankshaft to realize the revolution, and the cycloidal wheel feeds its rotation to the tie bar through the output disc as the input motion of the differential gear train, thereby, thus forming a closed planetary transmission.

RV reducer is characterized by a wide range of transmission ratio ($i = 31$–171), large loading capacity, high rigidity, motion precision, and transmission efficiency ($\eta = 0.85$–0.92). It is widely used in automation equipment such as industrial robot joints and high precision CNC machine tools. Compared with industrial robots, the current space robots have lower load and stiffness, so harmonic reducers with a simple structure can meet their requirements. Therefore, there are few RV reducers used in space robots. With the development of space robots and the maturity and reliability of RV reducer technology, the application of RV reducers in the field of space robots will have a good prospect.

7.3.2.3 Sensors

The joint of space robot is a complete servo control mechanical system, in which the sensors are important components. Common sensors include position sensors, speed sensors, force sensors, and current sensors. Since speed can be obtained by differentiating the position information, the position sensors are also used as speed sensors. In addition, in order to adapt to the high- and low-temperature environment in space, thermal control measures for joints are often needed and temperature sensors are required. Table 7.4 shows the main types of sensors used in joints.

The selection of sensors is largely related to the functional requirements of the robot joints. If a closed-loop torque control function is needed for the joint, a torque sensor is required in joint design; if closed-loop position control function is required in the joint, a position sensor is needed. The following focuses on the resolver and the one-dimensional torque sensor commonly used in space robot joint design.

(1) Resolver

The working principle of the resolver is similar to that of the common transformer, except that the position of the input and output windings of the ordinary transformer is relatively fixed, and the ratio of the output and input voltage is a constant. However, the input and output windings of the resolver are placed on the stator and the rotor, respectively, and due to their relative position change, the link flux between

Table 7.4 Main types of joint sensors

Position sensor	Electromagnetic type	Electromagnetic resolver Reluctance resolver Inductosyn
	Optical type	Incremental angle grating encoder Absolute angle grating encoder
	Magnetic induction type	Hall sensor
Speed sensor	Tachogenerator (used as position sensor)	
Current sensor	Hall current sensor	
Force sensor	One-dimensional torque sensor	
Temperature sensor	Thermistor	

them changes, and the output voltage also changes. The so-called resolver refers to a monopolar resolver sine and cosine rotary transformer. Two sets of two-phase orthogonal windings with the same parameters are arranged on the stator core and the rotor core. respectively. and perpendicular to each other in space, as shown in the one-speed part of Fig. 7.14. If an AC voltage $U_{1\sin\omega t}$ is applied to the rotor winding R1R3, and the rotor winding R2R4 is short-circuited, a current will flow through the rotor winding R1R3, and establish an alternating magnetic flux Φ in the resolver. The link flux between this magnetic flux and the cosine winding S1S3 of the stator is $\Phi\cos\theta$, the generated induced voltage is U_{S1S3}; and the link flux between this magnetic flux and the sine winding S2S4 of the stator is $\Phi\sin\theta$, and the generated induced voltage is U_{S2S4}.

Two-speed resolver is a kind of high precision resolver developed on the basis of one-speed resolver, that is, a set of multipolar winding ($p > n$, n is a positive integer) is added in addition to the one-speed winding. For one-speed resolver, the electrical angle and the mechanical angle are consistent. For two-speed resolver, the electrical angle is p times of the mechanical angle. For a multipolar resolver, the electrical angle is a number of poles times of the mechanical angle. The equalization of the magnetic circuit segmentation of the multipolar resolver and the series connection of circuits can reduce the influence of process factors and geometric inaccuracy on the precision, thereby achieving the objective of improving measurement accuracy. Under ideal conditions, the output voltage equations for two-speed resolver conform to the formula (7.8) to (7.11), and there is a strict correspondence between the rotor position angle and the output voltage

$$U_{\mathrm{S1S3}} = K_1 U_{\mathrm{R1R3}} \cos\left(\theta + \theta_{op}\right) \tag{7.7}$$

$$U_{\mathrm{S2S4}} = K_1 U_{\mathrm{R1R3}} \sin\left(\theta + \theta_{op}\right) \tag{7.8}$$

$$U_{\mathrm{S5S7}} = K_{\mathrm{p}} U_{\mathrm{R1R3}} \cos(p\theta) \tag{7.9}$$

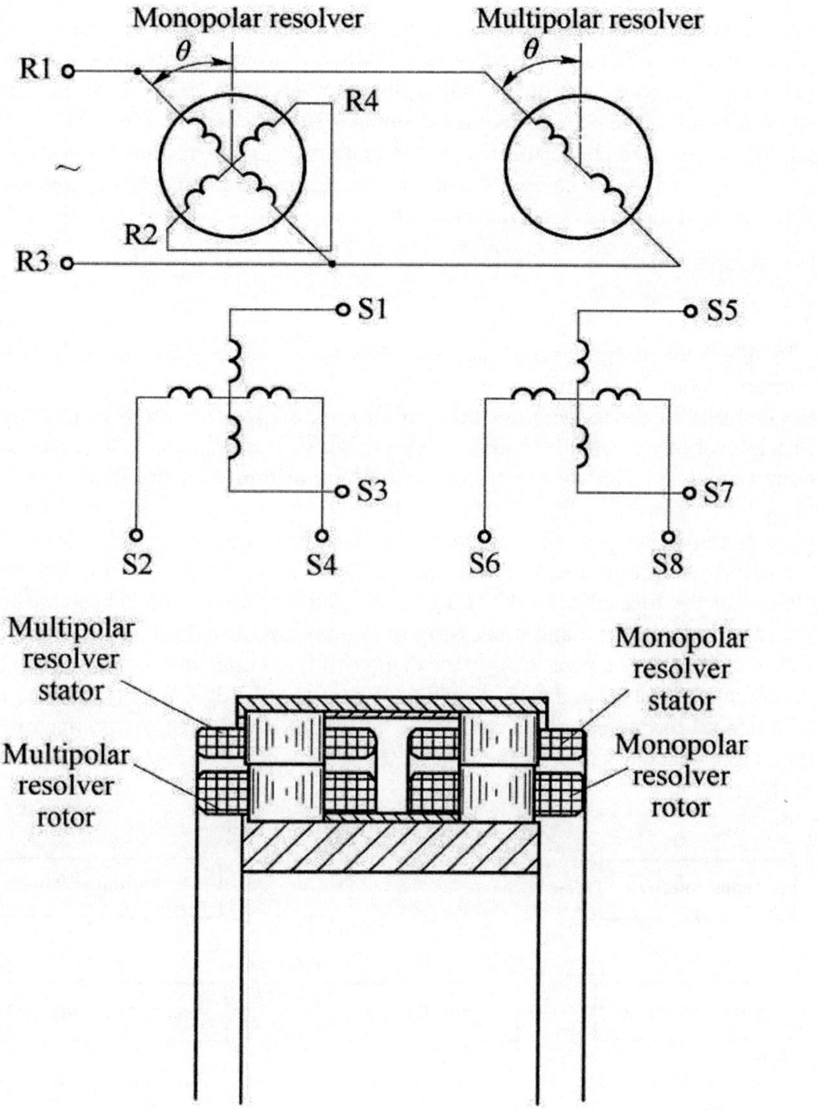

Fig. 7.14 Structure and principle of the resolver

$$U_{S6S8} = K_p U_{R1R3} \sin(p\theta), \tag{7.10}$$

where U_{S1S3} is the voltage between the stator terminals S1S3 (the rest are the same), in V; U_{R1R3} is the voltage between the rotor terminals R1R3, in V; θ is the mechanical angle of the rotor rotation in °; θ_{op} is the deviation between the zero position of the monopolar resolver and multipolar resolver, in °; K_1 is the transformation ratio of the monopolar resolver; K_p is the transformation ratio of the multipolar resolver; p is the number of pole-pairs of the multipolar resolver. Brushless resolver is an analog angle measuring device. It has the advantages of long service life, high reliability, good anti-interference performance, and can adapt to various harsh environmental conditions such as shock, vibration, high-temperature, low-temperature, and vacuum. It is a common angle sensor in space robot.

(2) Torque sensor

Torque sensors can be divided into various types according to the measurement principle, such as strain type, capacitor type, piezoelectric type, photoelectric type, magnetoelectric type, fiber type, wireless surface acoustic wave type. Considering the maturity of the technology, the strain torque sensors are more commonly used in space robots, in which the general elastic body is arranged in series with the joint output shaft in order to directly detect the output torque of the joint. Its working principle is shown in Fig. 7.15. The elastic body unit is deformed by the action of the external torque. The resistance value of the strain gauge is changed due to the deformation of the elastic body unit, and the mapping relationship between the change of the resistance of the strain gauge and the input torque is established. To facilitate the detection and processing of signals, the resistance of the strain gauge ($R1$, $R2$, $R3$, $R4$) is converted into the change of the output voltage value (U_e) by the processing of the related circuit such as Wheatstone bridge. Finally, the magnitude of the input torque acting on the elastic body unit is reflected by the magnitude of the output voltage value. In the figure, U_e is the supply voltage of the Wheatstone bridge circuit.

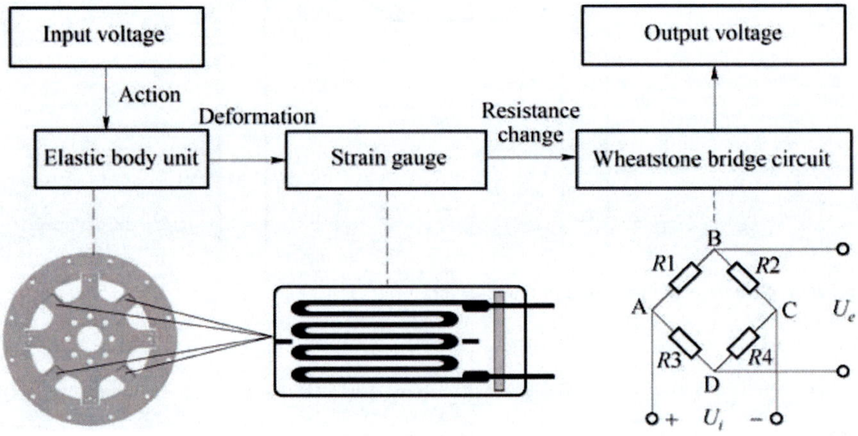

Fig. 7.15 Working principle of the torque sensor

7.3.3 *Joint Design*

7.3.3.1 Overall Design

The joint overall design mainly considers the joint function requirement and performance indices, the details are shown in Fig. 7.16.

1. Joint function configuration

The joint function requirement is mainly determined by the functional requirements of the space robot. A typical robot joint consists of a brake, a drive motor, a reducer, a position sensor, a force sensor, and an output shaft.

(1) The brake provides braking torque for the joint to ensure its braking function.
(2) The drive motor provides the power source for the joint; and the reducer can reduce the output speed and increase the output torque to ensure the matching between the two.
(3) The position sensor can be arranged at the drive motor end or the joint output end according to requirements for joint angle or speed measurement.
(4) The force sensor is generally arranged at the joint output to directly measure the external load of the joint and is often used to achieve compliance control of the joint.
(5) The output shaft provides an output interface to realize the load function and ensures the stiffness and strength of the joint.

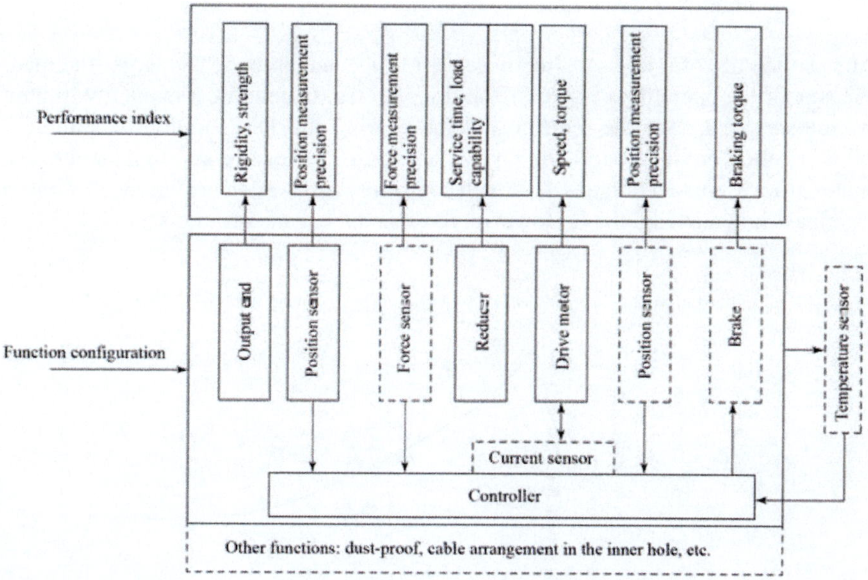

Fig. 7.16 Joint design

(6) The temperature sensors are usually arranged on components with specific temperature requirements to obtain their temperature data for thermal control.
(7) The current sensor is generally arranged inside the controller for measuring the current of the power (usually the winding of the drive motor) to realize the current closed-loop control or current limiting protection function of the motor.

2. Joint performance indices

(1) Joint load capacity

The load capacity of the joint mainly involves the output speed and torque, which directly affects the service life of the joint in actual operation.

The output speed and output torque are mainly determined by the matching of the drive motor and the reducer, and the relationships are as follows:

$$T_{\text{out}} = T_{\text{in}} \cdot i \cdot \eta \tag{7.11}$$

$$n_{\text{out}} = \frac{n_{\text{in}}}{i}, \tag{7.12}$$

where T_{out} is the output torque of the joint; n_{out} is the output speed of the joint; T_{in} is the torque of the drive motor; n_{in} is the speed of the motor; i is the transmission ratio of the reducer; η is the efficiency of the reducer.

In the actual design, the matching issue between the drive motor and the reducer (transmission ratio) is mainly considered and analyzed for the objective of meeting the envelope requirements and the pursuit of lightweight.

(2) Joint operating service life

The service life of a joint is directly related to the actual load. The larger the load, the shorter the operating service life, and vice versa. In general, it is mainly limited by the service life of the reducer and the bearing. In the condition of sufficient oil lubrication on the ground, the service life of the two can be approximated to the material-level contact fatigue life, which conforms with the contact fatigue *S-N* curve trend, as shown in Fig. 7.17. However, for space mechanisms, the service life of a

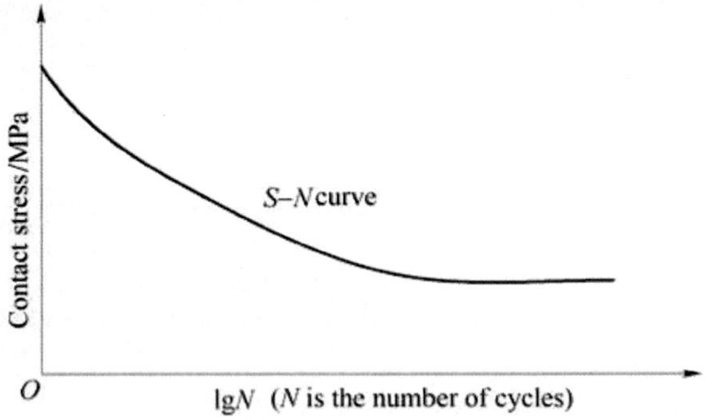

Fig. 7.17 Contact fatigue *S-N* curve

joint is often limited by the way of lubrication. The space lubrication methods mainly include grease lubrication and solid lubrication. Under normal conditions, the life of solid lubrication is shorter, and that of grease lubrication is longer. However, grease lubrication tends to increase the starting torque of the reducer and reduce the transmission efficiency under low-temperature. Therefore, in practice, various constraints should be considered comprehensively, and combined with the life constraints and characteristics of different lubrication methods, to finally determine the life index of the joint.

(3) Joint stiffness

The joint stiffness mainly includes torsional stiffness and bending stiffness. Torsional stiffness is determined by the stiffness of the output shaft and the reducer in series, it conforms to the following relationship:

$$\frac{1}{K} = \frac{1}{K_1} + \frac{1}{K_2}, \tag{7.13}$$

where K is the torsional stiffness of the joint output; K_1 is the torsional stiffness of the joint output shaft; K_2 is the torsional stiffness of the joint reducer.

The torsional stiffness of the joint output shaft is determined by the structure of the part. The torsional stiffness of the reducer is directly related to the transmission mode of the reducer. The influence coefficient of each stage is different. Taking the 3-stage transmission reducer as an example, it can be expressed as follows:

$$i = i_1 \cdot i_2 \cdot i_3, \tag{7.14}$$

$$\Delta = \frac{\Delta_1}{i_1 \cdot i_2 \cdot i_3} + \frac{\Delta_2}{i_2 \cdot i_3} + \frac{\Delta_3}{i_3}, \tag{7.15}$$

$$K_2 = \frac{T}{\Delta}, \tag{7.16}$$

where i_1 is the first stage gear ratio; Δ_1 is the input deformation of the first stage; i_2 is the second stage gear ratio; Δ_2 is the input deformation of the second stage; i_3 is the third stage gear ratio; Δ_3 is the input deformation of the third stage; T is the external load of the output shaft; Δ is total output deformation.

The bending stiffness of the joint is mainly determined by the bending stiffness and the bearing support stiffness of the output shaft, which also conforms to the relationship shown in Eq. (7.14).

(4) Joint accuracy

Joint accuracy includes backlash, transmission accuracy, positioning accuracy, force control accuracy, and speed stability.

The joint backlash is mainly caused by the engagement gap of the reducer. For open-loop systems, it will directly lead to accumulative errors; for closed-loop systems, it can be corrected by the output position feedback, but oscillation may occur due to over-correction.

The transmission accuracy of the joint is mainly measured by the transmission error of the reducer: the larger the error, the lower the accuracy. Transmission error refers to the difference between the actual rotation angle of the output shaft and the theoretical rotation angle when the input shaft of the reducer rotates in one direction. This is mainly caused by machining and assembly errors. For open-loop systems, it directly affects the joint's positioning accuracy; for closed-loop systems, it can be equivalent to the interference signal superimposed on the output shaft. When the error frequency is lower, it generally does not affect the positioning accuracy of the system.

The positioning accuracy and force control accuracy are the final control accuracy of the closed-loop systems, which are directly affected by the sensor accuracy and control algorithm, but the control accuracy cannot be better than the sensor accuracy.

Due to the high-speed rotation of the motor, the joint speed can be calculated by the position sensor at the motor end. Therefore, the speed stability at the motor end is mainly determined by the sensor accuracy and the algorithm of the control system. When the speed and the sensor accuracy are high, the speed stability is easier to achieve. In general, the higher the speed stability of the motor input, the higher the joint speed stability after reduction, but the speed fluctuation caused by the transmission error will be directly superimposed on the final joint output, thus affecting the final joint speed stability.

In short, various accuracy indices in the joint system interact with each other. The improvement of the sensor and the transmission accuracy is beneficial to the accuracy of the entire joint system. However, the system requirements and cost constraints should also be considered in the design to select the matching sensor accuracy and transmission accuracy.

3. Other functional designs

During joint design, other related functions, such as dust-proof and wiring layout, should be considered according to the environment and layout.

For the manipulator joints used for sampling on extraterrestrial bodies, the dust-proof function should be considered. A dust-proof plan can be designed according to the distribution of particles to be proofed. As shown in Fig. 7.18, when the proofed particle is small, the motor can use dynamics sealing on the gasket between the stator and the rotor to achieve the dust-proof function; if the proofed particle size is large, multiple gaps can be used to form a maze for dust protection.

Wiring of space robots is one of the important issues to be considered during joint design. If the centralized control mode is adopted, all components of the joint are powered and controlled by the same controller, so more wiring harnesses are required; if a distributed control mode is adopted, the drive is integrated into the joint body, the wiring between joints are only bus and power line, so less wiring harnesses are required. As shown in Fig. 7.19, the multiple joints adopt the way of cable arrangement in the inner hole, and each joint is distributed controlled.

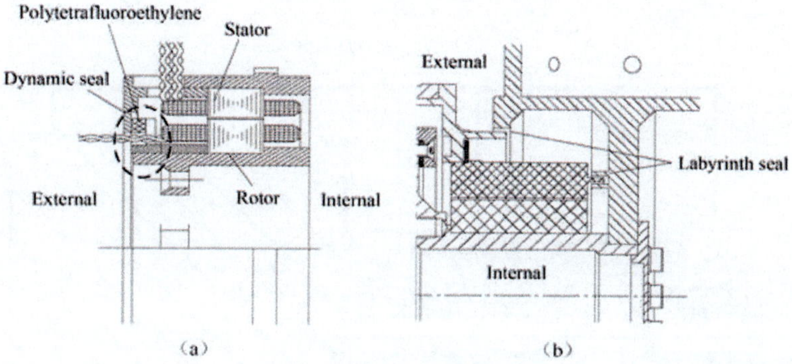

Fig. 7.18 Dust-proof design of joints **a** dynamics sealing dust-proof design; **b** Mazy dust-proof design

7.3.3.2 Torque Margin Design

Space robot is a space mechanism, and generally has static torque (force) margin and dynamics torque (force) margin requirements to ensure the realization of the motion function of the mechanism. The static torque (force) margin requirement is to ensure that the mechanism can produce motion; while the dynamics torque (force) margin requirement is to ensure that the resulting motion is sufficient to satisfy the need.

Countries have their own requirements for torque margin. For illustration, we take the US military standard (MIL-HDBK-83577) as an example.

The static torque (force) margin is defined as: the drive torque (force) minus the driving torque (force) required to generate the specified acceleration, then divided by the resistance torque (force) and then minus 1, expressed as a percentage as follows:

$$\text{Static torque(force) margin} = 100\% \times \left[\frac{\text{Drive torque(force)} - \text{Torque(force) required for generating acceleration}}{\text{Resistance torque(force)}} - 1\right] \tag{7.17}$$

See Table 7.5 for the required values of static torque (force) margin. Because it is difficult to determine the correct resistance torque or the resistance value, the required value is usually specified to be larger for the sake of reliability. When there is no sufficient confidence in determining the worst-case load or driving capacity, it may be more appropriate to consider a margin value higher than the requirements in Table 7.5.

When calculating the static torque (force) margin, the minimum drive torque (drive force) that may present and the maximum resistance torque (resistance) that may exist should be taken. The impact of voltage, temperature, motor, control parameters, and the worst-case conditions should be taken into account when determining the minimum drive torque (drive force); the factors of static friction, temperature

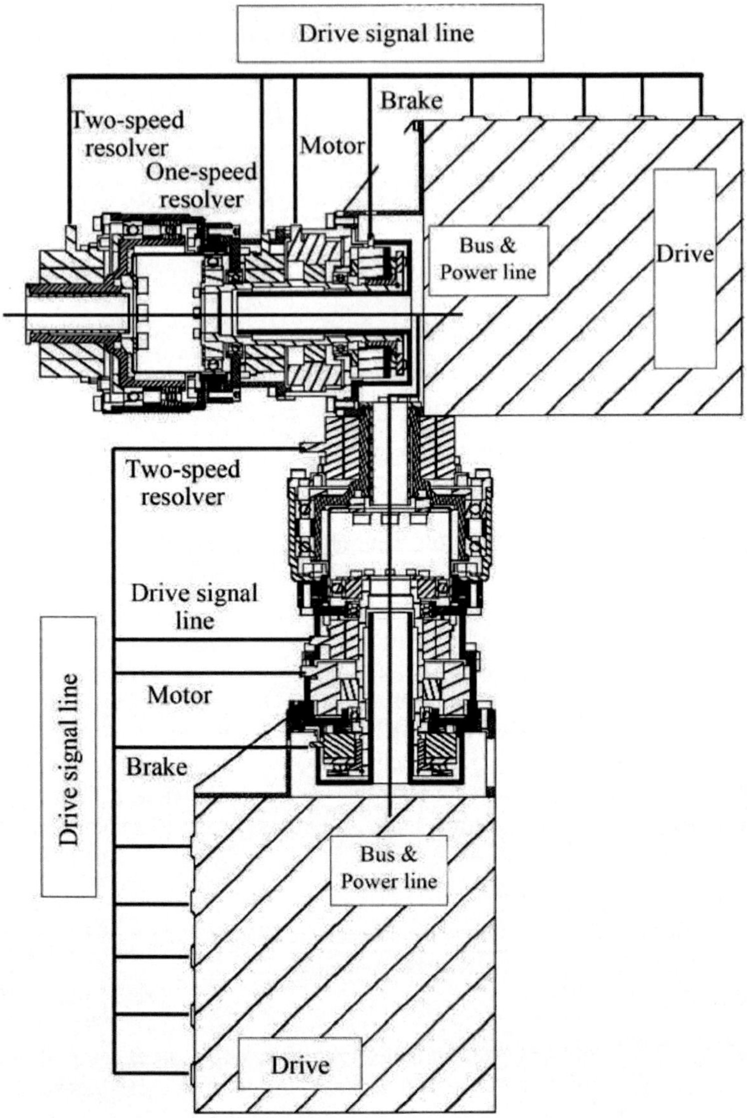

Fig. 7.19 Multiple joint wiring layout

Table 7.5 Required static torque (force) margin values

Phase	Required static margin value (%)
Program design	175
Preliminary design	150
Prototype design	125
Flight model design	100

variation, deformation, locking force, and harness load, as well as the degradation and loss of lubricating materials under vacuum or severe temperature conditions, should be considered in determining the maximum resistance torque (resistance).

The dynamics torque (force) margin is defined as: the drive torque (force) minus the resistance torque (force), then divided by the torque (force) used for generating acceleration, and then subtracted by 1, expressed as a percentage as follows:

$$\text{Dynamics torque(force)}margin = 100\% \times \left[\frac{\text{Drive } torque\text{(force)} - \text{Resistance torque(force)}}{\text{Torque(force)used for generating acceleration}} - 1\right] \tag{7.18}$$

From the point of view of pure mechanics, the right side of the above formula should be equal to zero. However, in practice, due to design, manufacturing, and environmental factors, it is not possible to correctly predict the inertia of the motion. Therefore, in order to fully guarantee that the mechanism can generate the acceleration required for the motion while overcoming the resistance torque (force), the dynamics torque margin is not zero. Because the inertia or mass is relatively easy to determine, the required value of the dynamics torque (force) margin can be set smaller, and generally defined as more than 25%.

7.3.3.3 Performance Analysis

In the design of space robot joints, its performance should be analyzed and reviewed. The main concerns include the following:

(1) Output capacity analysis: It mainly refers to the analysis of the output torque and speed of the joint. According to the output capacity requirements, the space environment, lubrication conditions, operation life should be considered, and corresponding motor and reduction ratio should be selected to match the relationship between output torque and output speed.
(2) Strength and stiffness analysis: The strength analysis refers to the analysis of stress and strain when the joint shells, output shafts, and bearings are subjected to load; the stiffness analysis refers to the deformation of the joints under load. The purpose of the analysis is to assess whether strength damage or excessive elastic deformation will occur to the joints under load.
(3) Thermal matching performance analysis: Compared with ground robots, space robots often work in harsher high- and low-temperature environments. Materials with matching thermal expansion coefficients should be selected according to the maximum temperature difference requirement, to ensure that the structure will not be damaged by larger thermal stress, and the gap of the transmission mechanism will not be too small in the high-temperature difference environment, resulting in the mechanism being stuck.

(4) Control performance analysis: The whole robot joint constitutes a servo control system. The structure and mechanism stiffness, gap, inertia, load, and damping will all affect the control performance of the system. A complete model with all parameters should be established to conduct simulation analysis.

7.4 End Effectors

7.4.1 Classification of End Effectors

The end effector (its moving parts is also known as the end effect mechanism) is a special space mechanism for target capture, tools use, and sample collection, and is usually installed in the operation end of a space robot. Because the types, methods, service life, and requirements of tasks that the space robots required to perform on-orbit are not exactly the same, it is necessary to carry out specific analysis and comprehensive design according to the robot's tasks. For example, if the task is to capture a target, the end effector should be designed as a gripping mechanism with a suitable interface shape and size; if the task is to complete processing and assembly, the corresponding end effector should be designed as a special operation tool or actuator (such as smart hands, screwdrivers, filling tools).

The end effectors can be classified into general end effector and special end effector according to the operation objects; multi-finger dexterous end effectors and simple operation end effectors according to the dexterity; mechanically connected and electromechanically connected end effectors according to whether it is electrically connected to the target; small, medium, and large end effectors according to on-orbit operation capability; cooperative and noncooperative target capture end effectors according to the characteristics of the target [2].

For different types of end effectors, their design principles, types of space operation tasks, space environment, and service life are different, but end effector technologies are similar. In the design process, one or more of the following characteristics are usually available:

(1) Large operational tolerance. Operational tolerance is one of the core technical indices of the end effectors. The greater the operational tolerance, the greater the probability that the end effector will operate successfully, and the lower the requirement of the system end positioning accuracy. For example, the Canadarm end effector utilizes the wire rope winding principle to achieve a large tolerance capture operation.
(2) Flexible capture operation. There is usually a collision contact force caused by the relative speed between the end effector and the operation targets, especially for noncooperative targets, high relative speed targets, and large inertia targets, which seriously affect the operational performance of the space robot and even jeopardizes the safety of the spacecraft system. Therefore, the impact force

is usually reduced through the flexible design of the end effector operating mechanism, and sometimes assisted by flexible control of the arm to further reduce the impact force.

(3) High operational accuracy. When the end effector performs an operation task on orbit, its positioning accuracy is determined by the motion control accuracy of the robot, and the operational accuracy is usually determined by the designed accuracy of the end effector itself. High accuracy end effectors can carry out on-orbit maintenance, connection, sampling, and filling operations precisely. The operational accuracy is achieved by both the mechanism transmission accuracy and the motion control accuracy of the end effector itself during design.

(4) Large load capacity. With the development and utilization of space resources, the space operation tasks required to be completed by end effectors are more and more complex. The end effectors are required to have the ability to operate large loads while meeting the requirements of miniaturization and lightweight.

(5) High connection stiffness. In order to avoid the relative movement between the end effector and the connected target in the process of driving the robot to connect the target, and ensure the reliable connection with the target, the end effector is usually designed with high connection stiffness for the on-orbit operation tasks with connection requirements.

(6) Intelligence and autonomy. Due to the actual space environment and specific tasks, it is difficult for astronauts or ground teleoperators to observe the actual position and state of the end effector in real time, and send operational instructions to the end effector in different operation states. This requires that the end effector has the ability of autonomous intelligent operation, which can comprehensively analyze and judge the feedback information from various sensors (visual, tactile, temperature, position, acceleration, etc.) during operation. In this way, the end effector can plan the operation control strategy in real time, accurately complete the required operation, and improve the operational safety of the robot system in the space environment.

(7) High rcliability. The end effector directly acts on the target, and the reliability of its operation is of great significance to the safety of astronauts and spacecraft and the successful accomplishment of the mission. In the design process, the reliability index of the end effector is usually obtained through ground verification and theoretical analysis and calculations, and then the weak links in the design are optimized to meet the requirements of on-orbit operation tasks for the end effector.

(8) Long service life. Due to the requirements of long-term and multiple operational tasks, the end effector needs a long service life. Usually, it is through simulation analysis, design optimization, and test verification in the ground vacuum and high/low-temperature environment to ensure that the end effector meets the task requirements.

7.4.2 *Composition of the End Effector*

In general, the specific structure of each type of end effector is determined by the specific tasks to be performed, but typically it includes the drive components, transmission components, actuator components, measurement components and control components, and other parts. According to the tasks and functions to be implemented, the end effector is composed of one or more of the above components, as shown in Fig. 7.20.

(1) Drive components

The drive components are the basic component of the end effector, which is usually composed of one or more drive sources (motor, spring drive unit, memory alloy unit, etc.) that provide the driving force or torque required for the end effector to perform actions. Among them, the motor has the features of long-term continuous, bidirectional driving, and excellent control performance, so it is most widely used in end effectors.

(2) Transmission components

The common transmission forms of the end effector include screw transmission, gear transmission, space/plane linkage transmission, tendon transmission, etc. The ball screw can be used to convert the rotary motion output by the drive unit into a linear motion, which has the advantages of smooth transmission, high transmission accuracy, and low frictional resistance. Gear transmission includes fixed-axis gear-train

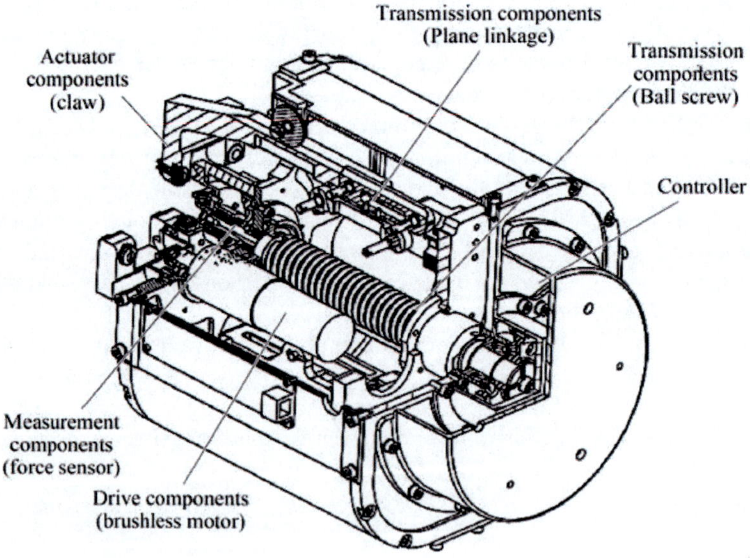

Fig. 7.20 Composition of the end effector

transmission, planetary transmission, harmonic transmission, etc., among which, the harmonic transmission has the advantages of small size, lightweight, and large transmission ratio. Based on the design principles of the series and parallel mechanisms, the space/plane linkage enables the end effector to perform planned actions along the designed trajectory according to actual needs. High-strength and high-toughness steel wire ropes and aramid fiber ropes are used to realize the smooth operation of the end effector by combining the tendon transmission with the drive unit.

(3) Actuator components

The actuator components are the core component of the end effector to finally realize the functional output. It usually combines with the transmission components to achieve the required functions according to on-orbit tasks. For sampling, the actuator components are mainly shovel, drill pipe, grinding wheel, etc.; for operation, the actuator components are mainly capture mechanism, winding mechanism, tool adapter, etc.

(4) Measurement components

In order to measure the working state of the end effector in real time and to feedback the end state information to the space robot system, a variety of sensors are used in the end effector. The measurement components usually consist of several sensing units, such as position, speed, force/torque, temperature, position switch. Common sensing units include resolver, Hall sensor, unidimensional force/torque sensor, six-dimensional force sensor, temperature sensor, space camera, radar, contact switch, proximity switch.

(5) Control components

The control components should have the ability to communicate with the host computer, control the drive motor, judge the working state, and control the temperature. The control components with noncooperative target operation requirements should also have autonomous decision-making capability. According to the task requirements, the control system should have a certain bandwidth to ensure a rapid response.

(6) Thermal control components

The thermal control components used in the end effector is similar to the joints in the form, generally includes active thermal control units and passive thermal control unit.

7.4.2.1 Drive Components

The end effector mostly uses a motor as the drive components, and its main types of application are the same as the joint drive source, refer to Sect. 7.3.2.1. In addition, spring drive components are often used in end effectors. The spring drive components (see Fig. 7.21) are typically composed of a drive spring and corresponding guiding and fixing devices, and has the advantages of small size, lightweight, reliable operation. The springs used in end effectors include helical springs, disc springs, leaf springs, etc.

(1) Coil springs

Coil springs used by the end effectors mainly include tension springs, compression springs and torsion springs. In order to ensure the reliable operation of the spring in orbit, it is necessary to appropriately adjust the allowable stress of the material according to the specific environment and the spring production and processing technology.

(1) If the failure of the spring may cause the failure of the end effector's task, its allowable stress should be reduced appropriately;
(2) The fatigue limit of the spring can be increased by prestressing treatment, and its allowable stress will be increased;
(3) When the operating temperature exceeds 60 °C, the shear modulus should be reduced properly.

In order to ensure that the load of the spring meets the requirements under specified deformation, the working deformation of the spring should be controlled within 20–80% of the maximum load deformation. When there is a strict requirement on the spring output force/torque, the deformation under the working load should be 30–70% of the deformation under the maximum load.

(2) Disc springs

Disc spring is a frustoconical compression spring stamped out of steel plate. It is mainly used for the position retaining and locking mechanism units in the end effector (see Fig. 7.22). It uses a guide rod or guide sleeve as the guide, and there should be a clearance between the guide and disc spring. It has the following application characteristics: large rigidity. Small deformation can output a large load, which is suitable for the mechanism with compact space and large output load. Properties of variable stiffness. When the disc spring is flattened, the ratio of deformation to the thickness is different, and its characteristic curve is also different. Combination of pairing and stack forms. When stacked, the more the number of disc springs for the

Fig. 7.21 Diagram of the spring drive component

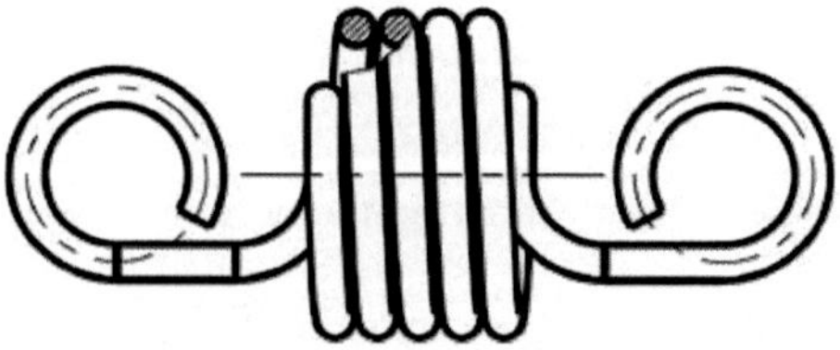

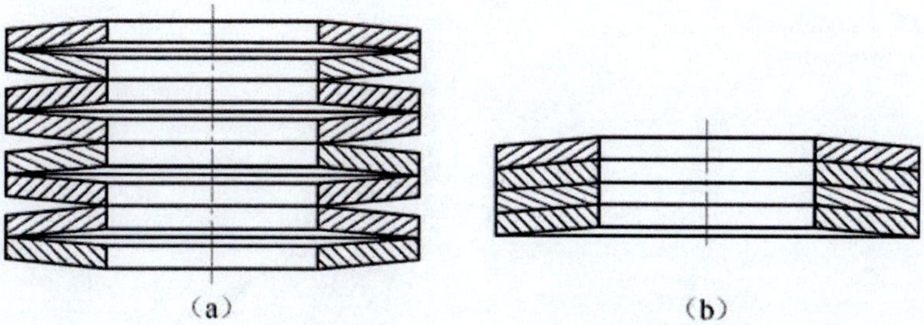

Fig. 7.22 Disc spring combinations **a** pairing combination; **b** stack combination

same amount of deformation, the larger the output load. When paired, the more the number of disc springs under the same load, the larger the deformation.

(3) Leaf springs

Leaf spring is made of a thin metal plate and uses the bending deformation of the plate as a spring. It is applicable for the occasions where the end effector is required to have a small load and a small spring stiffness and is usually used with the motor assembly. Leaf spring has many arcs and cross-sectional shapes in structure, which may cause stress concentration. When it is used intensively in orbit and the working time is long, the influence of stress concentration for the fatigue strength of the leaf spring must be considered.

7.4.2.2 Transmission Components

The end effector uses ball screw, gear, space/plane linkage, wire rope as the transmission components. The main application of the gear unit is the same as the transmission of the joint reducer. For more details, see Sect. 7.3.2.2. This section focuses on the screw transmission, space/plane linkage transmission, and tendon transmission.

(1) Screw transmission

The screw transmission refers to the transmission of motion and power through the screwing of nuts and screws. Generally, rotary motion is changed into linear motion. When the screw transmission is not self-locked, linear motion can also be changed into rotary motion.

The screw transmission can be divided into sliding screw, rolling screw, and static screw according to the friction property. Rolling ball screw transmission is widely used in end effector, its advantages are low frictional resistance, high transmission efficiency (generally above 90%), reversible transmission, and high transmission positioning accuracy.

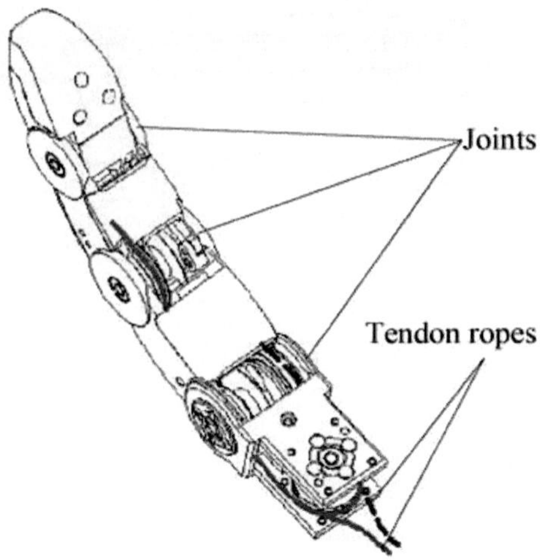

Fig. 7.23 Diagram of tendon transmission

(2) Space/plane linkage transmission

In the linkage mechanisms, the trajectory of each point of the linkage is a curve of different shapes. By appropriate design of the dimensions of each component, the linkage mechanism can realize different motion laws and motion requirements. In the design of the end effector, linkage mechanisms are often used to design the capture component of the end effector. By planning the motion trajectory of the linkage, a large capture tolerance can be obtained, and self-locking of the end effector is achieved through the dead-point position of the linkage mechanism.

(3) Tendon transmission

The tendon transmission (Fig. 7.23), is commonly used in space dexterous hand end effectors (such as NASA Robonaut2) and capture mechanisms (such as Canadarm II). For smart hands using tendon transmission, in order to reduce the volume and mass of the transmission mechanism, it is necessary to use the minimum number of tendon ropes to achieve the optimal control effect. According to the $N + 1$ theory, the minimum number of tendon ropes required for the control of N free tendon transmission smart hand is $N + 1$.

7.4.2.3 Actuator Components

The actuator components are the core units of the end effector for on-orbit operation. It is generally designed in conjunction with the operation target according to the task requirements.

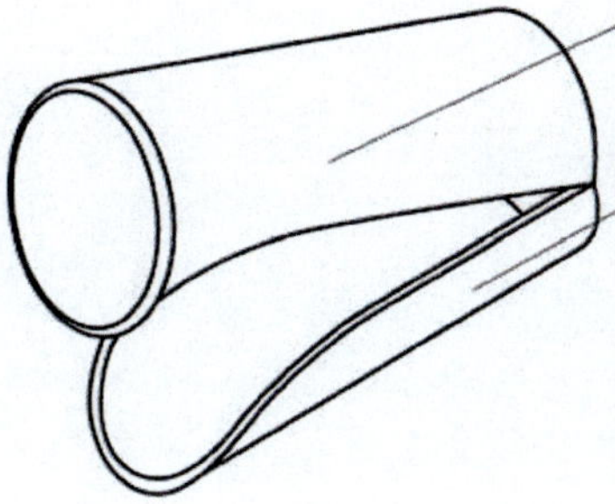

Fig. 7.24 Excavating shovel actuator

(1) Sampling actuator

It is an end effector designed for sampling tasks in deep space exploration. According to the physical properties of the sampling target, actuators such as excavating shovels (Fig. 7.24), drill rods (Fig. 7.25), and grinding wheels are often used as sampling tools.

(2) Operational actuators

For the end effectors with operational task requirements, their actuators are usually designed in conjunction with the specific types and interfaces of the operation targets. Most of them adopt the connection-separation mechanism and winding mechanism designs (such as the Canadarm I and II end effectors) to achieve on-orbit operations on a single target or multiple targets.

7.4.2.4 Measurement Components

In order to help the robot system control and feedback the state of the end effector itself, the measuring components are often designed in the end effectors. A unidimensional force sensor is arrange on the end effector to measure the drag force in real time, and a six-dimensional force sensor is used to measure the load forces and torques in various directions.

(1) Unidimensional force sensor

Unidimensional force sensors are commonly used in the end effectors to detect the contact force or the connection force of the operation target, their working principle is similar to that of the torque sensor. The elastomer is usually placed in the force path to directly detect the contact force of the end effector. See Fig. 7.15 for its structural form.

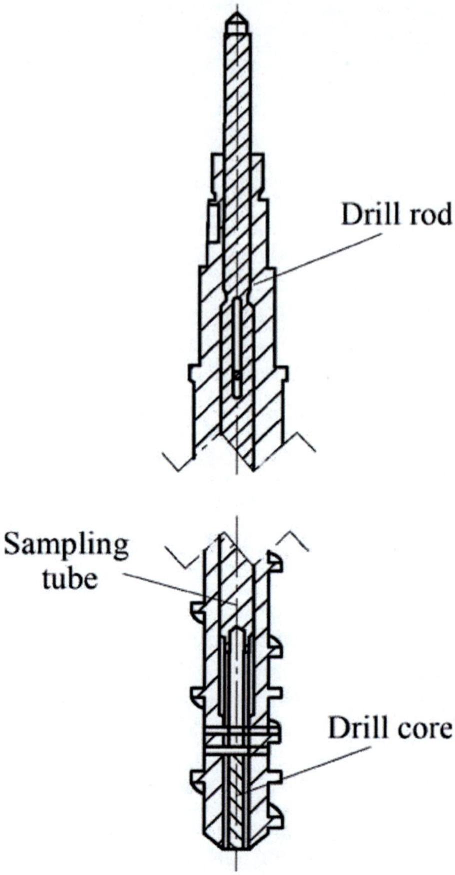

Fig. 7.25 Drilling and sampling actuator

(2) Six-dimensional force sensor

The six-dimensional force sensor mainly includes vertical beam type, beam type, composite beam type, barrel type, cylindrical type, and Stewart parallel structure. The form and arrangement of force sensitive components in the sensor structure directly affect its key features, such as sensitivity, stiffness, linearity, dynamics performance, and inter-dimensional coupling. The Stewart parallel mechanism has the advantages of high rigidity, stable structure, high load capacity, no error accumulation, easy solution, etc. It has become a widely used structure realization form of six-dimensional force sensors, as shown in Fig. 7.26.

(3) Switch unit

According to whether it is in contact with the measured object during operation, the switch components can be divided into two types: contact switch and proximity switch. The contact switch generates an on-off signal by means of the touch

Fig. 7.26 Six-dimensional force sensor based on Stewart parallel structure principle

sensing component to realize position detection, and the proximity switch relies on the electromagnetic signal generated when the motion mechanism approaches the sensing component to determine the position state. Proximity switches can be classified into inductive, capacitive, Hall, and photoelectric types depending on the principle. Among them, the Hall sensor has the advantages of small size, high precision, frequency bandwidth, noncontact, etc. It is suitable for applications where the installation space is small and the position measurement accuracy is high.

7.4.3 End Effector Design

Before starting the detailed design of the end effector mechanism, it is necessary to analyze the main performance parameters (such as capture tolerance, sampling volume, connection stiffness) of the end effector. In the program design stage, through mechanism analysis and mathematical simulation, the relationship between the design parameters and the mechanism performance is obtained. Through parameter optimization, the structural parameters that meet the system index are determined. The end effector design flow is shown in Fig. 7.27.

In the design of end effectors, the selection of materials, drive components, sensor components, transmission form, and the design of structures and mechanisms, and the analysis of index matching, strength of mechanical parts, stiffness of assemblies, etc., are basically similar to the joints, so we will not discuss it here. Refer to Sect. 7.3.3, for details. Because the capture tolerance is the primary technical indices of the end effector, this section focuses on program design according to the end effector capture tolerance requirements. In general, a larger capture tolerance is desired to meet reliability, envelope, mass, and power specifications. In the following paragraphs, we will take the typical end effector design program as an example to analyze the relevant parameters of the capture tolerance.

Figure 7.28 shows a rope winding end effector [3], its tolerance mainly depends on the length of the captured rod, the capture space and the on-orbit capture strategy of

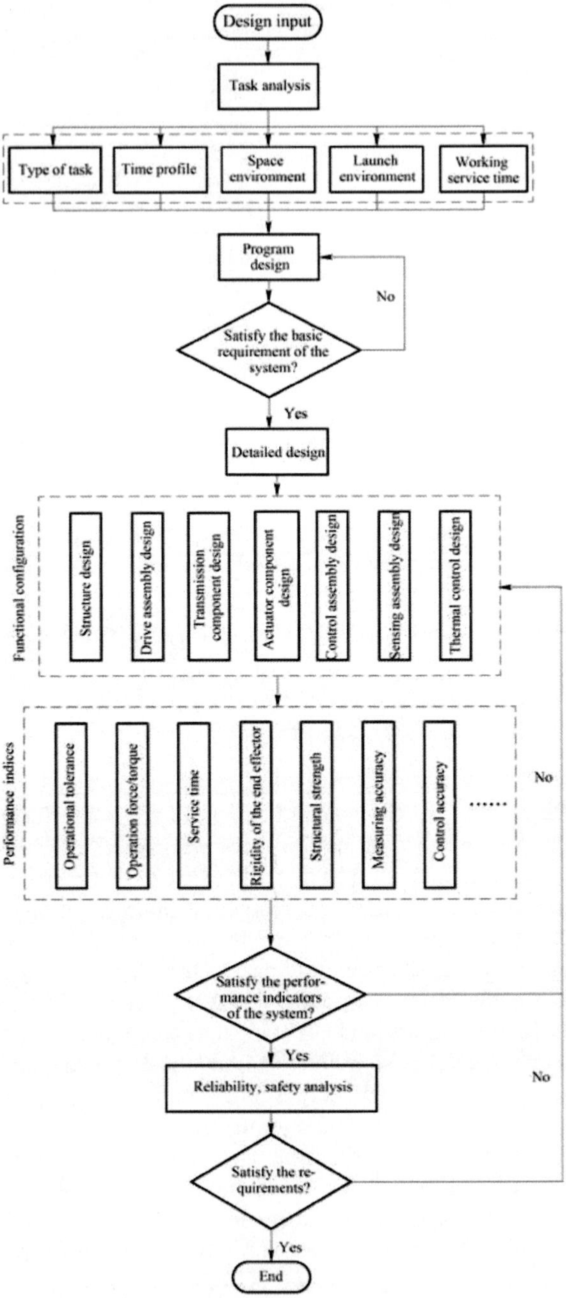

Fig. 7.27 End effector design flow

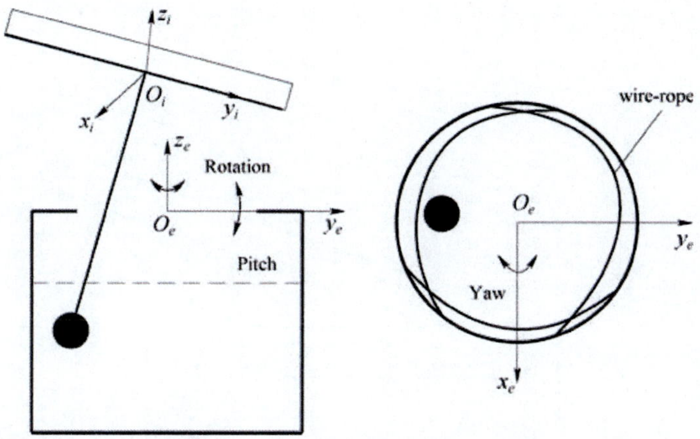

Fig. 7.28 Diagram of interface

the end effector, where the capture space is determined by the inner diameter and axial length of the end effector. During actual capture, a certain safety distance should be left between the interface coordinate $O_ix_iy_iz_i$ and the end effector coordinate $O_ex_ey_ez_e$ to ensure the safety of the robot arm system and the target.

The origin of the interface coordinate is O_i (0,0,0). The interface has radial deviation, axial deviation, and angular deviation relative to the end effector, which means that the deviation measurement coordinate has positional change relative to the end effector coordinate. Assuming that the variation of the deviation measurement coordinate with respect to the end effector coordinate is as follows: the radial deviations of the Ox and Oy axis along the end effector coordinate are Δx and Δy, respectively, and the axial deviation along the Oz axis is Δz, the angular deviation is φ (rotation), ϕ (yaw), and θ (pitch), respectively.

The coordinates of the captured rod endpoint a_0 in the interface coordinate are a_0 (0, 0, 1). After the coordinate transformation, the coordinates of O_i and a_0 in the end effector coordinate can be obtained. Thus, the end effector tolerance analysis model is established, as shown in Fig. 7.29, and the key points affecting the end effector tolerance performance are obtained. By analyzing the relationship between the key point coordinates and the parameters such as the diameter, the length of the end effector tolerance space, and the length of the captured rod, a quantitative analysis of the tolerance performance of the end effector is realized.

In Fig. 7.29, r is the capture space radius; D_i is the capture space diameter; l is the length of the captured rod; H_0 is the distance between the capture surface and the end surface of the end effector; L_i is the depth of capture space.

According to the coordinate definition of the end effector and its structural characteristics, the tolerance space of the end effector is a cylinder, and can be mathematically expressed as

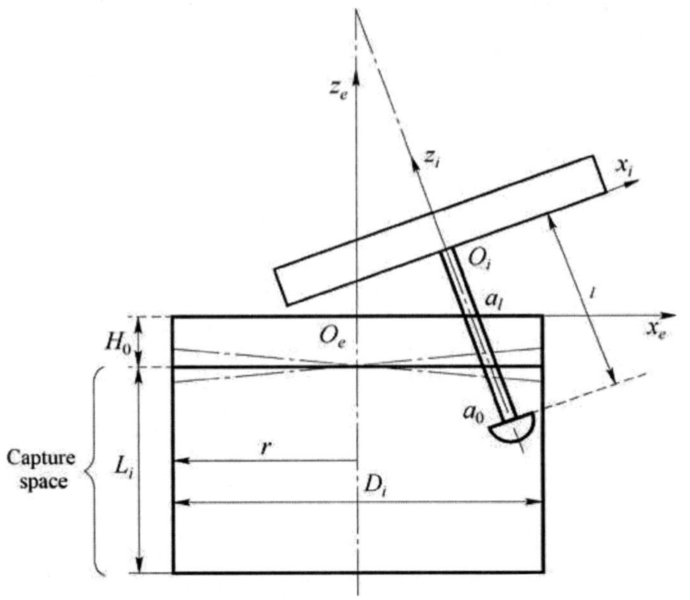

Fig. 7.29 Tolerance analysis model of rope winding end effector

$$\begin{cases} x^2 + y^2 = \left(\frac{D_i}{2}\right)^2 & -\frac{D_i}{2} \le x \le \frac{D_i}{2} \\ -L_i - H_0 \le z \le -H_0 & -\frac{D_i}{2} \le y \le \frac{D_i}{2} \end{cases}. \tag{7.19}$$

The position deviation of the interface coordinate T: $O_ix_iy_iz_i$ relative to the end effector coordinate G: $O_ex_ey_ez_e$ is (Δx, Δy, Δz), the attitude deviation is (ϕ, φ, θ); then according to the transformation matrix $R(\phi)R(\varphi)R(\theta)$ of the interface coordinate relative to the end effector coordinate, make out the coordinates of the end points O_i (0,0,0) and a_0 (0, 0, l) of the captured rod in the end effector coordinate (x_1, y_1, z_1) and (x_2, y_2, z_2); thus the linear equation of the axis line of the captured rod in the end effector coordinate is obtained as

$$\frac{x - x_1}{x_2 - x_1} = \frac{y - y_1}{y_2 - y_1} = \frac{z - z_1}{z_2 - z_1}. \tag{7.20}$$

The end effector end surface equation is

$$z = 0.$$

According to the linear equation of the captured rod axis under the position and orientation deviation, the coordinates $(x_3, y_3, 0)$ of the intersection of the captured rod axis line and the end surface of the end effector a_1 can be obtained. If the coordinates of a_0 and a_1 in the end effector coordinate meet the following conditions, respectively:

$$x_3^2 + y_3^2 \leq \left(\frac{D_i}{2}\right)^2 \tag{7.21}$$

$$\begin{cases} x_2^2 + y_2^2 \leq \left(\frac{D_i}{2}\right)^2 \\ -(L_i + H_0) \leq z_2 \leq -H_0 \end{cases}, \tag{7.22}$$

and the captured rod is within the tolerance space of the end effector, then the end effector can capture the target under this deviation state.

7.5 Mobile Mechanism

In order to accomplish a specific task, a robot usually needs to move its position to reach the task location. The mechanism that changes the position of the robot is called the mobile mechanism. The mobile mechanism is an important actuator component of the robot and typically consists of drive devices, transmission mechanism, position detection sensors, controller, and thermal control assembly. In addition to carrying its own weight load, the mobile mechanism also needs to carry various other payloads. If necessary, it should be able to change its own configuration to adapt to the load requirement. Therefore, it must have enough rigidity, stability, and adaptability to meet the load requirements of the robot.

Space motion mechanism is mainly used for planetary exploration, such as Lunar exploration robots and Mars exploration robots. In the future, on-orbit service missions of spacecraft will involve the operation of large-scale manufacturing, assembly, and maintenance, and the mobile robot working on spacecraft will be widely used. The design of the mobile robot is closely related to mission requirements. In general, the topography and geological conditions of the planet surface are full of uncertainties. Therefore, it is necessary to design different types of space mobile mechanisms for different detection areas. In the flat terrains, a wheeled mobile mechanism can be used; on rugged terrains, it is necessary to design a legged robot with a stronger climb obstacle and extrication capabilities. No matter what form is adopted, the mobile robots applied to the planet surfaces should have good adaptability to high- and low-temperature environment.

7.5.1 Legged Mobile Mechanism

Legged mobile mechanism is a kind of mobile mechanism inspired by animal skeleton structure. The common mobile mechanism has two-legged, four-legged, and six-legged forms. Compared with the wheeled or track mobile mechanisms, the legged mobile mechanism has better terrain environmental adaptability, because its legs can contact with the ground to realize discretized support and movement,

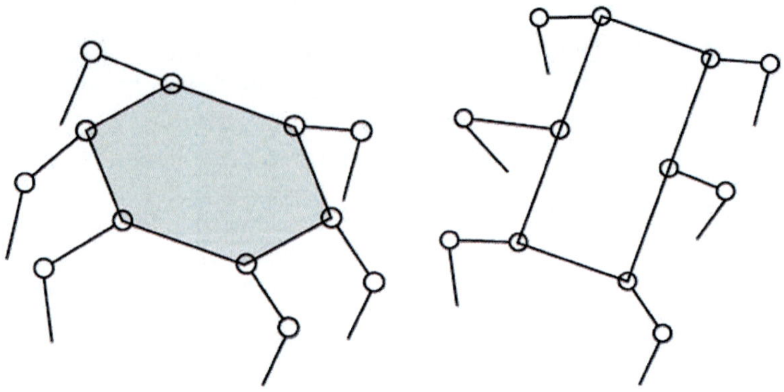

Fig. 7.30 Different arrangements of legs **a** hexagonal arrangement; **b** symmetric arrangement

especially under special conditions such as obstacle or slopes. In space applications, limited by poor stability of the two-legged mobile mechanism, presently, four-legged and six-legged mobile mechanisms are more used in design schemes. As the increase in the number of legs will increase the mass and control difficulty of the space robot, the number of legs should not be too much. Secondly, the six-legged robot is taken as an example to describe the legged mobile mechanism.

In order to facilitate modular management, six-legged robots are usually designed with the same structure for each leg. According to the design requirements, the six legs can be arranged on the robot body structure by multiple modes. For example, the hexagonal arrangement generally distributes the legs evenly around the body structure, and the symmetric arrangement generally aligns the moving legs symmetrically along the axis of the robot moving direction, both of which can help support the robot to move, as shown in Fig. 7.30. But the travel strategies of each arrangement are different.

Each leg is configured with multiple DOFs to increase dexterity and ground adaptability. According to the requirements, a leg is usually designed with a hip joint, a knee joint, an ankle joint, and a Local degree of freedom at the end, as shown in Fig. 7.31, and the number of joint can either be reduced or expended depending on the task.

There are various forms of design for the local degree of freedom at the end (shown in Fig. 7.32) according to different types of tasks. Three common forms are tapered, semi-cylindrical, and cylindrical ends. If it is rigid contact with the planet surface and has a flat contact surface, the three forms of ends correspond to the point contact, line contact, and surface contact, respectively. Among them, point contact can provide higher contact stress and improve the adhesion of the mobile mechanism.

Another widely used local degree of freedom is the wheel type in which a drivable wheel is fitted at the end of the moving leg, so it is also known as the leg-wheel hybrid mobile mechanism. The drivable wheel has two states: driving and locking. When the

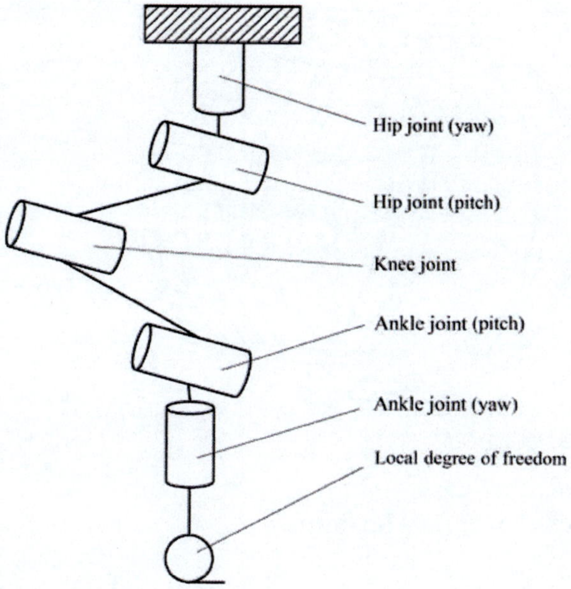

Fig. 7.31 DOF configuration of legged robot

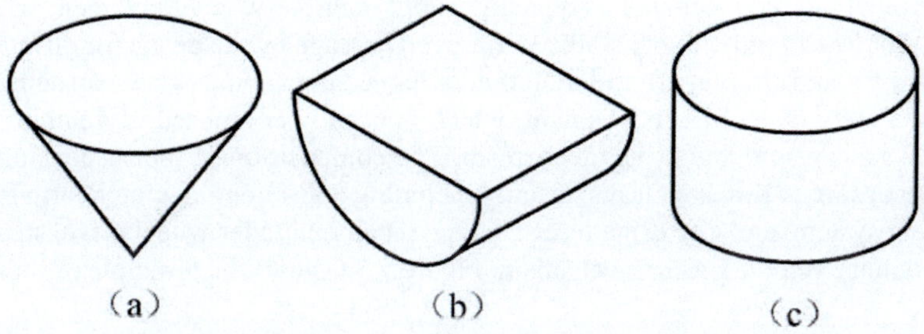

Fig. 7.32 Forms of local degree of freedom at the end **a** tapered; **b** semi-cylindrical; **c** cylindrical

six wheels are driving at the same time, the mobile mechanism can realize wheeled moving mode; when the six wheels are locked simultaneously, the mobile mechanism uses the legged moving mode. By designing the fin (pawl) structure on the surface of the drivable wheel (as shown in Fig. 7.33), the mobile mechanism can improve the adhesion to the ground, thus improve its passing capability.

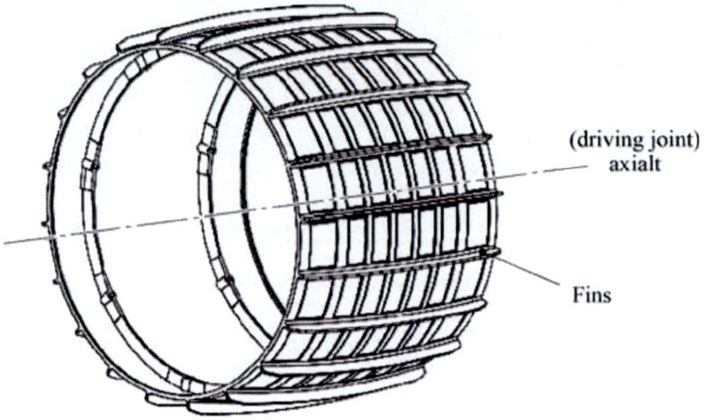

Fig. 7.33 Wheeled local DOF

7.5.2 Wheeled Mobile Mechanism

The wheeled mobile mechanism is composed of wheel, drive mechanism, suspension, and other parts. The mobile mechanism is mainly used for planetary surface roaming and exploration tasks. In order to improve their environment adaptability, the wheeled mobile mechanisms typically adopt multiple-wheel movement scheme, and is characterized by high stability, improved passage by the design of differential mechanism, and simpler control than that of legged ones. Four-wheel suspension is a simple form of mobile mechanism, which is usually composed of vehicle body, wheels, suspension, and drive mechanism. The composition of the suspension will vary according to different tasks, usually including a differential, clutch, arm boom, damping system, etc. The drive mechanism typically includes wheel drive, steering, and auxiliary vehicle lifting mechanism. Figure 7.34 shows the principle of a simpli-

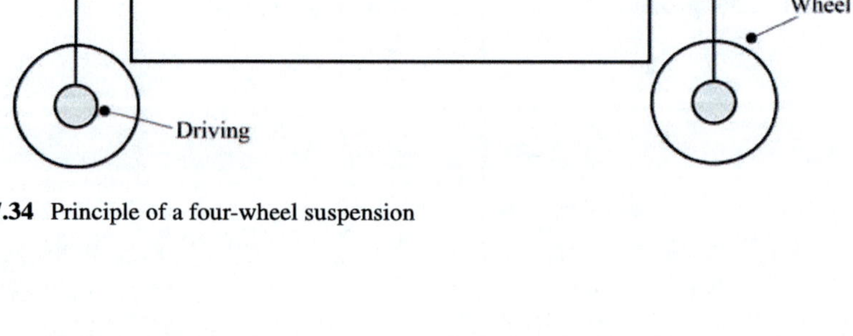

Fig. 7.34 Principle of a four-wheel suspension

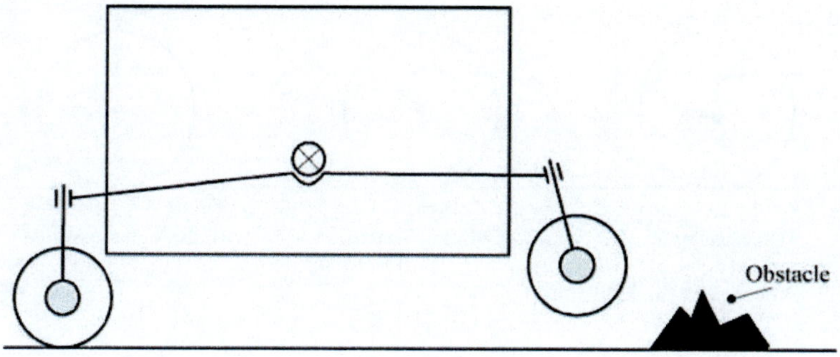

Fig. 7.35 Obstacle avoidance by a four-wheel suspension

fied four-wheel suspension, where there are 8 motors at 4 wheels, 4 of them are steering motors, 4 driving motors, and the suspension is equipped with a differential mechanism. Suspension can be divided into active suspension and passive suspension. The active suspension requires a power source, as shown in Fig. 7.35. When the vehicle encounters an obstacle, the active suspension can lift up one of the wheels to improve its trafficability; if it is a passive suspension, the whole vehicle will travel as usual without active adaptability.

In some special working conditions, the mobile mechanism needs to lift the vehicle. There are various vehicle lifting principles. Figure 7.36 shows a "double pendulum + slider mechanism" scheme, which changes the distance between the

Fig. 7.36 Wheel lifting principle

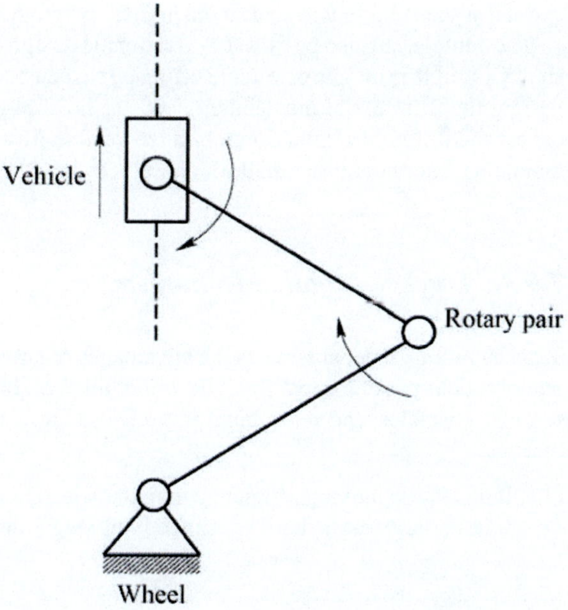

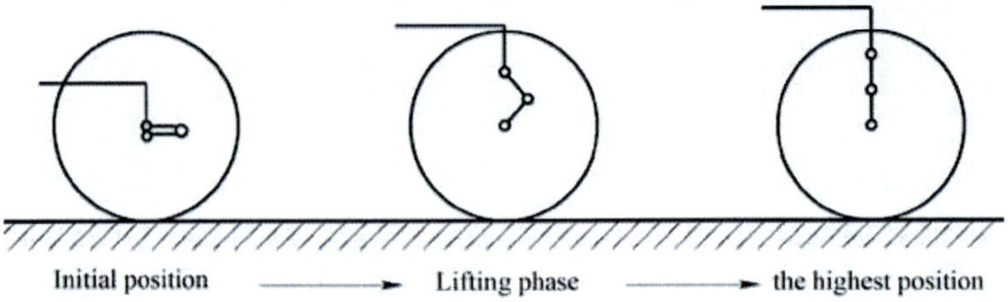

Fig. 7.37 Process of vehicle lifting

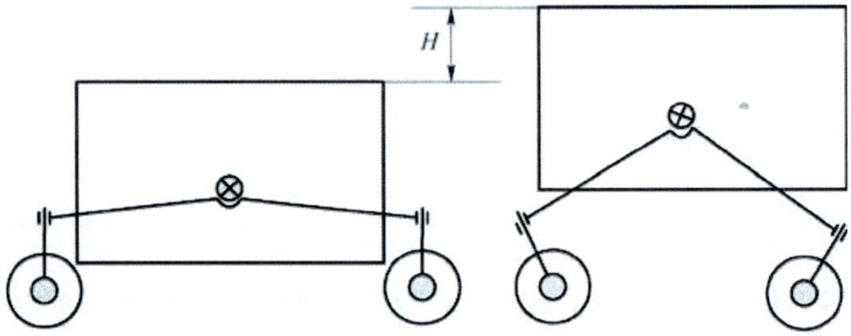

Fig. 7.38 Vehicle lifting by a suspension

vehicle body and the wheel by driving the revolute joint to raise and lower the vehicle. As shown in Fig. 7.37, before the vehicle is lifted, the two swing links are horizontal. With the lifting process, the angle between them is gradually increased, and when the vehicle is raised to the highest point, the swing links become vertical.

The vehicle can also be raised by a rationale design of the suspension. Figure 7.38 shows a simple principle of vehicle lifting by suspension control. In the actual product design, the differential mechanism needs to be configured with other mechanisms except the differential function to realize vehicle lifting and ensure that the vehicle maintains a horizontal or small tilt angle.

7.5.3 *Tracked Mobile Mechanism*

Tracked mobile mechanism has the advantages of strong adaptability to soft ground, and excellent passing capability and smoothness on harsh surfaces such as complex terrains, obstacles, and steps, but it is rarely used in space robots due to the following reasons:

(1) It has lower moving efficiency than the wheeled robots;
(2) It has a large mass, hard to realize lightweight design;

(3) The track is in direct contact with the ground, its form of movement will cause certain soil particles to enter the track and increase the wear of the mechanism;
(4) It may produce side slip during turning, which will increase the track wear.

7.6 Hold-Down and Release Mechanism

7.6.1 Function of the Hold-Down Release Mechanism

As the name implies, the main functions of the hold-down and release mechanism are hold-down and release. The hold-down function means that the mechanism presses the various parts of the space robot on the spacecraft body in a furled configuration, to ensure that the system can withstand large loads without damage during the launch and flight phases, at the same time satisfy the system fundamental frequency. The release function is that according to the mission requirements, after the spacecraft goes into orbit, the mechanism on spacecraft releases the constraints on the various parts of the robot smoothly to restores its normal operation.

Space robotics are complex in structure, including joints, end effectors, cameras, structures, and other components. Therefore, the hold-down and release mechanism is generally composed of several hold-down and release devices to ensure reliable fix in the furled state. As shown in Fig. 7.39, the space robot is tightly held down at the spacecraft platform by multiple hold-down release devices. When the space robot releases, it can adopt a single point release mode, which means, each hold-down release device is unlocked and released by a corresponding release device, or a multi-point linkage release mode, which means, a number of hold-down release devices are simultaneously unlocked by a single release device through a linkage mechanism. In practical applications, due to the envelope, weight, unlock timing, and reliability constraints, the hybrid mode of single point release and multiple points simultaneous release is often adopted.

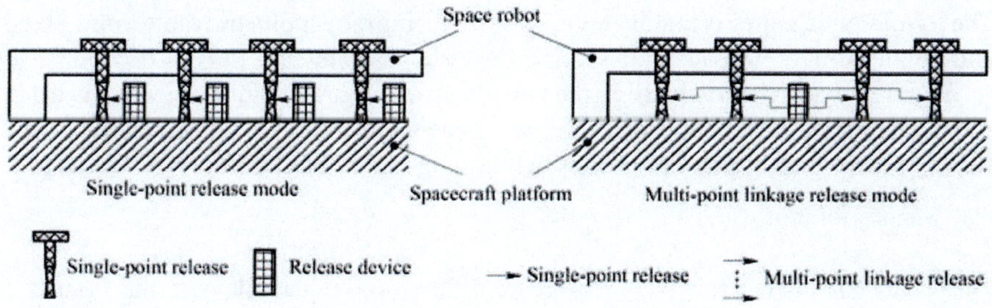

Fig. 7.39 Hold-down release modes of space robot

If a hold-down release mechanism has no release function, it is no different from the fastening connector, so the release device is the key to the hold-down release mechanism, and many release devices have the function of connection before they are released.

7.6.2 *Types of Release Devices*

The Release device is used to release the fastened constraints. Since the purpose of the release device is to eliminate the original connection relationship, the types of release device are closely related to the original connection form. There are a variety of release devices for spacecraft mechanisms, which can also be used on space robots. The release devices can be divided into pyrotechnic release device and non-pyrotechnic release device. See Table 7.6, for types and characteristics of release devices.

The pyrotechnic release device uses the energy generated by combustion or explosion to achieve the release function. It has the advantages of lightweight, small volume, large specific energy (energy/weight) and reliable performance, etc., but it also has the inevitable defaults such as a large impact and pollution. Compared with the pyrotechnic release device, the non-pyrotechnic release device has less impact, and generally does not produce harmful gas during operation, so the payload will neither be damaged by the impact, nor be polluted. In addition, the non-pyrotechnic release device can be reused in the ground test, and the test cost is low. The application research in this aspect has received more and more attention.

Due to the mature technology of the pyrotechnic release device, it has become the main device used for the release function of spacecraft mechanisms. The pyrotechnic release device used in space robot's hold-down release mechanism is mainly divided into two types: pyrotechnic cutter and separation nuts.

(1) Pyrotechnic cutter

The Pyrotechnic cutter is mainly composed of casing, shear pin, movable knife, fixed knife, nut, igniter, gunpowder, a sealing ring, etc., as shown in Fig. 7.40.

After receiving the instruction, the two electronic igniters ignite at the same time, and a large amount of high-pressure gas is generated in the closed cavity of the casing. The high-pressure gas pushes the moving knife to cut off the hold-down rod.

(2) Separation nut

The separation nut is a relatively mature hold-down and release device, characterized by the hold-down bolt which can be designed according to the requirements of the bearing capacity, and being able to meet the requirements of a large connecting load. It is composed of an initiating explosive actuating source and an actuator. Figure 7.41 shows the composition of a certain type of separation nut.

Table 7.6 Types and characteristics of release devices

Type	Name	Characteristics
Pyrotechnic release device	Explosive bolt	It has both connection and release function, characterized by large carrying capacity, small size, lightweight, simple structure, and reliable operation. The disadvantage is the large impact caused by the release
	Separation nut	It is used to release hold-down bolts, characterized by simple structure, reliable performance. It can be easily combined with other devices
	Pyrotechnic cutter	The release function is achieved by cutting off the hold-down rod; the cutter itself does not involved in the connection; no need to bear large loads
	Pyrotechnic pin pullers	The constraint of the target is released by pulling out the pin that fixes the target; the pyrotechnic Pin Pullers itself does not directly involved in the connection, so the size is usually small, and the typical diameter does not exceed 9.5 mm
	Pyrotechnic locker	It has three functions of connection, release, and separation; the impact during release separation is low, generally, no fragmentation occurs; with a complex structure, and high manufacturing cost
Non-pyrotechnic release device	Memory alloy actuating device	Using the shape change caused by the phase change of the memory alloy during heating to generate the force required for release; very low impact; long braking time; not suitable for synchronous release
	Paraffin actuating device	Using the expansion of paraffin from solid to liquid (up to 15% volume variation) to produce the mechanical energy for release; very low impact; long release time, 30 s or more

(continued)

Table 7.6 (continued)

Type	Name	Characteristics
	Thermal knife actuating device	The heating component of the thermal knife is energized and heated (up to 1000 °C), the tensioning rope is gradually weakened and finally broken; very low impact, without pollution; the knife body can be reused during ground testing

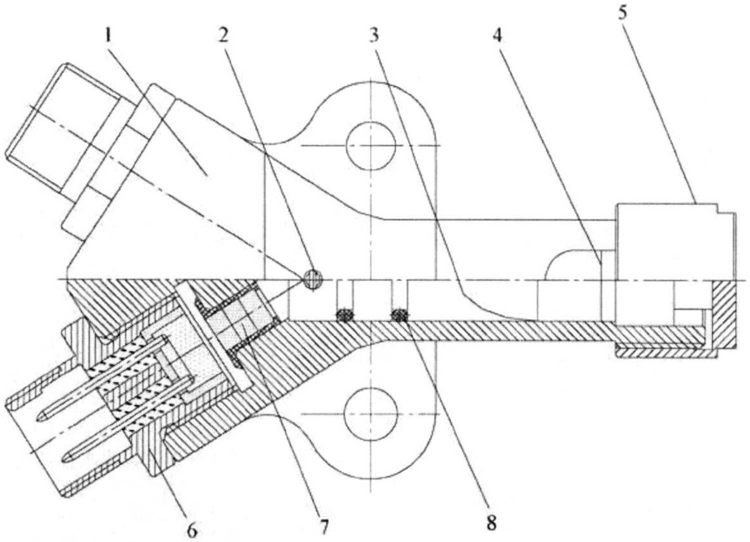

Fig. 7.40 Schematic of the pyrotechnic cutter structure 1 housing; 2 shear pin; 3 movable knife; 4 fixed knife; 5 nut; 6 igniter; 7 gunpowder; 8 sealing ring

During operation, the EED (Electro-Explosive Device) is electrically excited to ignite the main charge, and instantaneously generate a high-pressure gas, which pushes the sleeve down by a certain distance to release the radial constraint of the nut. At this time, the piston pushes the nut, the radial force expands the nut splits outward, releases the constraint on the thread of the hold-down bolt.

7.6.3 Design of the Hold-Down Release Mechanism

The hold-down and release mechanism is designed to meet the hold-down and release requirements and to minimize development costs.

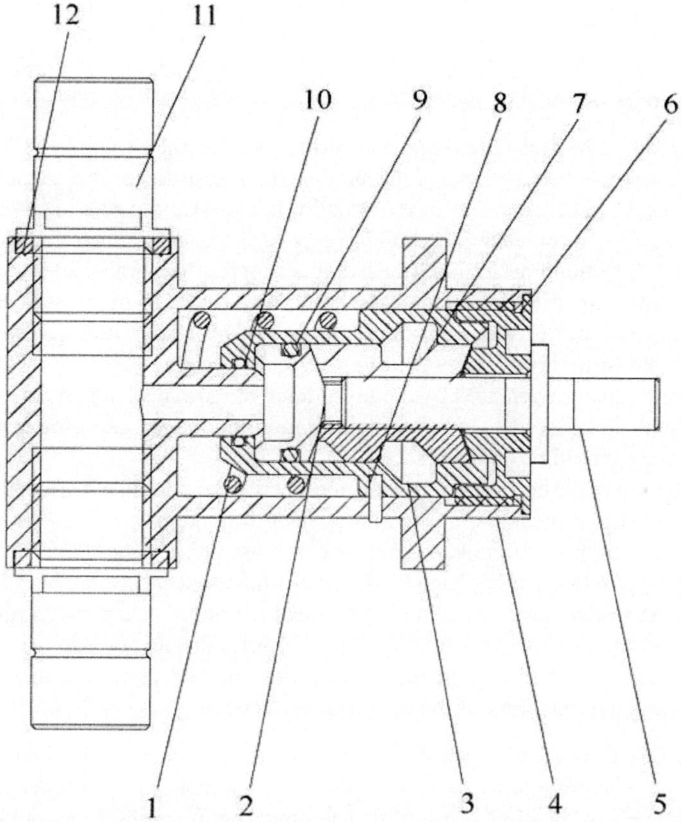

Fig. 7.41 Composition of a type of separation nut 1 spring; 2 piston; 3 sleeve; 4 shell; 5 hold-down bolt; 6 end cap; 7 nut Split; 8 shear pin; 9 sealing ring I; 10 sealing ring II; 11 EED; 12 sealing washer

Considering the design characteristics of the space robot hold-down release mechanism, the following principles should be followed:

(1) Inherit the existing design technology foundation, try to use mature technologies and materials;
(2) On the premise of satisfying the fundamental frequency of the product hold-down state, reduce the mechanical response of the product and ensure the minimized mass;
(3) Carry out the layout design of the hold-down points according to the principle of a reasonable load path, and optimize the number and position of the hold-down devices by analysis of the mechanical characteristics of the system;
(4) No remains should be produced after release, and no pollution should be caused to the environment and other equipment;
(5) Full consideration of the Processing and assembly technology;

(6) Economic considerations.

In the layout design of the hold-down and release mechanism, in order to make the layout more reasonable, the following aspects should be considered:

(1) The load path. The most important principles are the principle of force flow continuity and the principle of the shortest load path. According to the layout of each structural component of the robot, it is necessary to provide a reasonable load path between each supported component and the launch vehicle. On the premise of obtaining high stiffness and strength, the number and the mass of hold-down and release mechanism should be reduced as much as possible.
(2) The position of the hold-down and release devices should be set at the location where the structural rigidity is large.
(3) The parts that are released by the hold-down release mechanism may not interfere with or cause damage to surrounding components, and proper protective measures should be considered.
(4) The layout of the hold-down and release mechanism should not affect the robot's on-orbit deployment and normal movement of its joints.
(5) The spacecraft structure and the held-down assemblies should be separated as much as possible to avoid high load on the component caused by the deformation of the structure, and reduce the dynamics load on the spacecraft structure and other equipment from the held-down assemblies during release.
(6) The layout interface design should consider the deformation of the spacecraft body and the matching of the installation interface.

Figure 7.42 shows the layout of the hold-down and release mechanism of a space manipulator. The configuration of the manipulator adopts a "2 shoulder joints + 1 elbow joint + 1 wrist joint" design, that is, there are one yaw joint and one pitch joint at the shoulder, one pitch joint at the elbow, and one pitch joint at the wrist.

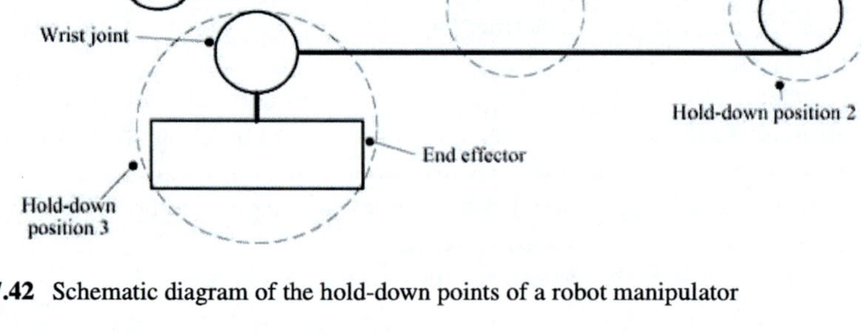

Fig. 7.42 Schematic diagram of the hold-down points of a robot manipulator

Since the shoulder yaw joint is directly mounted on the spacecraft body through the base, there is no need to set a hold-down release mechanism on the shoulder. So, one hold-down and release mechanism is set in the middle of the two arm booms, one hold-down and release mechanism is set at the elbow and the wrist. Through the mechanical analysis of the location of the hold-down release mechanism, the system fundamental frequency is calibrated to meet the requirements, and the strength of the robot manipulator is calibrated to meet the requirements under the launch load condition.

7.7 Lubrication for Space Mechanism

Lubrication failure is the most common failure in space mechanism. Different from ground-based lubrication, space lubrication needs to withstand complex space environments such as ultra-high vacuum, microgravity, high- and low-temperature alternating, redox medium, atomic oxygen, ultraviolet radiation, and other special environments. Under the condition of space vacuum, the disappearance of the gas lubrication effect between two relatively moving parts and the aggravation of lubricant/grease evaporation will affect the lubrication design.

The lubrication design of space robot should meet the following requirements:

(1) Its adaptability to the space environment should be satisfied, and verified by test if necessary.
(2) It should meet the minimum requirement for lubrication performance at the end of servicing life.
(3) It must not cause unacceptable damage to non-lubricated parts and other devices.

The common lubrication methods for space robots include solid lubrication, grease lubrication, and solid-grease composite lubrication.

7.7.1 Grease Lubrication

Grease lubrication combines the advantages of oil lubrication and solid lubrication and has the following functions. First, in the period from ground implementation to launch, the grease can isolate the air, water vapor, and corrosive gas so as to prevent the corrosion of friction surfaces and improve the storage life. Second, the grease can change the mechanical energy of shock vibration into hydraulic energy to a certain extent, thus playing a damping role. In addition to the advantages of ground-based grease, the space grease should also have low volatility, excellent adaptability to high- and low-temperature, and good space lubrication performance.

Perfluoropolyether-based grease has been widely used in space mechanism products owing to its excellent comprehensive performance. The Braycote 600EF, 601EF, and 602EF made by Castrol are all the grease based on Braycote 815Z. The perfluoropolyether (PFPE) lubricating oil has a pour point as low as −72 °C and excellent viscosity-temperature characteristics. The vacuum evaporation loss is small, and the friction and wear performance is good, which conforms to the standard MIL-G-27617. The grease lubrication used in space mechanism products should be supplemented by active temperature control to ensure the working temperature of grease.

7.7.2 *Solid Lubrication*

Solid lubricating film and coating are generally divided into solid lubrication films (generally no thicker than 5 μm) and solid lubrication coatings (generally thicker than 5 μm).

Sputtering-deposited MoS_2 composite film is a solid lubrication film commonly used in space. Its main characteristics are as follows. First, the vacuum friction coefficient is very low, the adhesion with the substrate is strong, the wear-resistant life is long, and the wear debris produced in the process of movement are few. Therefore, it applies to the precision parts working in a vacuum environment. Second, it will be oxidized in a humid atmosphere so that its performance will decline. Therefore, the humidity in its storage and use areas should be strictly controlled. In addition, the sputtering-deposited MoS_2 composite film can be used with grease.

High-temperature-cured MoS_2 inorganic bonded coating and medium-temperature-cured MoS_2 resin bonded coating are the solid lubrication coatings commonly used in space. The main characteristics of high-temperature-cured MoS_2 inorganic bonded coating are as follows. First, it has good anti-atomic-oxygen and anti-irradiation performance and high load-bearing capacity, which are suitable for the surface of non-precision parts directly exposed to space atomic oxygen environment. Second, it has lower flexibility and impact resistance and more wear debris than resin bonded coating. The main characteristics of medium-temperature-cured MoS2 resin bonded coating are as follows. Firstly, it has a low vacuum friction coefficient, a medium load-bearing capacity, and a medium working life. Secondly, it produces quite a lot of wear debris in the movement process, so it is not applicable to the surface of precision parts. The humidity resistance of medium-temperature-cured MoS_2 resin bonded coating is better than that of high-temperature-cured MoS_2 inorganic bonded coating. However, the humidity in their storage and use areas should be strictly controlled. In addition, neither coating is recommended for use with grease.

In the design and selection of solid lubrication film/coating, consideration should be given to the thermal expansion matching between the film/coating and the base material, the size matching between the thickness of the film/coating and the roughness of the metal substrate surface, the surface stress tolerance of the film/coating, and other factors. The selected solid lubrication film/coating must be process qualified and meet the application requirements.

7.7.3 Solid-Grease Composite Lubrication

Proper use of solid-grease combined lubrication can yield good synergistic effect, which is mainly reflected in two main points: (1) the grease covers the surface of solid lubrication film/coating, isolates the film/coating from the water and oxygen in the air, and avoids the degradation of lubricating performance caused by improper environmental conditions; and (2) a single solid lubrication material has a relatively large friction coefficient and a short wear-resisting life, but the addition of grease can effectively improve its service life. The solid-grease composite lubrication often used in space includes sputtering-deposited MoS_2 composite film + Braycote 601EF grease.

7.8 Test Verification

Test verification is an important process in the development of space robot mechanical system. It can check whether the system meets the design specifications and technical indices under expected conditions. This is the final test of the design result. In the design of space robot mechanical system, the first step is index analysis, calculation, and detailed design according to the input conditions and design scheme, then carry out preliminary conformity analysis on the design results. After the production and assembly are completed, a verification test is to be done to verify that the product can achieve the specified functions and performance under specified conditions.

Space robot mechanism test mainly includes mechanism function/performance test, environmental test, and reliability test. Function/performance test is the verification tests of the mechanism's functions and the specifications. These include the design indices broken down from the system to equipment levels, product design specifications, requirements, and inspection items identified during the product design process. Environmental test is to verify the product's capability to perform specified functions and comply with the specifications under defined space environmental conditions. Reliability tests include verification and evaluation of product reliability.

The function/performance test is usually performed separately before, after or during an environmental test. The criterion for the product to pass the environmental test is that the test value of function/performance indices meets the design requirement during or after the test. The environmental test is generally carried out in the detailed design stage. It mainly verifies the correctness and rationality of the product's design scheme and process plan, exposes defects in manufacturing quality, materials and processes in advance, verifies the mechanism's adaptability to the space environment, exposes potential defects, with the ultimate goal of improving mechanical reliability, and provides a baseline for the flight model products.

The core assemblies of the space robot mechanical system include the joints and the end effectors. Although the design schemes of the joints and the end effectors vary widely, the test items for their function/performance indices are similar. For the joints, the main test items include motion accuracy, backlash, torque characteristics, rotational speed, and torsional stiffness. For the end effectors, the main test items include capture tolerance, docking accuracy, operating force, operating time, and connection stiffness. Therefore, the test items and test methods of the space robot mechanical system are described by taking the above two core assemblies as examples.

7.8.1 Function/Performance Test for Space Robot Joints

7.8.1.1 Stiffness Test

Stiffness mainly affects the natural frequency of the robot joints, including the system's natural frequency determined by the inertia of the driven object and the torsional stiffness of the joint. Since the modal frequency affects the upper limit of the system bandwidth, higher stiffness joints enable the mechanical system to obtain higher servo performance.

Torsional stiffness of the joint is one of the important technical indices that determine the dynamics characteristics of the joint output. Depending on the composition of the joint structure, the torsional stiffness is co-determined by the components of the force transmission path, such as the torsional stiffness of the motor, the stiffness of the planetary transmission, the stiffness of the harmonic reducer, and the torsional stiffness of the connecting flange. When measuring joint stiffness, use the layout shown in Fig. 7.43. Hold the brake at the joint input end, apply moment loads to the joint output end, and record the rotation angles at the output shaft corresponding to each load, then the curve of the output shaft rotation angle change with the load torque value is obtained, i.e., the torsional stiffness curve of the joint. Generally, the joint torsional stiffness has a nonlinear feature, under low load conditions, the stiffness is relatively low due to factors such as backlash and transmission gap. When the load increase to the elastic range, the stiffness increases and tends to be stable.

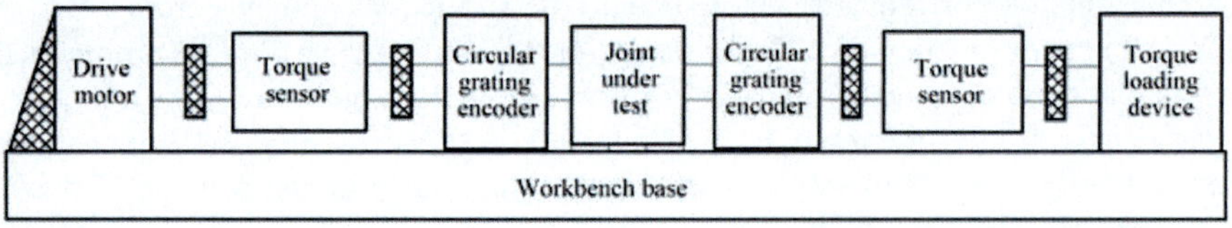

Fig. 7.43 Test platform for joints

7.8.1.2 Joint Transmission Error Test

(1) Transmission accuracy

The difference between the joint's actual motion angle and the given target angle is called the transmission accuracy of the joint.

During the test, use the test equipment shown in Fig. 7.43, compare the difference between the given motion curve and the measured joint angle output curve to obtain the transmission accuracy of the joint. Transmission accuracy includes no-load and loaded motion accuracy.

(2) Joint backlash

There are two joint backlash test methods: input test and output test. Simply put, the input test method is to input a forward and backward rotation angle on the input shaft, and measure the angular change at the first and last positions of the output end; the output test method is to load a positive and negative torque at the output end, and measure the angular change of the output shaft. Both methods can apply a certain load according to actual requirements.

7.8.1.3 Torque Property Test, Joint Output Property Test

Torque property refers to the output torque property of the joint. The output torque includes rated output torque, maximum output torque, continuous stall torque, and backdrive torque for brake release. Measure the corresponding speed for each output torque condition. The test layout can refer to Fig. 7.43, and the torque-speed curve is shown in Fig. 7.44.

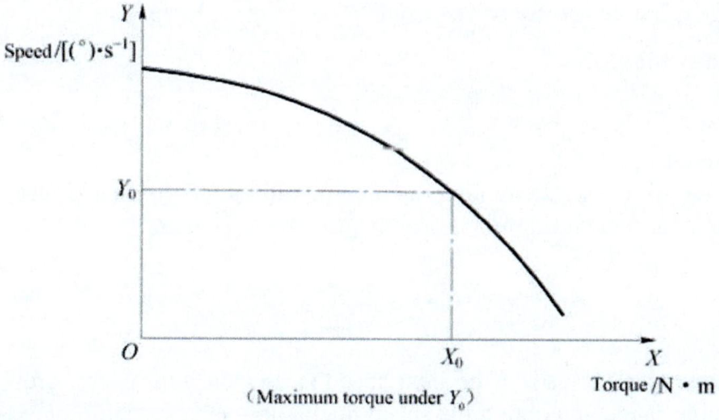

Fig. 7.44 Torque-speed curve

(1) Rated output torque

The rated output torque is the maximum output torque produced by the joint under rated operating conditions. At rated voltage, a torque is applied to the output end by a torque motor until the input current reaches the rated current value. After thermal balance is reached, the measured torque at the output shaft at this time is the rated output torque. The corresponding points at the rated output torque property curve are generally selected where the joint efficiency is higher.

(2) Maximum output torque

The maximum output torque is the maximum torque that the joint can provide during motion. The test method is as follows: under rated voltage, apply a torque to the output end, and measure the speed and torque on the output shaft. With the torque-speed property curve, the maximum output torque value at the specified speed is obtained.

(3) Stall torque

The torque motor provides a resistance torque on the output end, and measure the temperature of the joint motor with a temperature sensor, while ensuring that the joint motor is stalling for a long time when the stable temperature rise does not exceed the allowable value. The maximum torque measured at this time is the stall torque.

(4) backdrive torque

Change the joint motor from braking state to released state, gradually load the output of the joint under test in one direction until the input end of the joint rotates, record the reading of the torque sensor at the output end at this time; then, change the joint motor from braking state to released state again, gradually load the output end in the opposite direction until the joint input end rotates, and record the reading of the torque sensor at the output end; finally, take the larger value of the two measurements as the brake releases reverse drive torque of the joint under test.

(5) Speed stability

Speed stability mainly describes the speed fluctuations of the joint under different speeds and loads.

When the joint runs stably under certain conditions, let the maximum transient speed be $n_{\max}$ and the minimum transient speed be $n_{\min}$, then

$$\text{Speed fluctuation} = \frac{n_{\max} - n_{\min}}{n_{\max} + n_{\min}} \times 100\%. \tag{7.23}$$

Under the specific load, if the joint speed is divided into several groups within the adjustable range and obtain the speed fluctuation values at different speeds, the speed stability curve can be plotted, as shown in Fig. 7.45.

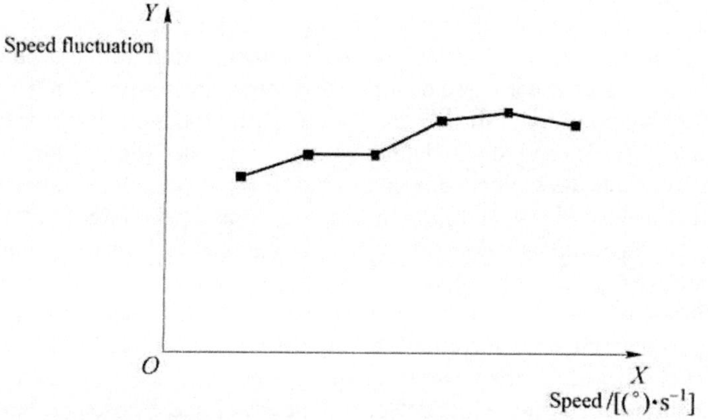

Fig. 7.45 Speed stability curve

7.8.2 Function/Performance Test of Space Robot End Effector

The test items of the end effector mainly include capture tolerances, operating forces, operating times, and docking stiffness. The operating pose tolerance of the end effector involves 6 DOFs, and more combined working conditions are available. Generally, according to the design characteristics of the end effector, the harsh working condition is selected for testing. The ground test platform for simulating poses between the end effector and the target to be operated includes a multi-DOF manipulator and a parallel mechanism. The simulation test platform can simulate the pose relationship and apply force/torque load on six DOFs, thus enabling the test verification of technical indices such as end capture tolerance, operation force, and operation time.

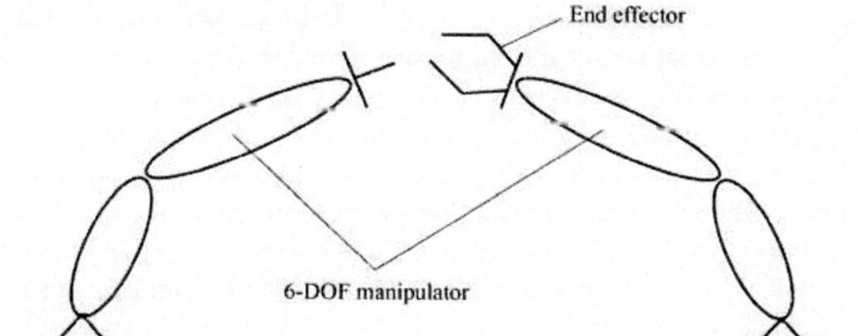

Fig. 7.46 End effector test platform (double-manipulator structure)

Figure 7.46 shows a 6-DOF manipulator end effector test platform. It consists of two 6-DOF manipulators, a six-dimensional force sensor, an end effector, and an operating target. The end effector and the operating target are connected to each manipulator by a six-dimensional force sensor, among which the manipulator connected to the end effector is used to simulate the motion characteristics of the end effector of the space robot; the manipulator connected to the target is used to simulate the motion, initial pose, mass, and motion characteristics of the target to be operated in space. The operational principle of the test platform is: when the end effector operates on the target, the manipulator connected to the end effector simulates the on-orbit motion of the space robot' end effector under load condition in real time according to the dynamics force detected by the six-dimensional force sensor, and the manipulator that fixes the target simulates the on-orbit motion of the target under forced condition according to the detected dynamics force. These two interact with each other to realize the ground semi-physical simulation verification of the end effector's on-orbit operation on the target. This test system has the advantages of large simulation motion space and simple system control.

The parallel mechanism test platform is shown in Fig. 7.47. It consists of 2 parallel mechanisms, a six-dimensional force sensor, an end effector, and an operating target. The end effector and the target are connected to the parallel mechanisms through the six-dimensional force sensor separately. The parallel mechanism connected to the end effector is used to simulate the motion characteristics of the space robot' end effector, and the other mechanism connected to the target is used to simulate the initial pose, mass, and motion characteristics of the real space target. During the test, the end effector operates on the target, and the parallel mechanism connected to the end effector simulates the on-orbit motion of the space robot' end effector under the load condition according to the operation force, and the other parallel mechanism that fixes the target simulates the on-orbit motion of the target under forced condition according to the reaction force, thus achieving the ground semi-physical simulation verification of the end effector's on-orbit operation on the target. The test system has the advantages of large connection rigidity and rapid dynamics response.

For end effectors with target connection requirements, usually, the connection stiffness needs to be tested separately. Figure 7.48 shows a connection stiffness test platform. The end effector and the target are assembled into a combination, one end of which is fixed on the rigid test platform, and the other end is connected to the loading tool beam. Loads are applied to the combination by six directions, the displacement measurement devices are arranged at proper locations on both ends of the combination. The deformation of the combination is measured under different loading states, and the connection stiffness between the end effector and the target combination is calculated according to the applied loads and the corresponding deformation.

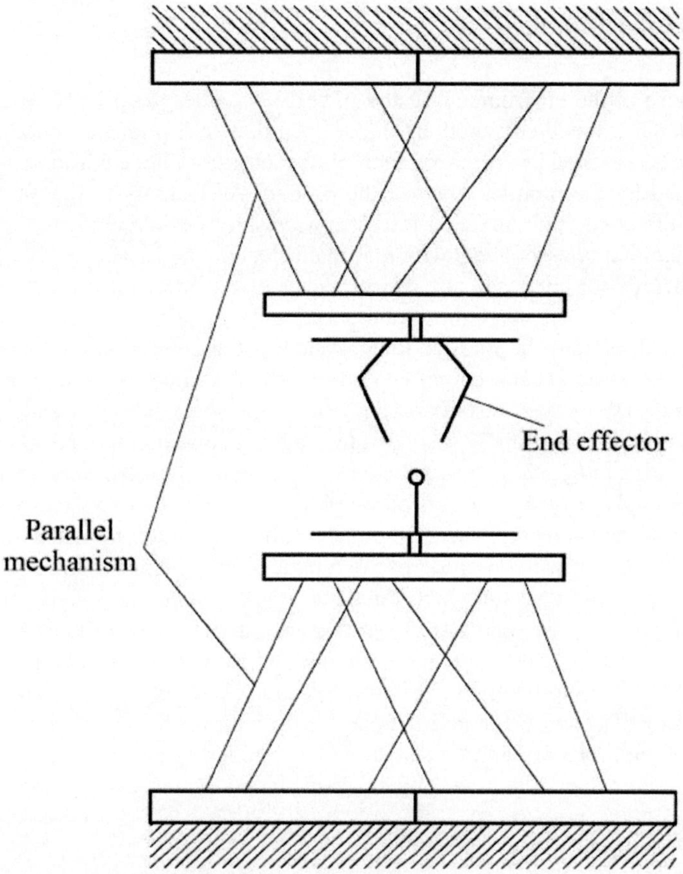

Fig. 7.47 End effector test platform (parallel mechanism)

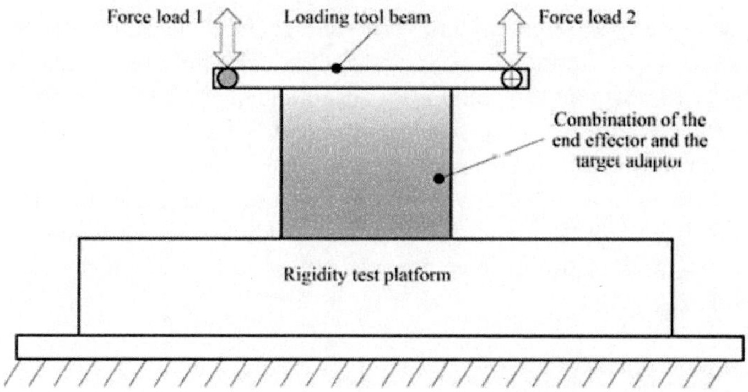

Fig. 7.48 Diagram of the connection stiffness test of the end effector and target complex

7.8.3 Environment Test

The purpose of the environment test is to verify whether the product is capable of performing the specified functions under specified environmental conditions and meeting the specified performance indices. According to the environmental profile experienced by the product, analyze the product's sensitivity to the environment, select an effective environmental test item accordingly, and complete the environmental adaptability assessment with reasonable test methods. The environment tests mainly include the mechanical environment tests (acceleration, shock, sinusoidal vibration, random vibration, micro-vibration, and acoustics), thermal environment tests (thermal vacuum, thermal cycle, high- and low-temperature storage), space environment tests (vacuum discharge, ESD, irradiation, atomic oxygen, space debris), and ground environment tests (storage, lifting, transportation, salt spray). During the environment tests, usually, a single environment condition will be applied to the product, but an integrated or special environment may be used if necessary.

The function/performance tests of the product should be conducted before, after or during the environment tests, and this test result is used as the criterion for the environment test. Usually, in order to test the function/performance indices of the space mechanism, ground test equipment can be used to simulate the on-orbit operational mode of the mechanism and test its performance under different modes. If the on-orbit vacuum, high- and low-temperature environment are also taken into account, the ground test equipment should be placed in such simulation equipment. If the ground test equipment is too large, it may be difficult for the simulation equipment to meet the requirements, and the test must be simplified by a comprehensive analysis of the characteristics of the space mechanism so that the test can be implemented.

7.8.4 Reliability Tests

Reliability test is a special test used to verify and evaluate the reliability of the space robot mechanisms, including environmental stress screening test, reliability growth test, reliability assessment test, reliability acceptance test, and service life test. The purpose of the reliability test is to find the defects of the mechanical products in the design and production process, and provide objective test data for quantitative evaluation of the product reliability. The test processes and results are also analyzed using the test data to verify that the mechanical products meet the system reliability index, and the product design and process are optimized to further improve the product reliability [4].

7.8.4.1 Environmental Stress Screening Test

Environmental stress screening test aims to reveal defective parts and processes, so as to improve the reliability of the product. Therefore, this test is generally carried out at the component level of mechanical products to prevent early failure during use. The conditions and procedures of the tests are focused on eliminating early failures in use. The faults found during the test must be disposed in a timely manner and appropriate improvements are taken to ensure that similar faults do not recur. In order to fully expose the possible defects during the test, the applied stress can be greater than the stress during use, but the corresponding evaluation standard or evaluation method must be developed to avoid possible evaluation errors that may impede a conclusion whether the results are qualified.

7.8.4.2 Reliability Growth Test

Reliability growth test refers to a series of tests conducted during the development of mechanical systems, especially in the initial stage of product development. The purpose of the tests is to reveal weak links of the product and find out the problems, so that measures can be taken in a timely manner to improve the design and the process and promote the reliability growth of the product. When the effectiveness of the improvement measures is verified, the product reliability will be improved. Reliability growth should be concerned at each stage of the product development, the closer the reliability growth test is to the actual on-orbit operation of the product, the higher the credibility of the test results, the more effective the improvement measures, and the more significant the reliability growth of the product.

7.8.4.3 Reliability Acceptance Test

The purpose of reliability acceptance tests is to evaluate the reliability of the delivered mechanical product. These tests should simulate the actual working conditions and provide an evaluation of the reliability based on the test results. For products such as joints, end effectors, mobile mechanisms, and hold-down mechanisms in mechanical systems, the test loads and boundary conditions should be clearly defined according to the actual on-orbit use characteristics, sufficient and reasonable test items and evaluation methods of measured data should be specified. Under the premise of ensuring the comprehensiveness and sufficiency of the examination, the number of tests can be reduced.

7.8.4.4 Reliability Assessment Test

Reliability assessment test is a special test used to verify that the reliability index (reliability degree) of the mechanism product meets the design requirements under the specified comprehensive environmental conditions. Reliability refers to the probability that a product successfully accomplishes the required functions under given conditions and within a specified period of time.

In the reliability assessment test of the space mechanical system products, the installation should simulate the real boundary conditions. For example, joints and end effectors in the mechanical system should simulate the actual installation interface and be applied with the corresponding load. In addition, the test conditions also simulate the real environmental conditions, including mechanical environment, temperature, vacuum, and irradiation environment. Scientific and rationale test methods should be determined according to the expected working environment and conditions, and a reasonable reliability evaluation method should be defined according to the function and performance characteristics of the mechanical system. Finally, through the analysis and evaluation of the objective test results, Finally, by analyzing and evaluating the objective test results, it is judged whether the reliability of the mechanism meets the design requirements.

With the advancement of reliability assessment methods, in order to reduce the test time and the sample size of test products, the success-or-failure test methods are no longer directly used to determine the reliability of the mechanical system products. The common method is to take a small number of test samples to test and verify the core performance indices of the product, and comparing the test results with the minimum value required for the product to complete the functions. Finally, the reliability of mechanical system products is obtained by using the corresponding statistical theory.

Stated above, among the reliability tests, the environmental stress screening test covers all products and includes all environmental stresses; the acceptance test generally uses a sampling test program and simulate the real tasks or service life profile to verify that the product can meet specified performance and reliability requirements under specified conditions; reliability growth test is to reveal the product faults or defects, and take measures to eliminate these faults or defects so as to improve product reliability; the reliability assessment test is to verify that the product's reliability index meets the requirements.

7.8.4.5 Service Life Test

For space robots with long service life requirements, if they do not inherit similar space robots that have undergone service life tests, it must undergo operational service life test and ground verification. Depending on the mechanical systems, long service life tests can be performed according to their actual on-orbit working hours and times of operations.

There are strict requirements on the unit under test (UUT) for service life test: it must truly reflect the technical state of the flight product, and has passed all specified tests to verify its design capabilities. The UUT used for service life test should have the same technical state as the flight product, and belong to the same batch of products and be randomly selected. In order to improve the success rate, the components that are most likely to fail (such as bearing components with lubricant) can be tested for service life.

The service life test of space robot mechanical systems often consumes a large amount of manpower and material resources. In order to shorten the test time, accelerated service life test may be carried out on the mechanisms, but before making such a decision, the conditions of real-time life test and accelerated service life test must be carefully considered to ensure the results of the accelerated service life test can truly reflect the results of the real-time life test, then make a proper choice according to specific conditions.

The real-time life test shall ensure that the total working time/number of work cycles in the test is not less than that of the regular ground test plus that of on-orbit operation, and appropriate margins shall be reserved according to the designed specifications and requirements.

References

1. L. Chen, *Spacecraft Structures and Mechanisms* (China Science and Technology Press, Beijing, 2005)
2. Y. Tan, L. Ren, H. Zhang, Development of large-scale end-effector for space station. China Mech. Eng. **13**, 1838–1845 (2014)
3. F. Feng, L. Tang, J. Xu, et al., A review of the end-effector of large space manipulator with capabilities of misalignment tolerance and soft capture. Sci. China Technol. Sci. **59**(11), 1621–1638 (2016)
4. D. Yu, J. Yang et al., *The Spacecraft Mechanism Technology* (China Science and Technology Press, Beijing, 2011)

Chapter 8
Space Robot Control System

8.1 Control System Overview

The space robot control system consists of command scheduling layer, motion planning layer, and execution control layer. Control commands and data are transmitted among these layers, as shown in Fig. 8.1.

8.1.1 Command Scheduling Layer

Typical execution equipment: operation console.

Basic functions:

(1) Providing external operation interfaces, and converting instructions from the astronaut or spacecraft platform into robot motion control commands;
(2) Sending ground commands to the robot;
(3) Autonomously monitoring and managing of the robot states.

8.1.2 Motion Planning Layer

Typical execution equipment: central control unit.

Basic functions:

(1) Receiving instructions from the command scheduling layer, calling the corresponding planning and calculation unit, then generating motion control commands, and transmitting the commands to the execution control layer;
(2) Collecting the motion state of the robot and feeding it back to the command scheduling layer;

Y. Wang, *Space Robotics*, Space Science and Technologies,
https://doi.org/10.1007/978-981-15-4902-1_8

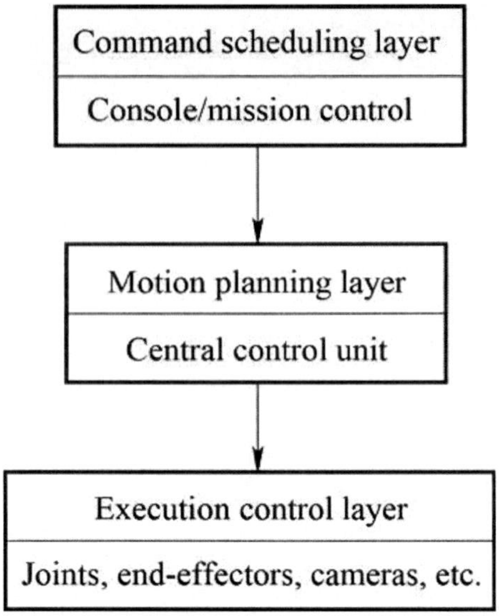

Fig. 8.1 Composition of space robot control system

(3) Monitoring the motion state of the robot to ensure the robot's safety, which includes collision detection, motion overrun detection, safety detection, etc.;
(4) Receiving measurement data from robot sensors for motion planning and state monitoring.

8.1.3 Execution Control Layer

Typical execution equipment: joint controllers, end-effector controllers, camera controllers, etc.

Basic functions:

(1) Receiving motion commands from the motion planning layer, executing the operation or forming a closed-loop control in the moving part through servo control strategy to ensure that the servo motion performance meets the requirements of the motion planning layer;
(2) Collecting and processing camera images and autonomously adjusting camera parameters to ensure the validity of measurement data and the stability of image transmission;
(3) Real-time monitoring of the state of the components, and according to the requirements of the motion planning layer, feeding back the operating state of the components, such as joint motion angle information and joint in-position state;

(4) Providing safety protection for the components' motion and ensuring safe operation of the robot through motor current detection and joint motion overrun detection.

8.2 Design of the Control System

8.2.1 Control System Architecture

The control system of a space robot typically has two forms: centralized control and distributed control. The centralized control system refers to a system where the functions of the command scheduling layer, the motion planning layer, and the execution control layer in the control system are realized in the same control equipment, and various electrical signals (such as motor current and joint angular displacement) for various execution mechanisms of the robot are input into the same control equipment; the control system and the robot execution mechanisms are connected by power supply cables and signal cables. A distributed control system refers to a system where the hardware platforms corresponding to the various task layers of the control system are independently designed, and the data and instructions are transmitted between the hardware equipment of the task layers by dedicated buses.

Compared with the distributed control system, due to the centralized design of the control system hardware, the centralized control system has obvious advantages in terms of equipment weight and power consumption. Because the data transmission between the task layers is not constrained by the bus transmission rate, the design of a control algorithm is more flexible. But for the manipulator, all the execution mechanisms are generally at the remote end, whose cables need to be connected to the control system through various mechanisms (such as joint), resulting in a large number of cables. If the internal wiring design of components is adopted, not only the size of components will be increased, but also the difficulty of the EMC design will be increased; if the external wiring design is adopted, there will be very strict requirements for cable bending radius, bundling, and protection, so as to reduce the risk of motion interference. In a word, the centralized control system is more suitable for robots with smaller DOFs and smaller body sizes. The typical structure of the centralized control system is shown in Fig. 8.2.

The distributed control system is designed with separate hardware platforms between different task layers and between different equipment of the same task layer, so its weight and power consumption will increase accordingly. The data transmission between various layers is generally achieved by buses according to the intersystem communication needs. In addition to the bus connection, there are also power, command, and telemetry connections between different task layers. At present, in the design of a space robot with a distributed control system, the modular and standardized design will often be used for equipment on each task layer. For example, in the design of joint, the factors such as the configuration and layout of the joint controller are all taken into consideration; the cable passes through the

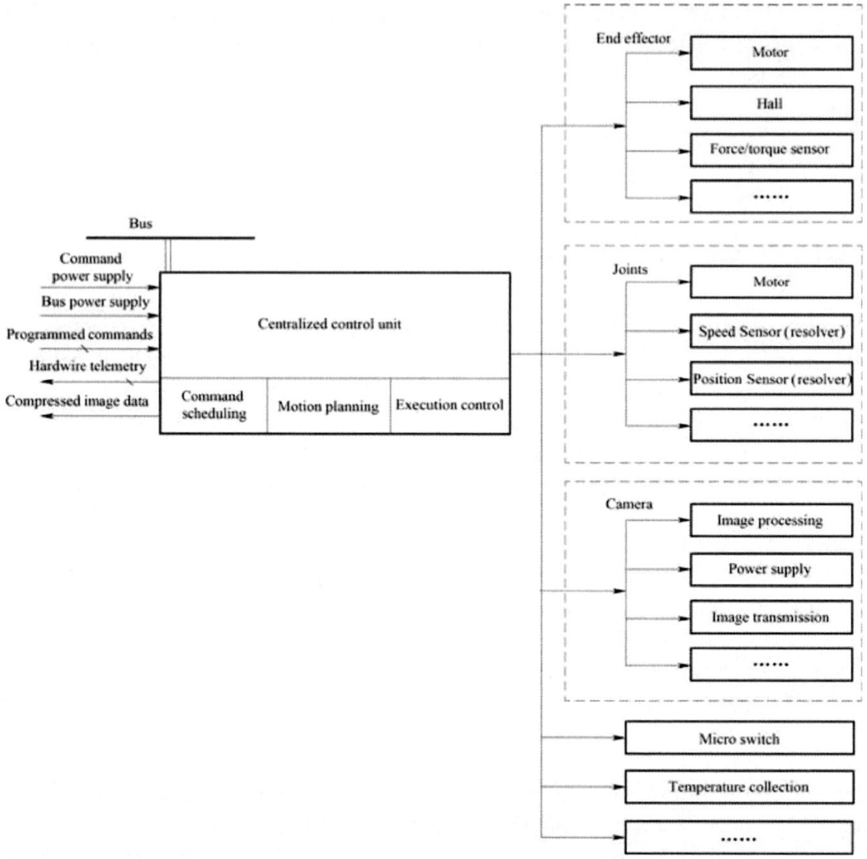

Fig. 8.2 Typical structure of a centralized control system

joint center to ensure that the whole joint is compact and has noninterference inside. Constrained by the high demand for performance and reliability for anti-irradiation devices, the miniaturization design of space robot control system products is still difficult to meet the requirements of small space robots. Therefore, the distributed control system is more suitable for space robots with larger DOFs and larger body sizes. The typical structure of a distributed control system is shown in Fig. 8.3.

8.2.2 Information Flow

The information flow design of a space robot control system is closely related to the design of the control system architecture, working mode design, working environment constraints and performance requirements, etc. The design of information

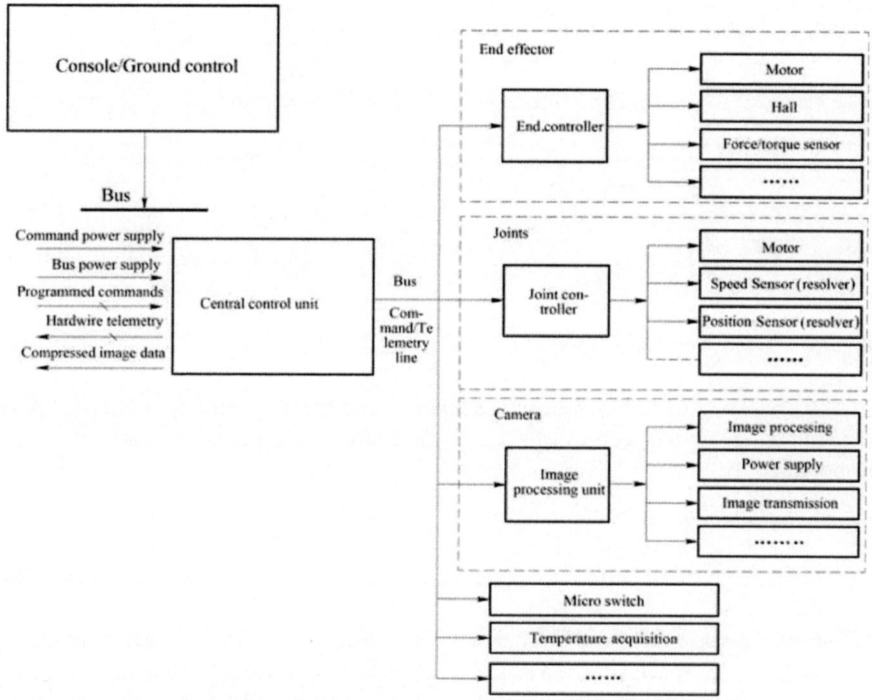

Fig. 8.3 Typical structure of a distributed control system

flow is mainly based on the communication bus selection, data type analysis, fault condition analysis, and working mode analysis.

8.2.2.1 Communication Bus Selection

In order to reduce the number of cables and the design complexity, standard buses are selected to connect the control units when the distributed architecture design is adopted in the control system. The buses that can be used in the space field mainly include the 485 bus, CAN bus, 1553B bus, IEEE 1394, and space-wire bus. Among them, the 485 bus, CAN bus, and 1553B bus are suitable for low-speed applications (transmission rate of less than 1 Mb/s), while the IEEE 1394 and space-wire bus are suitable for high-speed applications (transmission rate of more than 100 Mb/s). They can be selected according to specific design requirements. The comparison of the above buses is shown in Table 8.1.

Table 8.1 Comparison of common buses for space robots

	485 bus	CAN bus	1553B bus	IEEE 1394	Space-wire bus
Communication mode	Half duplex	Half duplex	Half duplex	Half duplex	Half duplex
Transmission rate	Maximum 10 Mb/s	1 Mb/s	1 Mb/s	32 Gb/s	400 Mb/s
Reliability	Low	Lower	High	High	High
Cost	Low	Lower	High	High	High

8.2.2.2 Data Type Analysis

The internally transmitted data in the space robot can be generally classified into the following types according to the transmission cycle period requirements and importance:

(1) **Regular telemetry data** used to characterize the state of the equipment typically includes power supply voltage, instruction counts, telemetry counts, etc., which is generally transmitted periodically, and the transmission period is generally 0.5–1 s.
(2) **Critical state data** used to characterize the important state of the equipment, such as joint speed/position state, force sensor output value, visual measurement value, over-current state, should be closely monitored by the host computer to take measures in time to prevent equipment damage when an abnormality occurs, which can be transmitted periodically or when an abnormality occurs, the transmission period is generally 10–50 ms.
(3) **Control data** used to transmit control commands, including joint speed/position command, motion stop command, end-effector capture/release command, mode switching command, etc., is with the highest priority, which is periodically transmitted or sent on command. The information system should be designed to ensure the reliability of the data transmission, so that the working state of the robot can be controlled under any circumstances.

8.2.2.3 Fault Condition Analysis

In the information flow design, each data transmission path should be analyzed so that the availability of the system under fault conditions must ensure that the system is "available in case of the first fault, safe in case of double faults". For some key data transmission paths, the reliability of information transmission must be ensured by redundant design, error detection/correction coding design, and automatic retransmission design.

8.2.2.4 Working Mode Analysis

Information flow is designed according to the working mode. Different working modes have different requirements for data transmission. When the robot is not working, the health condition of the robot is mainly monitored; during the task execution, the speed, position, and other motion information of each component of the robot are mainly monitored. In the case of a limited data transmission rate, it is necessary to reasonably plan the transmission of the above data, and the data transmission needs are minimized according to the function/performance requirements.

8.2.3 Working Modes of the Control System

The working modes of the space robot control system correspond to the working mode of the robot, and represent the working state of the robot. The advantage of setting the working mode is to enable the robot to quickly reach certain specific states and facilitate mode switching and management. For different types of robots and different mission requirements, working mode settings are not the same. Only the common working modes are given here for reference.

8.2.3.1 Power-OFF Mode

Power-OFF mode indicates that the robot is not powered ON. In this mode, all equipment is inoperative. For space robots, the power-OFF mode is usually used only during ground test or main/standby switching. This mode is rarely used during on-orbit operation, because the on-orbit state of the robot needs to be monitored, more importantly, the thermal control of the space robot requires real-time processing in the high and low temperature space environments. For space robots with distributed control architecture, some controllers can be powered OFF when one equipment runs for thermal control and overall condition monitoring.

8.2.3.2 Standby Mode

Standby mode refers to a state in which the robot is powered ON but does not perform any operation. In the standby mode, parameter settings and status monitoring can be performed on each controller. After the control system is powered ON and the software/hardware self-tests are completed, it automatically enters the standby mode. During long time on-orbit operation, the control system can also be set in the standby mode to maintain a basic working state.

8.2.3.3 Limp Mode

Limp mode means that the robot is powered ON, no operation is performed, and the mechanisms are in an unlocked and follow-up state. The limp mode is usually designed for the space robot in the orbital environment. In this mode, the robot is easy to realize the functions of joint back drive and follow-up.

8.2.3.4 Servo Preparation Mode

Servo preparation mode refers to a mode where the robot is powered ON and has completed the preparation before motion, including drive power supply powered ON and brake released, etc. All mechanisms of the robot are in a position-holding state, i.e., maintaining the current configuration and state.

8.2.3.5 Motion Mode

Motion mode refers to a state in which the robot is in motion and execution of an operation. In this mode, the robot receives commands and performs prescribed actions, and completes motion planning, end-effector operations, and dynamics captures. In the motion mode, various sub-modes can be designed depending on different control methods and the operating source:

(1) Preprogramming mode

Preprogrammed data is prestored in the internal memory of the robot. The stored data specifies the motion information of all driving components. The motion planning layer sends the preprogrammed data to each driving component in real time according to the control cycle, and controls the output trajectory of the robot or completes the operation. In the preprogramming mode, the working path and the operating sequence are fixed.

(2) Visual servo mode

Visual servo mode means that the robot autonomously performs motion planning, dynamics tracking, or target capture according to the targeted pose information provided by the visual system in real time.

(3) End/target parameter input control mode

End/target parameter input control mode is a control mode in which the operator inputs the targeted end position and attitude or the desired motion points of the robot into the control system, which automatically performs the trajectory planning.

In the end/target parameter input control mode, the targeted position and attitude of the robot ends or movement, motion velocity, and acceleration can be set and sent to the motion planning layer; the motion planning layer performs motion planning calculation according to the end-point targeted position and attitude or the desired motion point, and generates motion commands and sends them to each component of the execution control layer; the execution control layer completes the specified action.

(4) Handle control mode

Handle control mode refers to a mode in which the operator gives the robot end position and attitude or motion destination by handle commands according to the robot state information (telemetry, vision, etc.) to achieve the robot motion. In the handle control mode, the operation and control of the robot's motion trajectory, speed, acceleration, and other states can be realized.

(5) Single-component control mode

In the single-component control mode, the speed and position of an individual component of the robot (such as a single joint in the manipulator and a single wheel in the mobile robot) are controlled by parameter input or by means of a handle, teleoperation, etc.

(6) Multicomponent linkage mode

In the multicomponent linkage mode, multiple components (such as multiple joints in a space manipulator and multiple wheels in a mobile robot) can be simultaneously controlled.

(7) Force control mode

The force control mode includes force/torque-related control modes such as impedance control and hybrid force/position control, in which the functions of target operation and manipulator-assisted docking can be realized, and force control of the whole robot or individual components can be performed.

8.2.4 Control System Hardware

8.2.4.1 Control Hardware Circuit

Control system hardware design should go deep into the following items: selection of key components, reliability design requirements, and mechanical system matching design.

1. Selection of key components

The key components of the control system include processors, FPGA, communication interface chips, motor driver chips, etc. The main constraints on component selection include performance requirements, anti-irradiation requirements and reliability requirements, etc.

The aerospace field processors can be classified by a performance from low-to-high into MCU (such as 80C32), CPUs (such as TSC695 and TSC697), and DSP (such as TI's C6000 series), which can be selected for space robot control systems according to the requirements for specific applications. If the processor should perform path planning operations, the DSP should be selected; MCU or CPU may be selected for task management.

The FPGAs used in the aerospace field are mainly divided into SRAM-type FPGAs (such as Xilinx's Virtex series) and anti-fuse-type FPGAs (such as ACTEL's A54SX series). The space robot should preferably use anti-fuse FPGAs. If the anti-fuse FPGA cannot meet the requirements, SRAM-type FPGAs may be selected.

The commonly used communication interface chips include DS26C31/32 (485 bus), SJA1000 (CAN bus), and BU-61580/65170 (1553B bus), which can be selected according to the bus type.

There are many types of motor drive chips, which can be divided into separation solutions and integrated solutions. The separation solution adopts the "driver + MOSFET" design, and the integrated solution integrates the drive circuit, power circuit, and related protection circuits into one module. The separation solution has more components than the integrated solution, which is unfavorable to equipment miniaturization, so the integrated solution is preferred for space robots. Anti-irradiation performance and the quality grade requirements of the component should be considered in the process of solution selection.

2. Reliability design

Reliability design requirements include reliability requirements (such as anti-mechanical environment, derating, and thermal design) and quantitative requirements (such as 0.999/5 years). The main considerations are

(1) Anti-mechanical environment design: Analyze the mechanical environment of space robot during ground testing, launch, and on-orbit operation to obtain the mechanical environment of the controller, and carry out design and test accordingly.
(2) Derating design: All components used in space robots should meet Class I derating requirements.
(3) Thermal design: Analyze the thermal environment conditions of the space robot during long-term storage on orbit and mission execution, especially focusing on the processors, FPGAs, and power devices when performing thermal designs.
(4) Anti-irradiation design: For the total dose effect, the total dose radiation resistance capability of each total dose sensitive device (such as integrated circuit) should be confirmed according to the space environment analysis results on

specific orbits. If any device does not meet the requirements, radiation shielding reinforcement should be designed or components should be replaced; for the single-event effects, in addition to common Single-Event Upset (SEU) and Single-Event Latch-up (SEL) prevention designs, analysis on anti-SEB (Single-Event Burnout) and anti-SEGR (Single-Event Gate Rupture) effects should be carried out on the power devices (power MOSFETs), which are used on space robots in a large number.

(5) Failure Mode and Effect Analysis (FMEA): Comb the failure modes of the control system, analyze the effect of each failure mode, and design preventive measures accordingly, so as to avoid single point of failures as much as possible.

(6) Quantitative analysis of reliability: Predict the reliability of each control unit by part stress analysis (PSA) method, make design improvement to the low-reliability part, and determine the redundancy schemes (cold/hot backup, dual/triple backup) for relevant control units according to the reliability calculation results and the reliability requirement values.

3. Mechanical systems matching design

The design of the control system products of a space robot is closely coupled with the mechanical system of the space robot. The characteristics of a mechanical system, such as the power and rated speed of the motor assembly, the configuration of the sensor, and its output signal characteristics, all have an influence on the control system design. The cables of the control system are to be connected to the parts with relative motion between them, so the layout and routing of cables are particularly important. The matching design of the control system and the mechanical system should focus on the following considerations:

(1) Analysis of the limit working conditions: In the control system design, not only the rated operating state of the mechanical system but also its operating state under the limit conditions, such as the power demand of the driving module under stalled condition of the motor, the transmission system efficiency change at high and low temperatures, the relationship between the joint angle and the cable bending angle, should be considered.

(2) Mechanical system test analysis: After the control system and mechanical system are assembled, the complete robot system should undergo debugging, testing, and experimentation. During this process, the control system will be tested, disassembled, and the software will also be updated. Therefore, the control system should be designed under the consideration of the testability and supportability. If necessary, special test interfaces may be designed. For the software that needs to be determined only after the debugging test of the whole robot system is completed, it is recommended that the software be designed as on-orbit maintainable software when the hardware is designed, so as to facilitate software updates in the test and experiment phases.

(3) Analysis of cable routing: Constrained by the configuration of a space robot, the space for the control system cable routing is relatively small; the cables

also need to bear the bending and torsion forces caused by the movement of the parts; and the signals transmitted by the cables of the space robot contain multiple characteristic data, such as the power signals, sensor signals, and digital signals, so the EMC design is of great importance. Therefore, the cable should be designed in the early stages of the mechanical system design to reserve space for cable routing.

8.2.4.2 Control System EMC

The Electro-Magnetic Compatibility (EMC) of an electronic equipment mainly refers to the ability of the electronic equipment to operate normally in its environment according to design requirements, and does not constitute unacceptable Electro-Magnetic Interference (EMI) to anything else in that environment. Electronic equipment that meets EMC requirements will neither affect the normal operation of other equipment, nor cause performance degradation or failure due to the operation of other equipment.

The EMC design of space robot products should be carried out synchronously with its functional design. The EMC indices should be determined under the comprehensive consideration of the factors, such as life cycle and development cost-effectiveness. During the product design, the basic parameters of the EMC design should be explicitly defined, including the EMI values, EMI emission limits, EM susceptibility thresholds, EM susceptibility limits, safety margins, performance degradation criteria, performance failure criteria, etc.

In the development of space robots, space robot electronic equipment and their main electrical performance parameters must be listed separately to facilitate systematic frequency planning and analysis. For equipment that may cause interference or disturbance after analysis, appropriate protective measures should be taken in terms of grounding, cable layout, equipment layout, and working mode.

The electronic equipment of the space robot control system will use various types of power modules, relays, and other electronic equipment. For the servo drive equipment, a motor drive module will also be configured. These modules will generate EMI during operation. The following requirements should be considered in the design:

(1) Ensure that all electronic equipment of the space robot can work normally during the entire mission profile by carrying out reasonable equipment and system design;
(2) Ensure that the EMI conducted or radiated by all electronic equipment of the space robot during the entire mission profile will not affect the operation of other equipment by carrying out reasonable equipment and system design.

According to the design requirements of the EMC of the control system, the following aspects must be considered in the design of electronic equipment hardware and system solutions:

(1) Reasonable selection of components to avoid the use of EMC susceptible devices.

When designing electronic equipment, devices with good EMC characteristics, such as devices with slower rise/fall edges, analog devices with larger output values, and those with differential characteristics should be preferred.

(2) Carry out reasonable grounding design to avoid ground loop interference.

Analyze the ground signal of the system and classify them by power ground, digital ground, analog ground, etc., to avoid forming a ground loop and reduce interference to analog devices. In order to reduce the ground loop interference, the internal circuit board of the controller should be so designed such that the signal ground line and the chassis ground are insulated to form a floating ground. For the power, digital, and control system signal grounding, Single-Point Grounding (SPG) should be used to avoid forming a ground loop.

(3) Carry out reasonable PCB layout design.

With large ground planes, the grounding impedance of the printed circuits can be reduced, the interference and noise between the signal lines on the other side of the PCB can be effectively suppressed, the EM and static electricity can be shielded, and the external high-frequency radiated interference to the circuits can be reduced. When designing the PCBs, each PCB can be surrounded by a metal frame, and provided with internal circuit layers, including large area ground layers. The mutual induction between the printed lines on the other side can reduce by a complete ground plane, thus suppressing the electrostatic coupling noise formed by electrostatic induction.

When designing related motor-driven PCBs, the area enclosed by the motor current printed circuits should be minimized, and the motor current printed lines to reduce remanence and interference should be symmetrically arranged and widened.

In the PCB design, in order to reduce interference between signal lines, ground isolation and shielding measures should be taken for signal lines that are susceptible to interference.

A network ground wire system should be used to reduce the equivalent inductance of the PCB ground wires, and wide ground wires should be used as much as possible to minimize the internal impedance in the ground wire so that the circuits meet the EMC design requirements.

(4) Reduce bus voltage fluctuation caused by current ripple through filtering design

For the primary bus, an EMI filter is added before supplying power to the secondary power module; an SMD tantalum capacitor and a ceramic capacitor with different values are connected at the input of the main power supply in series and in parallel; the power supply noises in low, intermediate, and high frequency should be filtered; a ceramic chip capacitor is connected at the input of each chip in series, and placed near to the power supply of the chip; for integrated circuits, a capacitor with a smaller equivalent inductance should be selected, and each integrated circuit is filtered by a capacitor close to the power supply.

(5) Suppression by the housing

Using housing to suppress EMI is a common shielding method. The metal housings of the controllers are generally made of aluminum alloy, which has a good suppression effect on the electric field or plane wave. Metal materials with enough thickness can ensure good electric field shielding. Good sealing and effective treatment of the gaps, holes, and joints are of great importance for the housing.

(6) Reasonable wire routing and cable design.

The interference between cables can be prevented by reasonable wire routing, which includes that the cables should be treated by twisted paired or other methods to reduce interference to other circuits; signal lines and supply lines should be bundled and fixed separately, and the bundle can be covered with shielding material to reduce the influence of the supply line on the signal; the shielding and other treatment measures are taken for the cables of measuring electronic equipment; the shielding of the connectors should be taken into account by selecting connectors.

8.2.5 *Control System Software*

The control system software is divided into a system management part and an algorithm part. In the system management part, the scheduling and monitoring of the whole system are realized, and the functions realized by the software are connected and managed. It is the "nerve" of the control system operation. The algorithm part is the "brain" of the robot, because it controls the robot to perform the actions and complete the tasks. The roles and functions of the control system software are different. Chapters 2–5 give a detailed description of the basic control algorithms of the robot. This section only introduces the issues that should be noted during the software implementation process based on the requirements of the spacecraft software engineering and the characteristics of the space robot.

8.2.5.1 The Impact of Hardware Resource on Software Design

Constrained by space environment and component levels, the hardware resources of the space robot control system are not as rich as the ground controllers. Therefore, the influence of the controller hardware resources on the software design needs to be considered when implementing the control system software. Particularly, with the increase of the complexity of the control algorithm, the resources required by the software must be estimated and evaluated at the beginning of the software design to ensure that the task is executable and the algorithm is implementable. The control algorithm generally has time requirements. The calculation speed of the controller should ensure the cycle of the control algorithm. System interruption and branches should also be considered in detail to avoid confusion of the control cycle.

8.2.5.2 Software Development and Testing

The software development and testing mainly include requirement analysis, software design, software implementation, software test, system testing, third-party evaluation, and software acceptance.

In the requirement analysis, the tasks, functions, and interfaces on the system, subsystem, and equipment levels are analyzed; the compatibility with the hardware system is confirmed; the information flow is determined; and the software requirements are proposed by the task proposer (demand party). The task contractor (development party) completes the requirement specification description according to the software requirements; outline design and detailed design are carried out; development methods and tools are determined; software coding, static analysis, code review are performed according to requirements; unit test is carried out after software coding to ensure coverage of all software branches. Then, the task contractor designs assembly test examples to test the interface between the software unit and the components, and carries out confirmation tests to confirm that the test covers all software functions, performance, software/hardware interfaces, etc. After the confirmation tests, the system tests can be carried out to mainly check the different software configuration items, external inter faces and other functions. The software should be tested by a third party before acceptance and delivery, and a summary report about the development processes and documentations should be completed before the acceptance. In the software development, the demand party, the development party, and the test party should be different.

For the control system of a space robot, due to its inherent characteristics, in the software testing, the system task requirements should be considered comprehensively, and typical working conditions should be given as the software test cases. In the development software that is in close conjunction with hardware (such as motor drive), a semi-physical simulation method can be used to perform a case test.

8.2.5.3 Synchronization Design

The control system software has different software configuration items, and the synchronization design between software configuration items and modules inside the configuration items should be performed in the early stages of software development. For systems with high synchronization requirements, the time sequence should be designed to avoid malfunction due to timing confusion.

8.2.6 Control System Reliability and Safety Design

The stable and safe operation of space robots, astronauts, and peripheral equipment can be ensured by reliability and safety design, which is also crucial for the control system. The reliability and safety of control systems should be considered in the design stage.

8.2.6.1 Reliability Design

1. Reliability guarantee

The development and testing of the control system should be carried out in accordance with the development specifications established for the spacecraft project, and be strictly followed throughout the life cycle of the control system. The control system software development, testing, and maintenance should be managed according to the software engineering management method. The third-party software walkthrough and validation tests should be implemented to keep the development process and software state strictly controlled and the quality of the software ensured.

2. Redundancy and fault-tolerant design

The control system should be designed according to the redundancy concept, including hardware redundancy design, software redundancy design, and remote storage of important data. For space robots with redundant DOFs, fault-tolerant measures should be designed to enable the robot to continue the tasks or return to the service point for maintenance in the event of a component failure. Considering the space environment, the following design should also be considered for the control system:

(1) For SRAM-type FPGAs, single-event upset protection should be taken by means of Triple Mode Redundancy (TMR) and timed updating;
(2) EEPROM, SRAM, and other memories should be designed with EDAC protections;
(3) Initialization program and interrupt vector of the system should reside in the ROM area;
(4) A watchdog should be designed to recover the software abnormality.

3. Software margin design

The design and implementation of the software must ensure that the system resources occupied by the software have a certain margin for the reliability and safety of the system operation. Therefore, when designing the software, an optimized space-saving design should be adopted to leave a certain margin in the PROM and RAM spaces.

4. Software robustness design

The software robustness design includes

(1) Power-ON detection. When the software is called, the hardware should be tested to verify whether the system is properly started. If the startup fails, relevant telemetry information should be output and sent to the ground for processing.

(2) Interference from the space environment. Single-event upset in the space environment may cause hardware failures. The software should be able to automatically detect and correct single-bit errors in conjunction with hardware. Besides, it should be able to detect double-bit errors and take backup measures to ensure the continuous operation of the system.
(3) Hardware device failure. The software periodically detects the hardware and feeds the detection results back to the ground through telemetry parameters. For major failures, the software will reset or switch to a backup machine.
(4) Interference signals. In order to prevent misjudgment of certain important parameters caused by interference signals, the software must perform multiple consecutive interpretations of important telemetry data involving in the control. Only if the results of the multiple interpretations are consistent, the occurrence of an event can be confirmed.
(5) Wrong operation. When a wrong instruction or data is input on the ground, the software should be able to verify the legality of the instruction or data according to the checksum, instruction length, type of block, type of instruction unit, length of instruction unit, instruction content, etc. Only verified instructions and data can be executed.

8.2.6.2 Safety Design

1. Operational safety design

   (1) Interference and collision detection. The robot's interference collision includes the robot's own interference and the interference between the robot and the environment. Before the robot moves, the full mission should be simulated and possible interference and collision should be detected by the simulation software at an early stage. For detection of the robot's own interference and those in a known environment, the collision detection algorithm and real-time visualized simulation methods can be used to detect the motion state of the robot, and give collision warnings and handling measures in the states where a collision may occur. For unknown environments, a robot's interference can be monitored by a visual sensor and other sensors.
   (2) Anti-misoperation safety design. In the hardware design, the double-protection principle should be adopted for important switches and buttons to avoid the misfunctioning of the robot due to mis-operation. In the control system, corresponding logic should be set to automatically suspend those instructions that are noncompliant with the designed logic and that give advisory information.

2. Overrun and computational safety design

In the path planning and servo control, the intermediate variables and output results generated by the robot should be tested for overrun and rationality. The erroneous data generated by calculation should be eliminated, and the overrun data should be limited in amplitude, including

(1) Detecting the overrun and accuracy of the input and output data, such as the desired position and attitude of the end coordinate, and the desired position of the joint;
(2) Limiting the signal of the motion range and motion state of the actuating component;
(3) Detecting the voltage, current, and other states of the control system;
(4) Laying down current protection measures for the motion mechanisms, which should be able to autonomously power OFF and protect in case of overload;
(5) Avoiding possible singularities during the planning process;
(6) Timely transmitting feedback of the actuating components' faults to the higher control layer that should give handling measures to the robots.

3. Power-OFF design

    (1) Developing a hardwire emergency power-OFF strategy. In order to prevent equipment damage caused by program runaway, hardware failure, etc., an emergency power-OFF strategy should be set for the robot to directly stop the robot through hardwire.
    (2) Setting termination instructions. In order to prevent situations such as parameter input errors and task changes, the action termination instructions should be set to enable the robot to shift from motion state to the stop state; a buffer process should be set to make the robot gradually transit to the stop state.

Chapter 9
Space Robot Perception System

9.1 Overview of the Perception System

Space robots involve different operating objects and operating environments while performing missions. For example, for on-orbit servicing robots, the operating objects are mostly cooperative targets with the preset visual markers and dedicated operation interfaces, and the operating environment is a spacecraft surface structured environment, while for planetary exploration robots, the operating objects are mostly noncooperative targets without visual markers or dedicated operation interfaces, and the environment features are unknown and unstructured. Space robot perception refers to the process in which the space robot obtains the information of the operating objects and the operating environments by analyzing raw data generated by various sensors onboard.

Since space robots face different types of targets and environments during missions, a variety of perceptual capabilities should be available to comprehensively obtain all kinds of information about the external environment. The perception systems commonly used in space robots include visual perception system and force perception system. This chapter will focus on the description of the visual perception system. The force perception system is mainly composed of various force sensors (see Chap. 7 for details).

The space robot visual perception system generally consists of the following parts:

(1) **Optical assembly** for receiving target signal light energy and suppressing stray light is usually composed of several optical components (such as lenses, prisms, and mirrors), each of which consists of a plane, spherical, or aspherical medium with certain refractivity, and is assembled by structural components as required.
(2) **Electronic assembly** collects measurement data by converting optical signals received by the optical assembly into electrical signals using photoelectric detectors (such as CCD and CMOS) and supporting circuits, and locally processes the measurement data using microprocessor chips and supporting circuits to

Y. Wang, *Space Robotics*, Space Science and Technologies,
https://doi.org/10.1007/978-981-15-4902-1_9

realize the functions such as data compression, coding and output, target detection and recognition, and target measurement; in an active perception system, the electronic assembly also includes an energy projection device.

(3) **Structural assembly** is mainly composed of structural parts such as the housing of the optical assembly, the lens hood, the housing of the electronic assembly, and mounting interfaces with the robots. It is used for supporting and protecting the optical and electronic assemblies, and improving the adaptability of the perception system to the launch environment and space operation environment.

The main functions of the space robot visual perception system are to realize target detection, recognition, and measurement. Target detection and recognition refers to detecting, recognizing different targets in the operating environment, and establishing associations between them, that is, determining whether the target detected or recognized at different times, in different positions, and by different sensors is the same target or a new target. Target detection and recognition also includes the recognition of the actions and behaviors of the partners (human) to understand human intentions. Target measurement refers to tracking the target, estimating the information of relative position, attitude, linear velocity, angular velocity between the target and the sensor, and establishing a three-dimensional model of the target, where the position and attitude characterize the transformation relationship between the target coordinate system and the sensor coordinate system (the 6-DOF pose data, including 3 position translations and 3 attitude rotation angles).

9.2 Visual Perception System Design

The design of the space robot visual perception system focuses on meeting the requirements of space robot on-orbit applications, covering the complete process of system design, production design, processing, installation, debugging, system testing, and experimentation. The system design is mainly based on the technical requirements for the development of the perception system. If there are no separate technical requirements for the visual perception system, the task must be analyzed in combination with the actual application of the space robot, and the requirements for the robot are decomposed into the requirements for the visual perception system, so as to determine the functional requirements, performance indices, and the interface of the visual perception system, as a design input condition for the visual perception system. Note that functional decomposition and technical specification assignment is not a one-way process, and multiple iterations and the repeated process of "design-simulation analysis-test verification-design improvement" are often used. The basic technical indices of the visual perception system are shown in Table 9.1.

The system design mainly includes combing and overcoming technical difficulties and key points based on the task analysis results according to the technical requirements and design constraints; determining the composition, function requirements, and operation modes of the system, and decomposing the function requirements and

Table 9.1 Basic technical indices of the visual perception system

Types of indices	Names of indices
Physical parameters	Volume
	Mass
Electrical parameters	Input voltage
	Power consumption
Optical parameters	Band range
	Field of view (FOV)
	Distortion
	Stray light coefficient
Imaging parameters	Image resolution
	Image acquisition frame rate
	Signal-to-noise ratio
Measurement parameters	Measurement range
	Measurement accuracy
	Measurement refresh rate

performance indices of each production (assembly) in the system; designing intra- and extra-system mechanical, electrical, thermal, and information interfaces; determining test and experiment items, conditions and plans; software design; reliability, safety, maintainability, testability, supportability and environmental adaptability design and analysis; technical risk analysis, etc.

9.2.1 System Design

Space robot visual perception system falls into various types according to different criteria, and each type has its own distinct advantages, deficiencies, and scope of application, which should be selected and combined according to the actual needs of the space robot and the design constraints in the overall system design.

The first to be determined is the installation position of the visual sensor. There are two options: one is to install it on the robot end effector, i.e., the eye-in-hand-type visual perception system; the other is to install it somewhere in the robot's working area, i.e., the eye-to-hand-type visual perception system, as shown in Fig. 9.1. The eye-in-hand visual perception system refers to a visual sensor mounted at a fixed position on the end effector of the robot, and its pose will be changed with the movement of the end effector. The visual sensor in the eye-to-hand visual perception system is installed at a location within the robot's working area, and its mounting base can be configured with a mechanical device to have a certain degree of freedom (e.g., pan-tilt-unit). For the eye-in-hand visual perception system, the visual sensor can be placed closer to the target by adjusting the position of the end effector, thereby improving the target measurement accuracy. However, the motion of the visual sensor

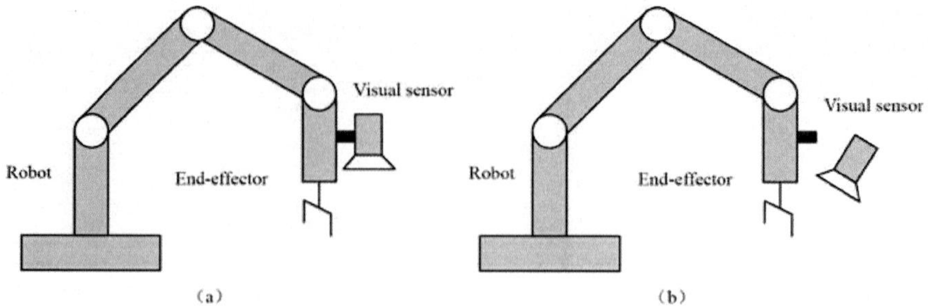

Fig. 9.1 Spatial distribution of visual perception systems. **a** Eye-in-hand visual perception system. **b** Eye-to-hand visual perception system

is likely to cause image blurring, and the target cannot be guaranteed to remain within the FOV, i.e., loss of target; in addition, since the visual sensor is mounted on the end effector, the load mass of the end effector is increased. The eye-to-hand visual perception system can monitor the robot globally and acquire the measurement data of the robot and its working area, but might be occluded during the movement of the robot; moreover, the target measurement accuracy is relatively low because the visual sensor is far away from the target.

The number of visual sensors at each installation location must also be determined, i.e., to check whether the visual perception system is a monocular, a binocular, or a multi-camera system. The monocular system uses a single monocular visual sensor for measurement, and it cannot directly obtain depth information. It completes the measurement usually with artificial visual markers with known target installation information, such as dimension, shape, distribution, and quantity, or by motion. With two or more monocular visual sensors at different spatial locations, the binocular/multi-camera system directly obtains depth information for measurement. Compared with the monocular system, the binocular/multi-camera system can effectively improve the measurement accuracy and the system reliability, that is, the measurement function is still available when one monocular cannot work properly. The disadvantage of the binocular/multi-camera system is that it is highly complex in computation and needs a large amount of computation.

Based on the selected visual sensors, an active or a passive perception system can be constructed. For example, the laser, structured light, and radar visual perception systems belong to the active perception system, and the visible light perception system belongs to the passive perception system. The active perception system achieves perception by transmitting energy from itself and receiving the reflected energy from the target. The passive perception system can capture scene images quickly, with high information acquisition efficiency, long observation distance, and larger amount of information contained in the observed data. The passive perception system is more suitable for understanding the scene and guiding the robot to complete the various tasks during movement, such as obstacle detection and avoidance, target recognition and tracking, scene modeling, and human behavior analysis, which acquires scene information by a noncontact and passive way (not transmitting energy to the environment during working) without causing interference to other sensors. It has the advantages of low power consumption, small size, lightweight, convenience, and dexterity; its disadvantages are that it is susceptible to light conditions and the measurement accuracy is lower than the active perception systems.

In addition, when designing a visual perception system, configuration optimization of visual sensors can also be considered to achieve the best measurement results. For example, we can only use one type of visual sensor (similar visual perception system), or combine several types of visual sensors (dissimilar visual perception system).

9.2.2 Information Flow Design

The typical information flow of the visual perception system is shown in Fig. 9.2. The space robot controller receives the telemetry data (including the target measurement results and the visual perception system status data) from the visual perception system to complete motion planning for the robot.

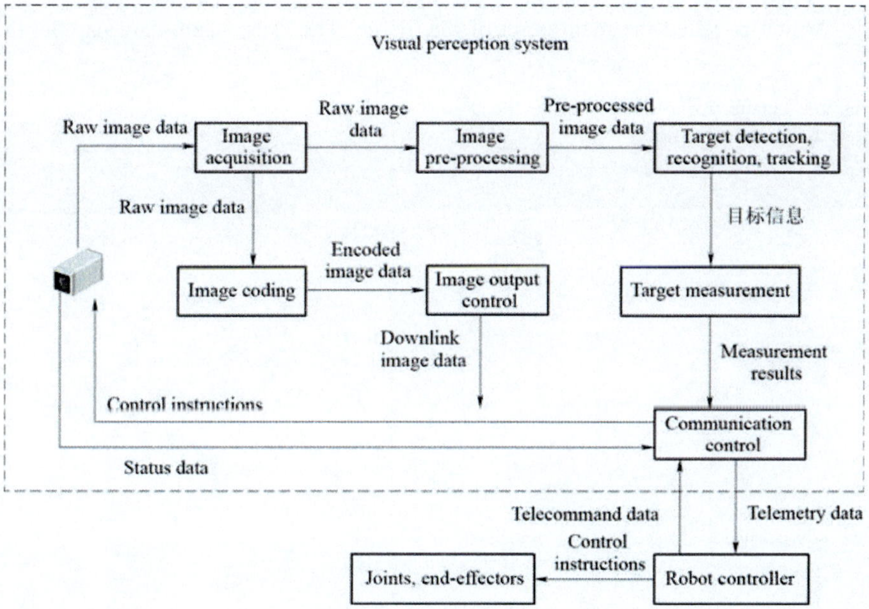

Fig. 9.2 Typical information flow of a perception system

9.3 Visible Light Visual Measurement

The two-dimensional image from a visible light camera is a perspective projection of the three-dimensional scene. This process can be expressed by the perspective projection geometric model shown in Fig. 9.3, which involves four coordinate systems: the pixel coordinate system, the image coordinate system, the camera coordinate system, and the world/target coordinate system [1, 2].

(1) **Pixel coordinate system**. The image is stored as an array in the computer. Each element in the array is called a pixel. The value of the pixel, that is, the brightness of the image points, is called grayscale (if it is a color image, the brightness of the pixels on the image will be presented by the intensity of the red, green, and blue colors). The pixel coordinate system O_{uv} is defined on the image, O is the origin, the coordinates of each pixel (u, v) are the numbers of column and row of the pixel in the array, and the pixel (0, 0) is located at the upper left corner of the image.

(2) **Image coordinate system**. Since the pixel coordinate system only indicates the number of columns and rows in which the pixel is located in the image, the physical position of the pixel in the image is not expressed in physical units, so an image coordinate system O_{1xy} expressed in physical units (for example, in millimeter units) is established. In the O_{1xy} coordinate system, the origin O_1 is defined as the intersection of the camera's optical axis and the image plane, which is called the main point of the image. The axes x and y are parallel to

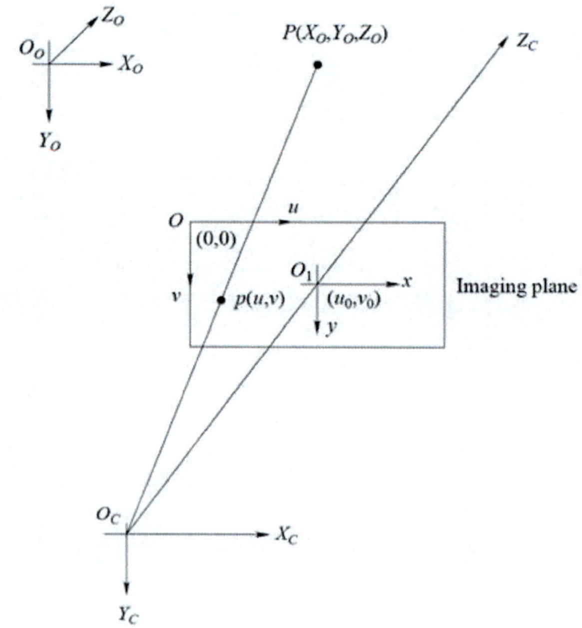

Fig. 9.3 Perspective projection geometric model

the axes u and v of the pixel coordinate system, respectively. If the coordinates of O_1 in the coordinate system O_{uv} are (u_0, v_0), and the physical dimensions of each pixel on the axis x and axis y directions are dx and dy, respectively, then the relationship between the pixel coordinate system and the image coordinate system is defined as

$$\begin{bmatrix} u \\ v \\ 1 \end{bmatrix} = \begin{bmatrix} \frac{1}{\mathrm{d}x} & \frac{-\cot\theta}{\mathrm{d}x} & u_0 \\ 0 & \frac{1}{\mathrm{d}y\sin\theta} & v_0 \\ 0 & 0 & 1 \end{bmatrix} \begin{bmatrix} x \\ y \\ 1 \end{bmatrix}, \tag{9.1}$$

where θ indicates the degree of non-perpendicularity between the axes u and v due to the manufacturing process of the camera (with increased improvement of the camera manufacturing process, the θ is very close to 90°).

(3) **Camera coordinate system**. The origin O_c of the camera coordinate system $O_cX_cY_cZ_c$ is called the camera's optical center, the axes X_c and Y_c are parallel to the axes x and y of the image coordinate system, respectively, the axis Z_c is the optical axis of the camera, which is perpendicular to the imaging plane. The intersection of the optical axis and the imaging plane is the origin of the image coordinate system O_1, and O_cO_1 is the focal length.
(4) **World/target coordinate system**. Since the camera can be randomly placed anywhere in a three-dimensional space, a coordinate system $O_oX_oY_oZ_o$ should be established to describe the pose of the camera, which is called the world/target coordinate system. Let the homogeneous coordinates of a point P in the camera coordinate system and the world/target coordinate system be $(X_C, Y_C, Z_C, 1)^T$ and $(X_0, Y_0, Z_0, 1)^T$, respectively, then there is the following relationship:

$$\begin{bmatrix} X_C \\ Y_C \\ Z_C \\ 1 \end{bmatrix} = \begin{bmatrix} R & t \\ 0^T & 1 \end{bmatrix} \begin{bmatrix} X_0 \\ Y_0 \\ Z_0 \\ 1 \end{bmatrix}, \tag{9.2}$$

where R is a 3×3 orthogonal rotation matrix, t is a 3×1 translation vector.

In visible light perception systems, the measurement methods can be divided into monocular and binocular (multi-camera) visual measurements according to the number of cameras. The specific measurement principles are described below for a cooperative target with artificial visual markers.

9.3.1 Monocular Visual Measurement

In the monocular visual measurement, one monocular camera is used to perform three-dimensional pose measurement, and it is mainly used for the measurement of cooperative targets with artificial visual markers.

The pose measurement of a cooperative target with artificial visual markers can be measured by a single frame image. The correspondence between the artificial visual markers and the image features can be established according to the camera model and the detection results of the image features which characterize the target geometry information in the image, such as points, lines, and circles, then the target pose can be calculated [3, 4].

The calculation formula of the target's three-dimensional pose by monocular visual measurement is

$$S_i \begin{bmatrix} u_i \\ v_i \\ 1 \end{bmatrix} = [A \; 0] \begin{bmatrix} R & t \\ 0^T & 1 \end{bmatrix} \begin{bmatrix} X_{oi} \\ Y_{oi} \\ Z_{oi} \\ 1 \end{bmatrix}, \tag{9.3}$$

where S_i is the nonzero scale factor, (u_i, v_i) is the corresponding image pixel coordinates of the visual marker center P_i, (X_{oi}, Y_{oi}, Z_{oi}) is the three-dimensional coordinates of the visual marker center P_i in the target coordinate system, A is the camera intrinsic matrix, R is a 3×3 orthogonal rotation matrix, t is a 3×1 translation vector, and R and t represent the three-dimensional pose of the target.

9.3.2 Binocular Visual Measurement

In the binocular visual measurement, two viewpoints (cameras) at different spatial positions are used for synchronous image capture, and the target pose is calculated according to the camera model and the correspondence solution (image feature matching) of the two images.

For cooperative targets with artificial visual markers, firstly, the coordinates of the visual markers in the binocular camera coordinate system are solved according to the camera model and the matching result of image features, then one-to-one correspondence between the coordinates of the visual marker in the target coordinate and binocular camera coordinate system is established, and the target pose is calculated [3, 4].

Taking the left camera coordinate system $O_cX_cY_cZ_c$ as the reference coordinate system, the intrinsic matrix of the left and right cameras is A_l and A_r, respectively, and the relative pose between the two cameras is R_c and t_c (R_c is a 3×3 orthogonal rotation matrix, t_c is a 3×1 translation vector), then the calculation formula of the three-dimensional pose of the target by binocular visual measurement is given by

$$S_{li}\begin{bmatrix} u_{li} \\ v_{li} \\ 1 \end{bmatrix} = \begin{bmatrix} A_l & 0 \end{bmatrix}\begin{bmatrix} I & 0 \\ 0^T & 1 \end{bmatrix}\begin{bmatrix} X_{Ci} \\ Y_{Ci} \\ Z_{Ci} \\ 1 \end{bmatrix}, \tag{9.4}$$

$$S_{ri}\begin{bmatrix} u_{ri} \\ v_{ri} \\ 1 \end{bmatrix} = \begin{bmatrix} A_r & 0 \end{bmatrix}\begin{bmatrix} R_C & t_C \\ 0^T & 1 \end{bmatrix}\begin{bmatrix} X_{Ci} \\ Y_{Ci} \\ Z_{Ci} \\ 1 \end{bmatrix}, \tag{9.5}$$

where S_{li}, S_{ri} are nonzero scale factors, (X_{ci}, Y_{ci}, Z_{ci}) is the three-dimensional coordinates of the visual marker center P_i in the left camera coordinate system, (u_{li}, v_{li}) and (u_{ri}, v_{ri}) are the image pixel coordinates of the visual marker center P_i corresponding to the left and right cameras, respectively.

The transformation relationship between the left camera coordinate system $O_cX_cY_cZ_c$ and the target coordinate system $O_oX_oY_oZ_o$ can be expressed as

$$\begin{bmatrix} X_{Ci} \\ Y_{Ci} \\ Z_{Ci} \\ 1 \end{bmatrix} = \begin{bmatrix} R & t \\ 0^T & 1 \end{bmatrix}\begin{bmatrix} X_{Oi} \\ Y_{Oi} \\ Z_{Oi} \\ 1 \end{bmatrix}, \tag{9.6}$$

where (X_{ci}, Y_{ci}, Z_{ci}) is the three-dimensional coordinates of the visual marker center P_i in the left camera coordinate system, (X_{oi}, Y_{oi}, Z_{oi}) is the three-dimensional coordinates of the visual marker center point P_i in the target coordinate system, R is a 3×3 orthogonal rotation matrix, t is a 3×1 translation vector, and R and t represent the three-dimensional pose of the target.

9.4 Laser Visual Measurement

In laser visual measurement, a ranging sensor is used to directly measure the distance from the target to the sensor itself in the robot's working environment. According to the working principles, the ranging sensor can be divided into two types: time-of-flight (TOF)-based ranging sensors and triangulation-based ranging sensors [5].

9.4.1 TOF-Based Ranging

The TOF-based ranging sensor consists of a transmitter and a receiver. The transmitter is used for projecting collimated beams (i.e., a laser) to the target, and the receiver is used to detect a component of the light that is reflected by the target and substantially coaxial with the emitted beams, then to measure the distance. The first distance measurement method is to directly measure the time t from the emission of the beam to the moment when the beam reflected by the target returns to the receiver, then calculate the distance d based on the beam propagation speed v:

$$d = \frac{vt}{2}. \tag{9.7}$$

The second method is to measure the differential frequency between the Frequency-Modulated Continuous Wave (FMCW) and the reflected wave received. The transmitter emits an equal amplitude sinusoidal signal whose frequency changes in time with a certain rule. The received signal reflected by the measured target is multiplied by the transmitted signal, and only the low-frequency signal therein is taken to obtain the differential frequency signal Δf, which is proportional to the distance, thereby the distance d can be calculated by

$$d = \frac{\Delta f v}{4 f_m f_w}, \tag{9.8}$$

where f_m is the frequency of the modulated signal, and f_w is the bandwidth of the modulated frequency.

The third method is called phase-shift measurement, which measures the distance by using the phase difference between the transmitted signal and the reflected signal. As shown in Fig. 9.4, the transmitted beam with a wavelength of λ is divided into two beams. One beam (called the reference beam) reaches the receiver through distance L, the other beam is reflected by the target (called the reflected beam) and reaches the receiver through distance d, there will be a phase-shift θ between the reference beam and the reflected beam. The distance d is calculated as

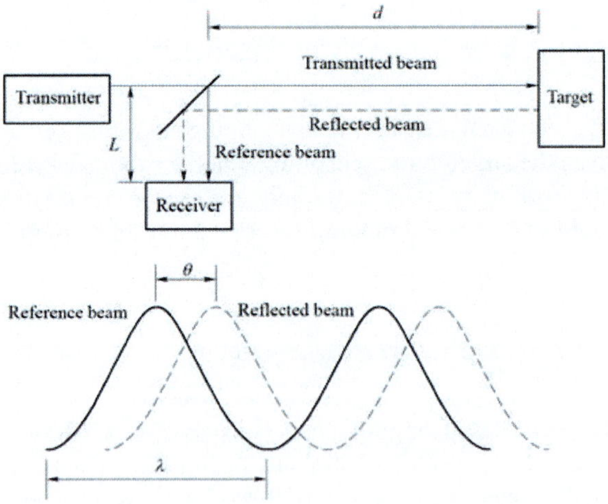

Fig. 9.4 Principle of phase-shift measurement

$$d = \frac{\theta}{4\pi}\lambda. \tag{9.9}$$

It can be concluded from Eq. (9.9) that the reference beam and the reflected beam will be aligned with each other when $\theta = 2k\pi$ $(k = 0, 1, 2, \ldots)$, that is, the two beams cannot be distinguished only based on the measured phase shift. Therefore, when using this method, it is required that $\theta < 360°$ or $2d < \lambda$. Since the laser wavelength is very short, in practical applications, the transmitted beam amplitude can be modulated by a wave with a much longer wavelength. Then the modulated beam is projected onto the target, and the reflected beam is demodulated and compared to the reference beam to determine the phase shift, then to calculate the distance.

9.4.2 Triangulation-Based Ranging

The triangulation-based ranging sensor determines the distance of the target by using geometric information. The sensor uses a transmitter to project known light patterns (e.g., points, lines, or textures) onto the target. The reflection of the known pattern is received by the receiver, and the distance is measured by using known geometric information and triangulation principles. The most basic triangulation principle is shown in Fig. 9.5, where the distance between the transmitter and the receiver (baseline length) is known as B, the angle between the transmitted beam and the baseline is α, and the distance d is calculated as

$$d = B \tan \alpha \tag{9.10}$$

Fig. 9.5 Principle of triangulation ranging

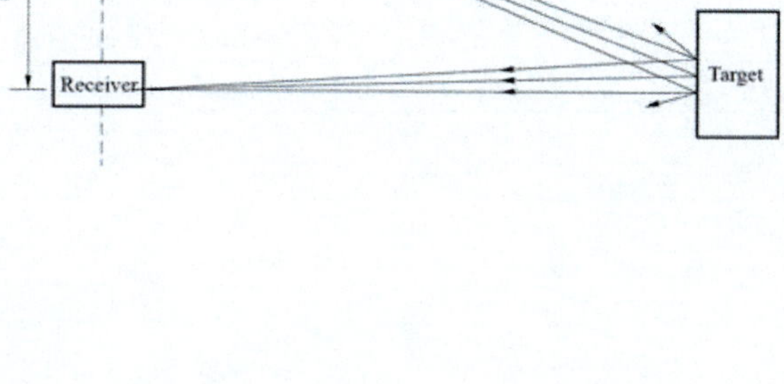

9.5 Test Verification

Ground test is an indispensable step in the development of the space robot visual perception system. Only a comprehensive and sufficient test can determine whether the visual perception system meets the requirements. The ground test includes three major categories: function/performance test, special test, and environment test. In the function/performance test, the function and technical indices of the performance of the perception system are mainly tested; the special test is a targeted evaluation test aiming at the special task requirements and working characteristics of the perception system; in the environment test, the perception system's ability is assessed to sustain environmental stress and its adaptability, the environments include the ground storage, transportation, loading and unloading environment, the launch environment, the space working environment, and returning environment. The specific test matrix is shown in Table 9.2.

Table 9.2 Space robot visual perception system test matrix

No.	Test category	Test items
1	Function/performance test	Mechanical interface test
2		Thermal interface test
3		Electrical interface test
4		Power consumption test
5		Communication test
6		Target detection and recognition capability test
7		Target parameter measurement capability test
8		Target tracking capability test
9		Environmental map construction capability test
10	Special test	Typical task verification test
11		Reliability test
12		Life test
13		Maintainability test
14	Environment test	Acceleration test
15		Vibration test (including sine vibration test and random vibration test)
16		Shock test
17		Thermal balancing test
18		Thermal cycle test

(continued)

Table 9.2 (continued)

No.	Test category	Test items
19		Thermal vacuum test
20		Burn-in test
21		Vacuum discharging test
22		Electromagnetic compatibility test
23		Damp heat test
24		Salt fog test
25		Sand and dust test

References

1. S. Ma, Z. Zhang, *Computer Vision——Computation Theory and Algorithm Basis* (Science Press, Beijing, 1998)
2. N. Zheng, *Computer Vision and Pattern Recognition* (National Defense Industry Press, Beijing, 1998)
3. Y. Jia, Machine Vision (Science Press, Beijing, 2000)
4. G. Zhang, Vision Measurement (Science Press, Beijing, 2008)
5. R. Szeliski, Computer Vision: Algorithms and Applications (Springer, London, 2011)

Chapter 10
Space Robot Teleoperation System

10.1 Teleoperation System Overview

Teleoperation system is an interactive tool between human and space robot. The process is: the operator acquires the information of the space robot and environment through the teleoperation system, judges its working state and gives the follow-up motion instructions, and sends them to the space robot through the teleoperation system; the space robot performs the desired motion. Therefore, as the control equipment of the space robot, the teleoperation system has two tasks: one is to sense and feedback the state of the space robot; the other is to convert the operator's operational intention or action into instructions.

There are many ways to sense the state of space robots. The most commonly used way for humans to obtain external information is vision, tactile, and auditory sensing. Therefore, it is preferred to obtain the operating state feedback of space robots by operator's accustomed natural perception, so that it has a sense of immersion and realizes the ideal human–computer interaction.

The purpose of human–robot interaction is to convert the operator's intention or action into operational instructions. The way of instruction generation is closely related to the intelligence level of space robot, the configuration of the actuator mechanism, and the properties of the tasks. Common methods include spatial mapping, semantic (voice) instructions, etc.

Two issues need to be considered in the design of the teleoperation system of space robot: one is time delay, the other is communication bandwidth.

In the Earth orbit, the moon orbit or the moon surface, there is an operation delay of 3 to 10 s (mainly determined by the information system and processing mode, the effect of space transmission delay is less important), which seriously affects the stability of the robot operation. For static tasks (position shift, static cooperative target capture, transfer, release, etc.), robots can operate well by remote control, but for the continuous operations of dynamics targets, virtual predictive display, shared control, and other technologies are required to reliably accomplish the tasks. In more remote planetary exploration missions (such as Mars exploration), the round-trip time of

Y. Wang, *Space Robotics*, Space Science and Technologies,
https://doi.org/10.1007/978-981-15-4902-1_10

information can be as long as tens of minutes. The operation efficiency of remote control based on telemetry information is too low to meet the mission requirements; under this long delay conditions, the risk of teleoperation is very large. Even if virtual predictive display technology is used, the operations can only be completed within the controllable (known environmental conditions) range. It is urgent to improve the autonomy of the space robot and use the teleoperation based on intelligent agent technology to complete the tasks. Therefore, time delay is the key problem to be considered in the design of the teleoperation system.

The increase of spatial distance will not only cause time delay, but also reduce the available bandwidth of information transmission system. The reduction of communication system bandwidth has a relatively small impact on the transmission of remote control information (it can be solved by advanced semantic instructions or autonomous control technology), but it has a greater impact on the telemetry information. Most modern deep space detectors use optical and radar imaging equipment to detect environment information, which requires a large bandwidth to complete information transmission. The operating mode of space robot teleoperation system depends on its available bandwidth: only enough bandwidth can support the interactive operation; if the bandwidth is not enough, it can only operate according to a certain period to ensure the safety of the space robot. This mean the teleoperation system accumulates information by consuming certain time. Therefore, the communication bandwidth is the second key issue to be considered in the design of the space robot teleoperation system.

10.2 Teleoperation System Design

The space robot teleoperation process typically includes the following elements:

(1) Operator. The operator is the core of the teleoperation system. Through the human–robot interaction interface, the operator can sense the state of space robot and environment, make operational decisions, expresses operational intentions through action, and converts them into operational instructions via the human–robot interaction interface.
(2) Human–robot interaction interface. Its functions include: receiving information of the space robot and its environment, physically reproducing or simulating the state of the space robot and the environment on the ground; converting the operator's motion into an electrical signal by using the inertial assembly, the space electromagnetic detectors, the motion mechanism, and supporting circuits, to realize the acquisition of input data; using a computer to perform mapping process of the input data according to a certain relationship to generate the sequence of instructions capable of driving the space robot. Commonly used devices include handle, hand controller, space pose sensor, space mouse, data gloves, etc.

(3) Space-ground communication link terminal. Space-ground communication link terminal is responsible for receiving and analyzing the telemetry data of the space robot, and distributing it to the human–robot interaction interface and related equipment and units. In addition, according to the predetermined protocol, the operation instruction sequences generated by the human–robot interaction interface are packaged and uplinked.
(4) Space robots. Space robot is a task execution mechanism. It receives and executes instructions from the space-ground communication link terminal to realize operations on the environmental objects, perceives the information about the interaction between robots and environment, and sends the information to the operation monitoring interface.
(5) Environment. Environment is the general term of operating environment and action object of the space robot. Different from the general control system, the teleoperation system must consider the action object of the system in the realization process.

In a sense, the human–robot interaction interface mentioned here is the teleoperation system, which also includes the environmental reproduction unit (including a visual reproduction module, a force reproduction module, etc.), instruction input and generation unit, etc.

The teleoperation system of space robot generally has the following main functions:

(1) Status feedback: it can accurately and appropriately reflect the relative position relationship between environment and robot, the interaction process, and the operating status of the robot on the ground.
(2) Instruction generation: It can detect and recognize the operation instructions or operation actions of the operator, convert (map) it into a sequence of instructions that the space robot can recognize and execute, and drive the space robot to operate and complete tasks.

The performance indices of the teleoperation system are closely related to key indices such as the control cycle, and affects the usability and availability of the system. The general technical indices of the teleoperation system are shown in Table 10.1.

10.2.1 System Design

The design of the space robot teleoperation system mainly includes the following aspects:

Table 10.1 Basic technical indices of the teleoperation system

Type of indices	Name of indices
Shape parameters	Volume
	Mass
Electric parameters	Input voltage
	Power consumption
Adaptability to delay	Delay
Output frequency	Instruction output frequency
Video parameters	Refresh rate
	Image resolution
Measurement parameters	End positioning accuracy
	Angular position accuracy
Force feedback parameters	Force feedback range
	Force feedback accuracy
	Force feedback stiffness (gradient)

10.2.1.1 Determine the Teleoperation System Scheme

The main factors that affect the teleoperation system scheme are delay and bandwidth. Delay and bandwidth are closely related to the mission of space robots. Divided by the boundary of the Earth–Moon system, teleoperation system has different application solutions inside and outside the Earth–Moon system, only due to the delay and bandwidth limitation caused by distance.

With the increase of teleoperation distance, the available bandwidth of the system decreases and the delay increases. Teleoperation needs to reproduce the state of the space robot and the environment on the ground for further decision-making and operation. Therefore, for the application tasks with short delay and wider bandwidth, the teleoperation system can be constructed by using various bilateral teleoperation technologies; for application tasks with longer delay and narrower bandwidth, the teleoperation system can only be constructed by using "command-execution-waiting for feedback" telecontrol technology and long-range intelligent agent teleoperation technology. Teleoperation based on virtual predictive display may be considered only in static tasks. Specifically, for space applications in the Earth–Moon system, telecontrol and bilateral teleoperation technologies can be considered. For tasks with high dynamics requirements, technologies such as virtual predictive simulation and shared control can be introduced into the bilateral teleoperation system to reduce and control the impact of the delay. For space applications outside the Earth–Moon system, teleoperation based on intelligent agent technology is required. If the environment is static and known (a credible virtual environment can be established), the bilateral teleoperation technology or the telecontrol technology verified on the ground can be adopted. However, for the space missions outside the Earth–Moon system, regardless of the teleoperation based on the intelligent agent or the bilateral teleoperation, due to the limitation of bandwidth, only the way of "time accumulation for information accumulation" can be used, that is, "command-execution-waiting for feedback". Only after the system completes the image data [Digital Elevation Model (DEM)] transmission and the reconstruction of environment and system models at the master end (on ground), further decision-making and operation can be performed.

10.2.1.2 Interactive Mode Design

After determining the teleoperation system of space robot, the specific task is considered in coordination with its structural characteristics to determine its mapping mode and select the interactive devices (the telecontrol system does not need to consider mapping and interactive devices). There are generally two kinds of space robot teleoperation tasks: one is displacement operation, the other is in situ operation. In displacement operation, the space robot is regarded as a point with spatial coordinates, and its position is changed by driving this point. In the in situ operation, the spatial position of the space robot body does not change, but coordinates the operation of one or more manipulators to complete the fine task operation in the reachable space. In order to complete tasks of extraterrestrial planet surface exploration and maintenance/repair of large mechanisms, space robots need to switch between the two kinds of tasks. Space robots and manipulators usually have two drive modes. One is the end mapping (point mapping) mode, which maps the control quantity as the targeted pose of the robot or the manipulator end, the control quantity of the specific joints is autonomously solved by the robot controller; The other is joint mapping, that is, the operating equipment directly outputs the joint angles, and after being processed by the robot controller, drives the corresponding joints to move to the specified angles. Space robots have many kinds of structures, including wheeled, multi-legged walking robots, etc. When designing the displacement function of the teleoperation subsystem, space robot is usually regarded as a single point. When climbing over an obstacle, the multi-legged walking robot can switch to in situ operation, so that each manipulator can operate independently. During in situ operation of the space robot, in order to improve the operating efficiency and simplify the operating system design, usually end mapping method is used to operate the single manipulators, but for the manipulators with fewer joints (less than or equal to 6 joint DOFs), joint mapping drive method may be adopted, especially for gripping mechanisms such as the grippers. The operating devices used in the end mapping (point mapping) generally include handle, hand controller, space mouse, pose tracking sensor, while those used in the joint mapping generally include pose tracking sensors, gesture-based devices, and data gloves.

When a space robot is teleoperated by a control device such as a handle (also a hand controller or a pose sensor), the displacement and in situ operations are generally combined. For different operation modes, the corresponding measurements and mapping control are also different. The switch is used for mode shift between the displacement and in situ operations. The displacement operation usually needs to move a large distance, so the position (angular displacement)—speed mapping is adopted for the operation device; the in situ operation belongs to fine operation in a small-distance, so the position-position mapping is adopted for the operation device. When the teleoperation subsystem designed with mapping mode switching function is switched, if the current pose of the master operating device is inconsistent with the current pose of the slave space robot, it may cause the pose to hop and damage the device. The general handling method is: whether switching from displacement operation to in situ operation (i.e., from position-speed mapping to position-position mapping), or from in situ operation to displacement operation (i.e., from position-position mapping to position-speed mapping), the master operating equipment is required to reset to zero, the system is restored to stop before switching; or the output of the master operating equipment is zero from the moment of switching to the moment the master equipment reaches zero, and the normal output is restored after switching.

10.2.1.3 Ergonomic Design

In the design of space robot and its teleoperation system, in order to adopt an appropriate implementation scheme, ergonomics should always be considered, and ergonomic requirements should be given priority when necessary.

Teleoperation system should adopt the perception mode suitable for the operator as far as possible to enhance the sense of immersion and substitution and improve the operating efficiency.

The polarity design of the teleoperation system should be consistent with the sensory cognition direction of the operator and the polarity of the robot's motion coordinate, so that the operator can replace the robot to achieve human–robot consistency and prevent misoperation.

10.2.1.4 Stability and Safety Design

When designing a teleoperation system, the stability and safety of the system must be considered at the same time, including:

The existence of delay may lead to system instability. Use of virtual predictive technology may isolate the delay from the operation loop. However, if the delay is too large, it will cause an excessive difference between the actual state and the predicted state, which will bring safety risks. Therefore, verification and appropriate correction are needed.

The existing teleoperation system adopts computer control systems, and the control quantity is the output of discrete quantity. These discrete quantities can be described as absolute or incremental values. Theoretically, the effect of the two is the same, but when used in the practical application system, the effect is still somewhat different. For the system with fewer DOFs (less than or equal to 6), the spatial solution is unique, singularity is not likely to occur, so absolute value output can be used; for systems with more than 6 DOFs, the spatial solution is not unique, so incremental output may be used to effectively monitor the increment value and limit the increment according to the executive capability of the manipulator joints, thus to improve the stability and safety of the system.

10.2.2 Typical Teleoperation System

The motion of the space robot is controlled by the operator through commands. The computer system of the robot analyzes the received commands, controls the execution components such as joints to move and complete the operation on the specific target, and collects signals with the sensor and feeds back to the operator. The above process is the most basic operation information flow of space robot. Many operation technologies are formed and developed according to the application environment and system characteristics.

The original teleoperation systems adopt telecontrol technology, that is, the operator generates the operation command sequence according to the tasks, and each command controls the robot to complete a state setting or a state change. The operator sends a command, waits for state feedback to confirm the execution state, and then sends the next command. In this way, the operator drives the robot to complete tasks according to the sequence of designed actions. Telecontrol operation is actually open-loop control, in which humans play a decisive role in the telecontrol chain, so it cannot adapt to the highly flexible real-time operation task.

In order to deal with the real-time teleoperation task, bilateral teleoperation technology (interactive teleoperation technology) is developed. The principle is to build a robot system on the master end, and map it to the far-end (slave end) robot. When operating on the master robot, it simultaneously drives the slave robot to operate, and the telemetry information is reproduced on the master end, so that the operator can feel the environmental action of the far-end robot and the interaction between the operation and the perception, so that the operator can play a decision-making role in the control loop.

With the increase of the telecontrol distance of the space robot, the delay is more and more large, and the stability of the bilateral teleoperation system is getting worse. Therefore, a teleoperation based on virtual predictive display technology is developed. The principle is: a computer simulation system is added to the master end of the bilateral teleoperation system to simulate and display the relationship between the virtual manipulator and the virtual environment. The operator realizes a visual closed-loop of the virtual image and isolates the delay from the operation loop. The teleoperation based on virtual predictive display is essentially an open-loop control (the operator's visual closed-loop is only for the virtual simulation image and video), but it can well cope with the application environment, with a delay of several seconds to tens of seconds.

As space missions become more and more complex, space robots are required to perform multiple tasks at the same time. To this end, sensors are provided on the slave robot to enable partial functional autonomy (e.g., autonomous dynamics tracking of the target), so that the operator can focus on more important operations, thus developing the shared control technology. Shared control refers to the combination of the robot's autonomous precision control and the operator's macro decision-making ability to achieve high intelligence and high-precision control of the robot. The shared control teleoperation system is essentially a dual-loop control system. The inner loop is the closed-loop control for the space robot sensors to achieve the target position tracking function; the outer loop is a human-in-the-loop control to enable the operator to play his decision-making role. Inner loop control creates conditions for outer loop control.

When the space robot performs the planetary exploration missions outside the Earth–Moon system, the delay time may reach up to tens of minutes. In this case, the advantages of the bilateral teleoperation technology totally become invalid. Therefore, teleoperation Based on Intelligent Agent is gradually developed, which greatly improves the autonomy of the far-end robot: after receiving the advanced semantic instructions from the teleoperation system, the space robot autonomously plans the task flow and independently completes the environment perception, path planning, obstacle avoidance/transfer, and exploration operations. This is an important development trend of the teleoperation system for deep space exploration robot in the future.

10.2.2.1 Telecontrol System

Telecontrol is the most basic operation mode, as shown in Fig. 10.1. It sends a single command or command sequence from the master end (operator), directly controls the space manipulator to complete specific operations, and determines whether the command is executed successfully by telemetry commands. Due to the need to send commands and wait for the confirmation of the execution result, its operation efficiency is relatively low, so it is more suitable for the static working environment. The telecontrol's task planning function is originally completed at the master end, the operator completes path planning and sends commands one by one to drive the space robot segmentally complete the motion and reach the specified target or position. Now, with the advance of technology, sometime the path and task planning functions are placed at the slave (space robot) computer, the ground sends target commands to the space robot, who autonomously completes the path or task flow planning and executes it, and then feeds back the execution process and the result to the operator for confirmation. The telecontrol system is more suitable for space

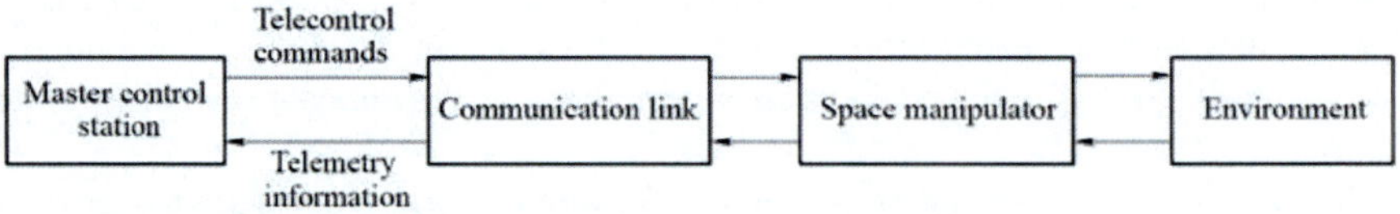

Fig. 10.1 Telecontrol system information flow

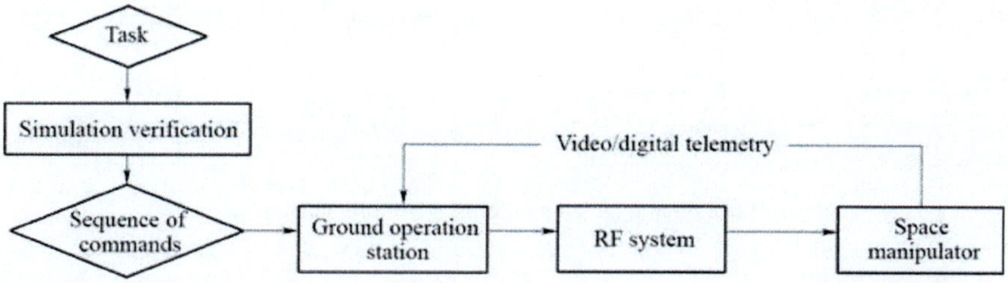

Fig. 10.2 Schematic of manipulator teleoperation

applications that are not sensitive to delay (within the range of near-Earth orbit), and has the advantages of low cost and reliable performance.

The teleoperation mode of a space operation robot is a typical telecontrol mode, as shown in Fig. 10.2. The robot is a 4-DOF manipulator controlled and driven by a central controller mounted on the lander. The controller has the trajectory planning function, but not the ability of task planning. The manipulator can be operated by single joint control mode, multi-joints linkage control mode, preprogramming motion mode, and visual servo control modes. In order to ensure safety, the system adds a ground simulation verification module. First, the ground simulation verification is carried out according to the task, and obtain a series of commands can be executed. Then, using the same operation mode as the traditional satellite to send the planned motion commands which have been verified one by one with ground equipment, and drive the manipulator to complete the tasks such as sample position transfer, sample acquisition, sample placement, and sample transfer. The space robot judges the execution results of the commands through the image information and telemetry parameters provided by the camera installed on the manipulator.

10.2.2.2 Bilateral Teleoperation System

(1) Ideal bilateral teleoperation system

Since traditional teleoperation cannot meet the requirements of continuous operation tasks such as on-orbit maintenance and fuel replenishment (due to poor real time and interactivity), a bilateral teleoperation technology based on telecontrol is developed, as shown in Fig. 10.3. The purpose is to establish a robot system (mapping robot) on

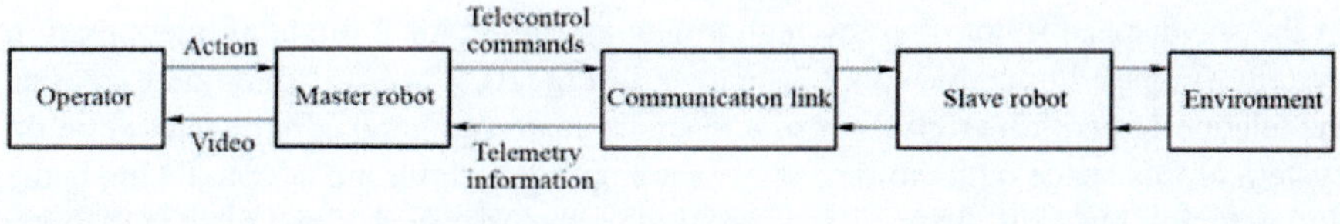

Fig. 10.3 Structure of bilateral teleoperation system

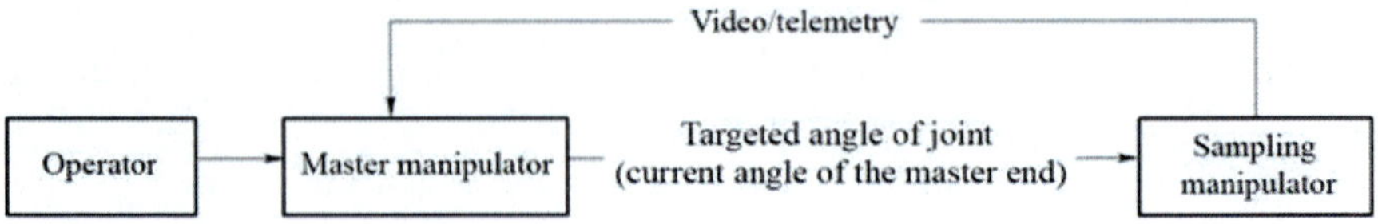

Fig. 10.4 Schematic of ideal bilateral teleoperation system for a manipulator

the master end, where the master robot is controlled by the operator at the master end, and generates a command flow to be transmitted to the slave (far-end) space robot, who reproduces the master actions and, at the same time, feeds back the video and telemetry information to the operator. The operator then adjusts the operation of the master robot according to the feedback information, forms a closed-loop control, and completes continuous operation tasks. In addition, a bilateral teleoperation system improves the operation efficiency.

The most ideal bilateral teleoperation system is to deploy the same space robot on the master and slave ends, build the same environment, and maintain the bilateral state consistency. The operator controls the master robot to work, and the commands are simultaneously distributed to the slave robot. The operator directly observes the relative position and interaction between the robot and the environment according to the master site and forms a human-in-the-loop control. Meanwhile, the slave robot completely reproduces the operation of the master and completes the operation tasks. The ideal bilateral teleoperation system used on a sampling manipulator in China is shown in Fig. 10.4. In order to operate the manipulator, a similar configuration manipulator (master) is equipped in the cockpit to collect the angles of the master manipulator joints with sensors; the teleoperation system sends the collected positions of the master manipulator joints to the manipulator outside the cockpit as a target quantity; the manipulator outside the cockpit generates control commands according to the targeted angle and the current joint angle to control the joint angle and move them to the position corresponding to the master, thereby reproducing the motion process and state of the master. The operator can observe the motion of the manipulator and its relative relationship with the environment through camera images, so as to realize the human-in-the-loop control operation. The ideal bilateral teleoperation system is suitable for tasks with a small delay, clear image, and human-in-the-loop control.

In theory, as long as there is enough investment, the ideal bilateral teleoperation system is feasible; but in order to save money and improve the mission flexibility, the commonly used bilateral teleoperation system only uses devices such as the handle, hand controller, space mouse, and space pose sensor at the master end to map the pose of the slave space robot, thereby realizing a simplified ideal bilateral teleoperation system. A space humanoid robot, as shown in Fig. 10.5, uses this method to realize the teleoperation of the robot's arms and smart dexterous hands. In the teleoperation system of this space humanoid robot, two mapping methods are adopted. One is the end mapping used for the arms motion, that is, the operator's wrist pose is mapped to the end pose of the arms of the humanoid robot; the other is that the operator's

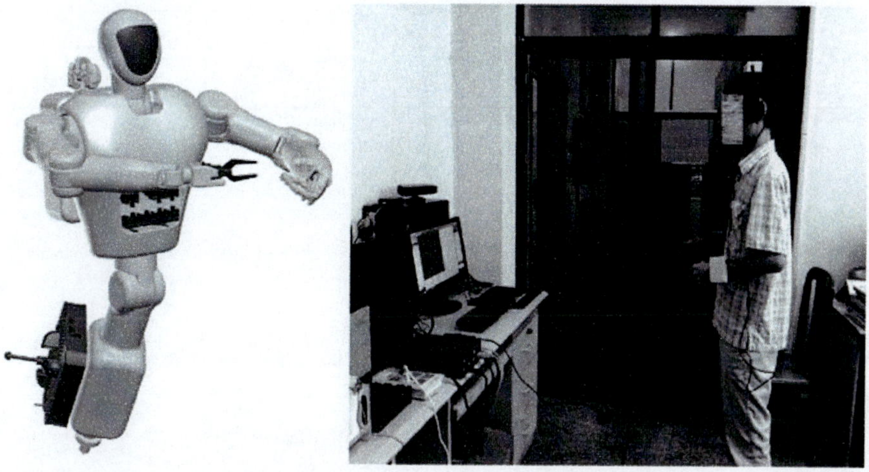

Fig. 10.5 Schematic of the bilateral teleoperation system of space humanoid robot

right hand finger joint angular position is mapped to the finger joint angle of the robot's smart dexterous hand. The system uses KINECT to identify the operator's bone information, and further calculates the angles of most joints, but it cannot detect the left and right rotation of the head, as well as the rotation of the arm and the wrist (9 angles), so an attitude sensor is added in these positions to identify the motion of these angles. Data gloves are used to sense and recognize the motion angles of the finger, and finally, form the instructions to drive the motion of a humanoid robot.

Gesture interaction technology for space humanoid robot is further developed on the basis of visual sensors. The operator faces the space humanoid robot, and uses his own body to flexibly operate the robot (smart dexterous hand) motion, and works according to the visual images, as shown in Fig. 10.6.

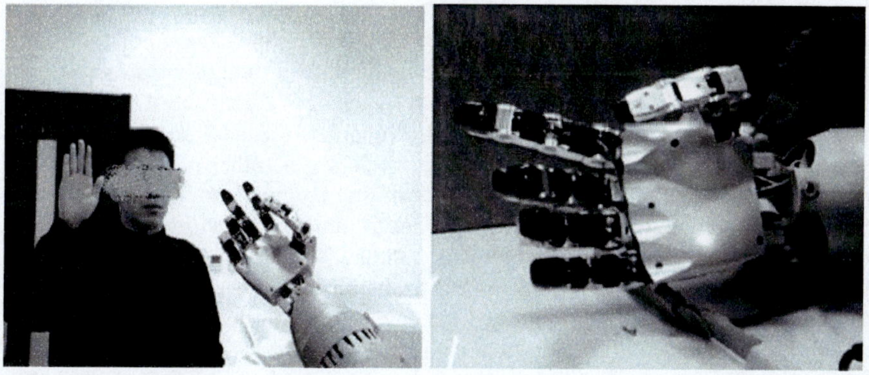

Fig. 10.6 Robot astronaut gesture control teleoperation

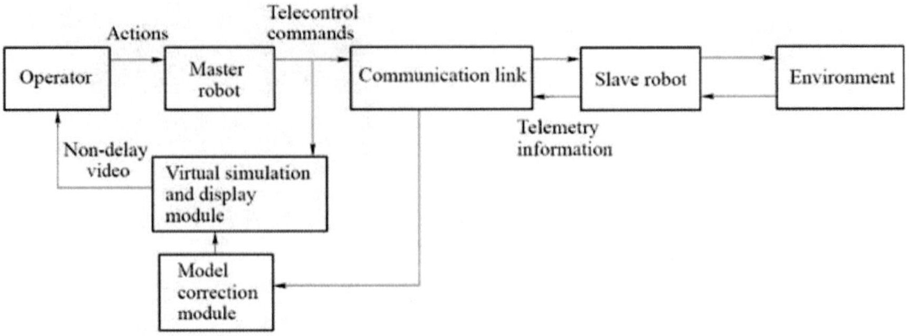

Fig. 10.7 Construction of bilateral teleoperation based on the virtual predictive display

(2) Teleoperation based on the virtual predictive display

When the distance between the master and the slave ends is very close (the delay can be ignored, forming a visual closed-loop directly), and the bilateral teleoperation system can work very well. Studies show that when the delay is less than 0.3 s, the operator does not feel the delay effect. When the space manipulator is applied to the near-Earth orbit and the Moon surface, the delay can reach up to 3–10 s, which may lead to the instability of simplified bilateral teleoperation system. In addition, because the distance is too far, the operator cannot directly form a visual closed-loop. Although it is still a man-in-the-loop, it lacks the necessary and intuitive decision-making information, so a virtual environment is introduced at the master end, enabling the operator to use the virtual environment to form a visual closed-loop, which promotes the birth of the teleoperation based on virtual predictive display, as shown in Fig. 10.7.

The teleoperation based on virtual predictive display replaces the master space robot of the ideal bilateral teleoperation system with the handle/hand controller and virtual robot. The working environment of the slave robot is also replaced by a three-dimensional computer model, and the virtual master robot is made to work in a virtual working environment of the slave space robot, at the same time, a mapping relationship between the handle/hand controller and the virtual robot is established. In this way, the control commands generated by the operator's control devices such as the handle/hand controller will be sent through two routes, one is sent to the master virtual reality computer system without delay, the other is sent to the slave space robot via the communication link. The master computer solutes the operation control command and drives the virtual robot to move in the virtual environment, which is displayed on the screen by using a three-dimensional display technology. The operator faces the virtual image, and forms a visual closed-loop according to the relationship between the robot and the environment in the virtual images. The control handle/hand controller drives the master virtual robot to perform operational tasks in the virtual scene. The master visual closed-loop operation commands are also distributed to the slave space robot, so that the slave space robot can reproduce the action of the master virtual simulation. From the perspective of the whole system, the action image generated by the master virtual reality precedes the execution action of the slave, which is similar to that the master is predicting the action of the slave, so it is called the teleoperation system based on the virtual predictive display. The master operator forms a visual closed-loop in the virtual scene and isolates the space-ground link delay outside the operation loop, which can better cope with the space delay. However, there are always errors between the model and the real physical system. In order to eliminate the accumulation of modeling error and calculation error, the model correction function is usually added in the virtual prediction system to correct the model error by using telemetry data, to ensure the consistency of the pose of master and slave robot and its mutual states with the environment.

A ground demonstration system for the teleoperation based on virtual predictive display is established on a large space manipulator. The teleoperation verification system uses a small cooperative manipulator as the control object, its master end uses a force feedback hand controller to map the end pose of the manipulator, and the task operation object is a small square wooden block. It is required to drive the manipulator by teleoperation to perform the operations of grasping, transferring, release of the small wooden block. The master virtual simulation computer simulates a small cooperative manipulator, a desktop, and a small wooden block. The visual sensor uses a Kinect depth camera. This system is used to verify the feasibility of teleoperation based on virtual predictive display technology and model correction technology under large delay conditions. The principle of the demonstration system is shown in Fig. 10.8.

After the teleoperation demonstration system is built, the operator faces the virtual environment simulated by the virtual predictive simulation computer and operates. He uses the hand controller to drive the virtual manipulator to complete the action of virtually grasping the small wooden block and traversing it for a certain distance and release. After 20s of delay, the slave small cooperative manipulator successfully

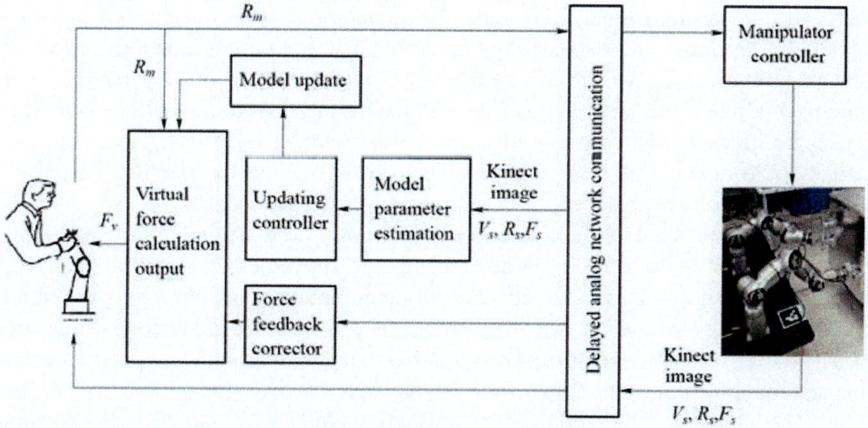

Fig. 10.8 Block diagram of the teleoperation demonstration system

reproduces the grasping, traversing, and releasing of the small wooden block, which verifies that the teleoperation based on virtual predictive display technology can effectively cope with teleoperation tasks with large delay.

(3) Virtual reality interactive teleoperation system

In the teleoperation based on virtual predictive display, in order to overcome the influence of delay and improve the operation safety and efficiency, it is necessary to improve two abilities: one is to minimize the transmission delay of the video images of the far-end space robot, the other is to maximize the visual depth of the master virtual prediction system. Since the bilateral teleoperation system is mostly used for contact operation, when the space robot interacts with the environmental target, especially when misoperation occurs, it is necessary to reposition the target as soon as possible, so as to take measures to repeat the operations or to avoid it. A small video delay can improve operational efficiency and protect the safety of the space robot.

The teleoperation based on virtual predictive display realizes the visual closed-loop between the operator and the virtual environment at the master, improves the visual depth of the virtual prediction system, which can improve the positioning ability of the operator, thereby improving the operational accuracy. In order to increase the visual depth and improve the operator's visual telepresence, a lot of new technologies have been developed, such as Virtual Reality (VR) technology and Augmented Reality (AR) technology at the master end of the teleoperation system.

VR is a computer-generated, interactive, three-dimensional virtual environment in which the computer system collects the operator's natural actions (head rotation, eye movement trajectory tracking, hand and finger movements, motion displacement, etc.), processes the operator's motion data, responds to the user's actions in real time and feeds back to the operator's perception (visual, auditory, motion, force, etc.). By introducing VR technology into the teleoperation based on virtual predictive display, the operator's wrist pose can be mapped to the space robot manipulator pose, and the head pose or the eyeball trajectory can be mapped to the space robot's line of sight (LOS) motion. In the master end three-dimensional virtual scene, the operator can use the arm to drive the digital model to operate the object. By moving and rotating his head, the operator can easily change the operator's LOS, viewpoint, and angle of view, and conveniently observe the relative relationship between the virtual environment and the operation object (virtual), so as to effectively improve the operator's telepresence, depth of field, and operational accuracy.

The equipment used for the collection of VR human body natural motion includes: hand controller, pose tracker, bone tracker, eye movement tracker, data glove, smart helmet, smart glasses, stereo projection equipment, and stereo glasses. Hand controller and pose tracker are high-precision pose tracking devices, which are usually used to map the end pose of robot manipulator, and through computer calculation and mapping conversion, to form a driving signal to drive the joint motion of the robot. The hand controller and the pose tracker generally adopt the position-position mapping mode to realize in situ operations. However, when the spatial position transfer or the position shift must be completed, the position-speed mapping mode can also be used, but attention must be paid to the switch between the two modes.

Gesture-based tracking devices can be used to track the joint motion attitude above the waist of the human body and drive the upper limb and the waist motion of the virtual model, but its accuracy is relatively low. Moreover, data gloves are usually used to drive multi-DOF robot hands.

Smart helmets and glasses are generally composed of a stereoscopic display and a head pose tracker. They can not only display the virtual video image, but also dynamically use the motion information of the user's head to drive the virtual environment scene change, so as to realize the dynamics closed-loop response to the changes of operator's LOS, and improve the operator's telepresence.

The teleoperation system of NASA's space robot R2 is a typical teleoperation system using VR technology. Since the R2 robot is a humanoid robot, its teleoperation system is similar to the ideal bilateral teleoperation system. R2's teleoperation system uses a position tracking device to obtain the 6-DOF pose data of the operator's wrist to drive the R2 robot's arms motion and to obtain the operator's waist motion information to drive the R2 robot's lumbar joint motion; it uses data gloves to drive the dexterous hands to complete grasping, inserting/extracting, screwing/pressing actions. The smart helmet has two functions: one is to track the motion of the operator's head to drive the neck joint motion of R2 and synchronize the motion of the R2 head with the operator; the other is to display the binocular visual image of the robot, enabling the operator to have a strong sense of substitution (immersion).

AR technology uses computer technology to add virtual environment and objects to the real environment image and video information collected by the operator, and then collects the interactive information of the operator and the environment (including the virtual environment) by the computer processing system to respond to the user's action, and displays the combined virtual and real image information on a three-dimensional terminal display. At present, AR technology is more used to superimpose important parameter information or feedback important prompt information on robot visual images. In the internal console display of a large space manipulator, in order to facilitate the astronaut operation, a camera measurement pose is superimposed on the video images of the manipulator end camera and the docking camera. When the astronaut operates the manipulator, he can obtain the image and digital data information without switching his LOS.

(4) Force and tactile feedback teleoperation system

When space robots perform fine operations such as inserting/extracting the orbital replacement unit (ORU), it is difficult to complete the alignment and insertion/extraction operations only with high-precision visual images; for fragile and vulnerable equipment, it is also difficult to complete the clamping and transfer tasks only with visual images. In such a task, the operator's perception of interactive force becomes extremely important. The operator can judge the position and direction of the stuck according to the magnitude and direction of the force, and actively adjust and correct to successfully complete the task. Therefore, force and tactile feedback devices have been developed in the VR technology, such as force feedback handles, exoskeleton force feedback data gloves, full-arm force feedback hand controllers, to enable the operator to feel the force imposed on the slave space robot at the master end.

The force and tactile feedback teleoperation system adds a virtual force calculation module (dynamic modeling) to the master computer virtual environment on the teleoperation based on the virtual predictive display. It uses the elastic contact model or the collision model to calculate the force or moment imposed on the virtual model according to the contact state of the virtual object model and the environmental target, then uses a force feedback device at the master end to apply the simulation calculated force to the operator, so that he can better perceive the force imposed on the real manipulator during operation. In the force feedback teleoperation system, the force fed back to the operator is also calculated at the master end, which presents earlier than the telemetry force value of the real robot (time difference = telemetry downlink delay + command uplink delay). It is also a virtual predictive value. In order to improve the feedback accuracy, it is required to correct the model periodically or occasionally. In addition, the force feedback system usually works in combination with the visual feedback system to keep the vision synchronized with the force and the tactile, so that the operator can comprehensively perceive the environmental state of the slave robot and improve the sense of immersion and operation efficiency.

Generally, in order to achieve better interaction effect during application, not only geometric and kinematic models are established in the master virtual environment to generate visual feedback, but also a dynamics environment model is constructed to provide virtual predictive force feedback, and if necessary, tactile feedback may be generated to form a multi-sensor interactive teleoperation system.

10.2.2.3 Shared Teleoperation System

Space robots, especially free-flying space robots, have difficulty to complete hovering target capture and multi-arm multi-DOF cooperative operation by using telecontrol or bilateral teleoperation. Therefore, it is necessary to improve the intelligence of space robots so that it can automatically use sensor devices to complete the basic motion control, and free the operator who can then complete the operation and judgment of critical actions. The control realized by the shared effects of human and the robot is generally referred to as shared control.

Unlike the teleoperation based on virtual predictive display, the shared teleoperation system (Fig. 10.9) has expanded its function on the slave end on the basis of the original bilateral teleoperation. It introduces space robot sensors (such as visual measurement information) to realize the functions of position tracking, visual direction tracking at the far-end, and changes the dynamics operation into a relatively static operation, which frees the operator and enables him to concentrate on critical operations (such as target capture). Shared control is essentially a dual-loop control, its inner loop is the position and LOS tracking loop of the far-end space robot, and the outer loop is the man-in-the-loop.

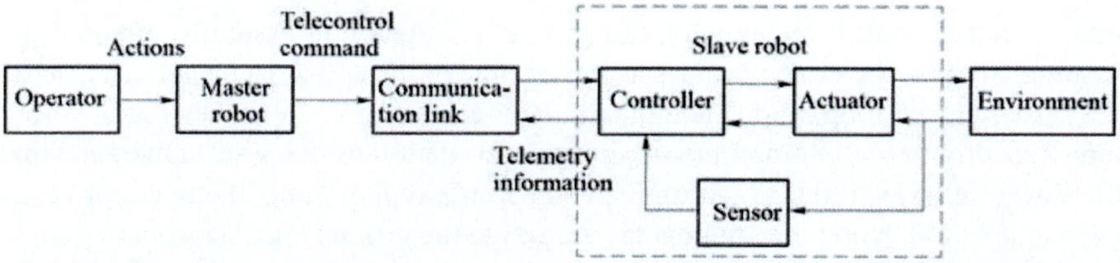

Fig. 10.9 Schematic of shared teleoperation system

The shared teleoperation system has less influence on the human-in-the-loop when the delay is small, and the influence increases with the increased delay. Therefore, shared control needs to be combined with virtual prediction. The process flow is shown in Fig. 10.10. It should be pointed out that due to the introduction of shared control, in the original virtual predictive simulation system, it also need to add a virtual shared control simulation loop to realize the position, LOS tracking functions at the master space robot, so that it is not affected by the system delay, and provide the operator with a relatively static image to capture appropriate dynamics operating opportunities.

There are two modes of space manipulator: automatic control and manual control. Its teleoperation system is a comprehensive system with telecontrol and bilateral interactive teleoperation capability, and can also use shared teleoperation technology during the task of the capture of hovering spacecraft. In the automatic control mode, the operator can use the keys on the console interface to input parameters and send commands to control the operation of the manipulator; in the manual control mode, the console is equipped with two 3-DOF force feedback handles for mapping the end pose of the manipulator, and use position-speed mapping mode to drive the operation of the manipulator end. When performing hovering target capture task, the manipulator is required to be able to steadily track the marker of the target, and the operator needs to operate the end to complete the capture action. The manipulator

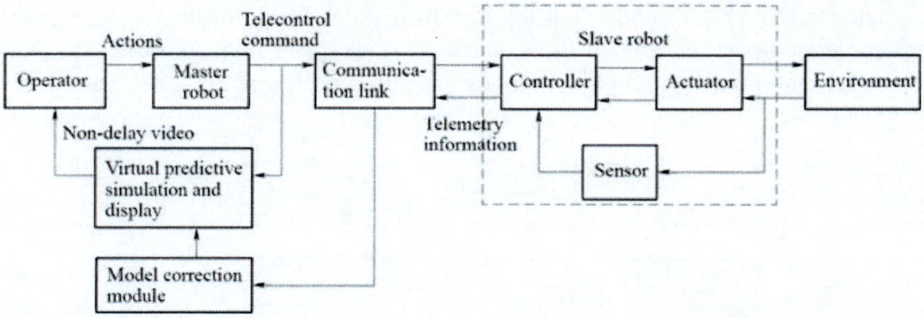

Fig. 10.10 Shared control introduced into teleoperation based on virtual prediction

adopts shared control technology and ground teleoperation system to operate the manipulator into the effective measurement range of the end wrist camera, and starts the visual closed-loop mode. The manipulator tracks and approaches the hovering target according to the visual measurement information of the wrist camera, so that the hovering target remains within the end capture range. Then, the operator sends the capture command to the manipulator through the ground teleoperation system to execute the automatic capture.

10.2.2.4 Teleoperation Based on Intelligent Agent

Space robots face two main problems: delay and bandwidth. The longer the working distance, the greater the impact of delay and bandwidth. For space robots operating outside the Earth–Moon system, the telecontrol and bilateral teleoperation are inefficient and have serious safety hazards. In the contact operations, the collision between robot and environment is inevitable. If the relocation of objects and the chain reactions thus generated cannot be properly handled in time after the collision, it may endanger the safety of the robot. Under such application conditions, intelligent agent control can play an important role. The teleoperation system based on intelligent agent greatly enhances the intelligence level of the space robot, enabling it to complete path planning, task planning, autonomous execution, and obstacle avoidance protection according to the instructions, and ultimately accomplish a relatively independent task. Although the concept of teleoperation based on the intelligent agent is relatively new, it is developed on the basis of the telecontrol system, and its working mode is similar to the telecontrol. As shown in Fig. 10.11, the teleoperation system based on intelligent agent first needs to reinforce the functions of the slave robot. The slave robot needs to be equipped with powerful measurement sensors, such as high-definition cameras and laser radars, to complete the image acquisition and DEM acquisition from the surrounding environment. The slave on-board computer performs image processing, including image segmentation, edge extraction, DEM data fusion, and environment model reconstruction, and finally obtains the latest environment model and target information. Based on the reconstructed environment model acquired by the sensors, the slave space robot completes the path planning, task planning, and autonomous execution according to task instructions.

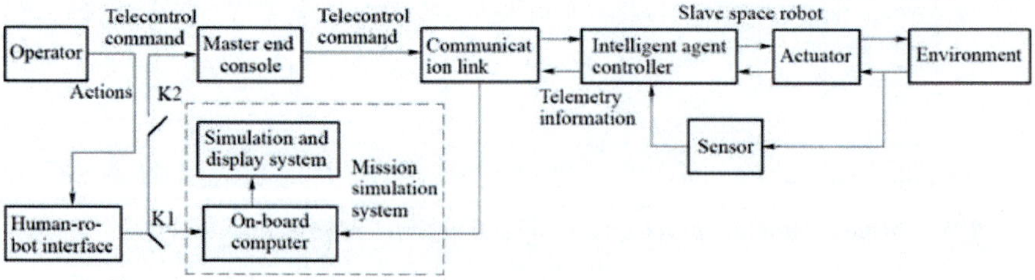

Fig. 10.11 Diagram of teleoperation based on intelligent agent

In addition, in order to ensure the completion of the task and meet the requirements of human intervention, it is necessary to configure the mission simulation system, which consists of a backup on-board computer and a simulation display system. The mission simulation system uses the space environment model transmitted back by the slave space robot (or uses the returned camera images and the environment model generated by the laser radar DEM data) to receive the task instructions given by the operator. The ground control computer performs task planning and sends instructions. The simulation display system drives the simulated space robot motion according to the instructions and displays it on the 3D display (ground rehearsal). After the entire execution process is evaluated by ground experts to be safe and reliable, task instructions are sent to the space robot for implementation. Upon receiving the instructions, the space robot autonomously performs the motion path and the task flow planning, and executes it independently. At the same time, the intelligent agent system needs to perceive and avoid the danger autonomously. When contact operations occur at the slave end, the space robot needs to update the image of the operation target, the DEM data, and the environment model in time, to get ready for the next operation. In the missions on celestial bodies outside the Earth–Moon system, such as on the Mars, the master mission simulation system of the teleoperation system based on intelligent agent, in addition to mission rehearsal and function evaluation, is also used to receive telemetry and reproduce the task execution process at the slave end, but it cannot participate in the decision-making of human-in-the-loop.

The US Mars rovers Spirit, Opportunity, and Curiosity all use teleoperation systems based on the intelligent agent. Similar solutions have been adopted in the teleoperation system of China's Mars exploration mission. The working process is as follows: the Mars rover uses cameras or other devices to acquire the surrounding terrain images, the on-board computer autonomously processes the image information, completes the environment modeling, and transmits the image information back to the teleoperation system; after receiving the image data, the ground teleoperation system performs data processing and modeling, the ground mission simulation computer plans the motion path according to the task requirements, and sends the operation instructions to the Mars rover after simulation verification; after receiving the commands, the Mars rover will autonomously plan the path for motion transfer. After arriving at the mission area, the rover transmits image information again. After the ground system receives the new image information, it updates the environment model and simulates the in situ operation task. After mission simulation is verified, the task operation instructions are sent to the Mars rover, who autonomously plans the task operation process according to the instructions and completes the sampling, analysis, and other tasks.

Chapter 11
Space Robot System Verification

11.1 Space Robot System Verification Methods

Each spacecraft has to undergo comprehensive and rigorous ground verification prior to launch, and space robots are no exception. There are two ways for the ground verification for general spacecraft: simulation and physical testing, and the latter is the most commonly adopted method. Physical test of spacecraft at assembly-level is generally easier to conduct with full coverage of test conditions, because the ground simulation environment in which the products actually work can be easily set up. However, due to the size and weight (including payload weight) of the space robot system, as well as the challenge of simulating the on-orbit microgravity (or low gravity) environment in which the space robot systems are supposed to work, system-level ground verification test for space robot systems is very difficult to carry out. So, system verification of space robots is more difficult than that of general space mechanisms.

Since the complexity of the robot systems and the particular challenges in conducting the physical verification tests on ground, system simulation is an indispensable part of the development process of all space robots. The most convenient and effective method for a space robot verification is to establish a simulation model according to its design or prototype, by which the task verification is performed based on the physical laws and system properties [1, 2].

In summary, for the space robot system-level verification, three methods are mainly used: physical test verification, semi-physical simulation verification, and mathematical simulation verification [3]. In the past decades, physical test and mathematical simulation verifications have been extensively used in the development of space robots. However, due to the difficulties in the construction of the physical test environment and the inaccuracy in mathematical simulation verification, for the purpose of both reducing the difficulty of building the verification environment and ensuring the credibility of the verification results, a semi-physical simulation verification method combining the advantages of physical test and simulation verification has been rapidly developed and becomes another system-level verification method.

Y. Wang, *Space Robotics*, Space Science and Technologies,
https://doi.org/10.1007/978-981-15-4902-1_11

MDA Canada adopted the semi-physical simulation as the verification method during the development of a Special Purpose Dexterous Manipulator (SPDM) [4, 5]. Compared with the physical test and mathematical simulation verification, the semi-physical simulation verification has the comprehensive advantages in verification cost, efficiency, accuracy, and effectiveness.

In summary, in the development of space robots, task verification is often performed by the combined methods of mathematical simulation, assembly-level physical test, system-level physical test, and semi-physical simulation, as shown in Fig. 11.1. The main principles to be followed in the design of a specific verification scheme include:

(1) The ground test should cover all functions and performance indices of the space robot, and the task verification should cover all on-orbit tasks.

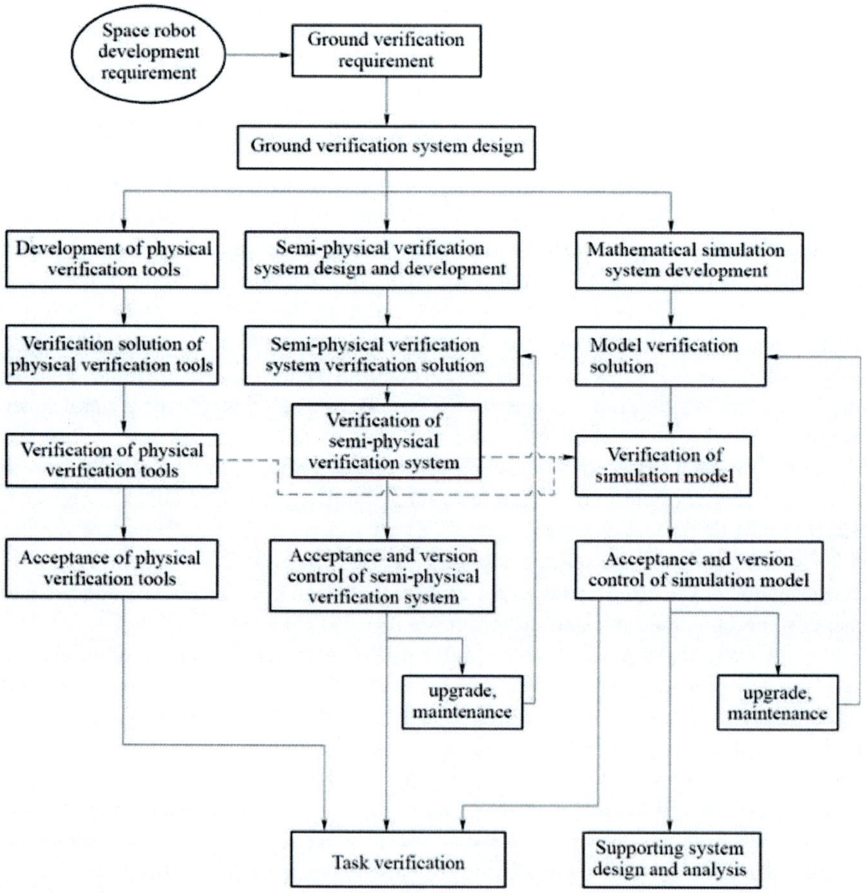

Fig. 11.1 Space robot task verification methods

(2) For the ground physical test, hardware systems should be used. For those system-level tests that cannot be performed with the physical test, they can be decomposed to component-level physical tests; for system-level full-load tests that cannot be carried out with a physical test method, a combination of system-level tests with reduced load and component-level tests with full load should be considered.
(3) In the task verification, the physical test verification method should preferably be adopted. For a verification task, such as large load operations, that cannot be carried out by a three-dimensional physical test, it can be done by a combination of two-dimensional physical test verification and mathematical simulation; for a task that is completely impossible to perform physical test verification, a combination of assembly-level physical verification and mathematical simulations can be used to check whether the space robot meets the specified requirements.
(4) All on-orbit tasks should be verified by mathematical simulation. The correctness and accuracy of the simulation model should be validated by physical tests both at the component-level and system-level. For the processes of on-orbit tasks that cannot be mathematically modeled or enough accurately modeled, such as a contact between objects of complicated shapes, the physical form of the objects can be used in the task verification as a substitute of the corresponding mathematical model. This method is called a semi-physical simulation verification method for task verification.

11.2 Necessity of Simulation Verification

As mentioned above, mathematical simulation is an important means for task verification of space robot systems. Simulation work runs through the entire process of space robot development. The credibility of mathematical simulation lies in that the simulation model is built based on the space robot's design and its physical prototype. In the entire development process, it is continuously modified and improved according to the test data to ensure that it can represent the actual characteristics of the space robot system.

Next, we will use the verification of a large space manipulator as an example to illustrate the necessity of simulation verification. In the development of the manipulator, the ground task verification covering all working conditions is an essential prerequisite for ensuring the reliable on-orbit operations of the manipulator and successful completion of the tasks. However, the ground task verification of the manipulator is extremely difficult, mainly embodied in the following:

(1) The manipulator has a large geometric dimension (the arm span exceeds 10 m) and payload (the maximum design payload is up to 25t). In the ground gravity environment, the manipulator is designed to work in the on-orbit microgravity environment, so the system-level verification tests on physical hardware has many restrictions and difficulties. In fact, it can't even lift itself off under the ground gravity condition by only its own driving ability, let alone to manipulate

a 25t payload. The zero-gravity simulation method based on the air-bearing system can realize the microgravity environment in the plane but cannot verify the three-dimensional motion behaviors of the space robot. Thus, the validity and coverage of the testing cannot be guaranteed.

(2) The manipulator is a complex system, with diversified tasks and long service life, which makes ground integration testing a huge workload.

As a general-purpose operating tool, the manipulator is designed to carry out various tasks. The variety of tasks, complexity of working conditions, and the operating life would be significantly increased compared with traditional space mechanisms. In order to reliably and efficiently carry out the tasks, the task execution procedures should repeatedly be verified by using different methods under different boundary conditions and different control parameters. The ground test and verification conditions of the manipulator are so massive that the time and economic cost of a comprehensive physical testing are unbearable, even if a zero-gravity test environment can be created on the ground.

(3) In order to ensure the reliability and safety of the manipulator's on-orbit task, a large number of fault tests should be performed and some task-level verification should be performed even in man-in-the-loop mode. So, this manipulator is very difficult to be verified by the physical test.

In order to ensure the safety and efficiency of the operation, and prevent faults from operation errors, wrong instructions, or safety hazards from program runaway, software errors, single-event fault, etc., the safety of the manipulator must be fully verified. Since the fault verification takes a long time, the fault injection test requires a huge workload, and there are certain risks that are not appropriate to be checked by physical tests. In addition, the same problems also exist in the verification of man-in-the-loop tasks.

(4) Possible improvised manipulator's tasks during on-orbit operation are impossible to be verified on the ground in advance.

Taking on-orbit injection as an example, the manipulators currently used in the International Space Station (such as SSRMS and SPDM) have many tasks, which are not considered in their design and construction stage. By the time of these tasks, the manipulators had operated on-orbit for many years, and there were no facilities and equipment on the ground that can be used for physical verification of these tasks. Therefore, such tasks could not be verified in advance.

In summary, the verification of space robots in the ground environment faces the challenges of "incomplete test" or "unable to test" due to the difficulties in simulating the space environment with high fidelity, which make it very difficult to ensure the validity and coverage of testing and verification. The most feasible and efficient method still depends on simulation. By establishing a simulation model based on product design and its prototype, and correcting the model with the measured data obtained from the model verification tests, a mathematical model that can fully reflect the physical properties and characteristics of the space robot can be achieved. On this basis, combined with some physical prototypes, a credible space robot ground verification system can be constructed to fully verify the product design and final product performance, and tackle the difficulties faced by the space robot task verification mentioned above.

11.3 Example of Space Robot Simulation Verification

SimMechanics is particularly suitable for task analysis and algorithm design in the early stage of product design due to its relatively easy modeling, fast simulation speed, and friendly user interface. The main simulation processes of SimMechanics include model construction, controller setup, model configuration, simulation, and debugging.

Now, we will take a space manipulator design as an example to introduce the process of system task verification using SimMechanics. As shown in Fig. 11.2, a space manipulator is mounted on a service satellite (floating base) to track and capture a target satellite, and to operate the target to complete a docking task. The manipulator has a "3+1+3" seven-DOF configuration.

11.3.1 Model Establishment

SimMechanics provides two modeling methods: one is to directly use the modules provided by the standard module library, the other is to import modules from CAD models. Direct modeling is relatively simple and easy, which is suitable for the modeling process with simple shapes and a few modules. The standard module library provides common modules such as rigid bodies, joints, coordinate system transformation, and configurations. Various multi-rigid body dynamical systems can be set up by properly connecting these modules. When the model is complex or requires high visualization, the simulation model can be generated directly from the CAD software assembly models. The SimMechanics Link is required to be installed for this process.

The satellites and the robot in this example are all automatically imported models. SimMechanics supports importing models from PTC Creo (Pro/Engineer), SolidWorks, and Autodesk Inventor. The following preprocessing work must be done before importing:

(1) Establish an entity assembly model (not support shell model import).
(2) Set mass property parameters for each entity.
(3) Correctly establish the constraint relationship between various rigid bodies.

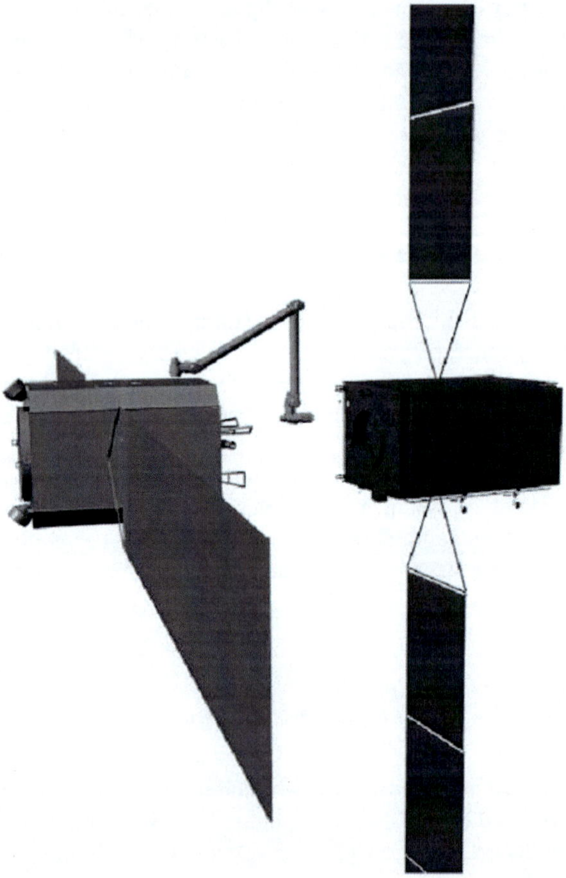

Fig. 11.2 Example of SimMechanics-based simulation

In order to make the simulation model neat and clean, multiple parts on the same rigid body can be combined into one part and re-assigned with quality parameters. The model import flow is shown in Fig. 11.3.

SimMechanics Link can automatically convert the entities in the CAD model into rigid bodies with inertial parameters and dimensional parameters in the simulation model, and transform the constraint relationships into the joints in the simulation model. In this example, the simulation model is generated from the PTC Creo assembly. After subsystem encapsulation, the resulting simulation model is shown in Fig. 11.4.

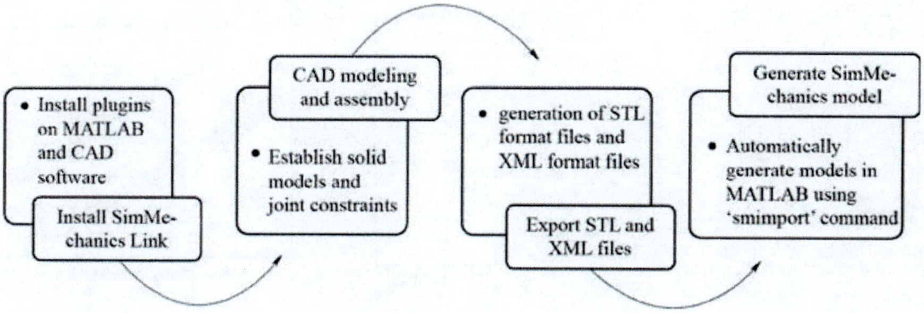

Fig. 11.3 Model import flow from cad software

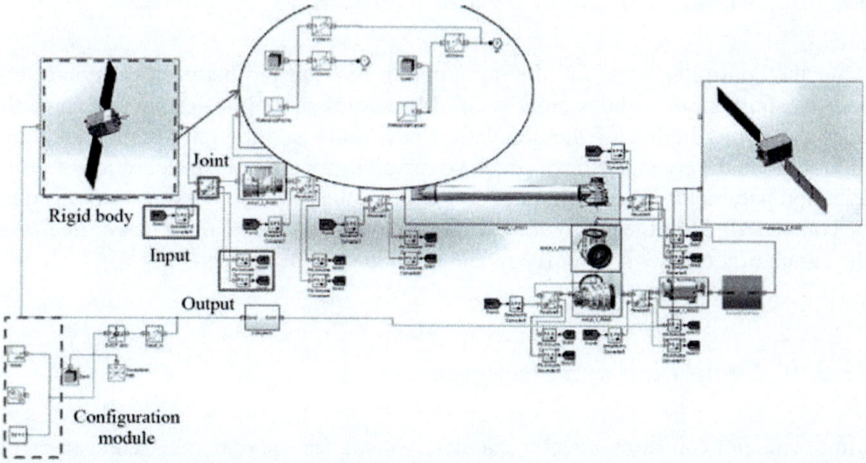

Fig. 11.4 Simulation model set up with SimMechanics

11.3.2 Controller Setup

The controller is set up by using the impedance control algorithm based on feedback linearization in Chap. 5, as shown in Fig. 11.5. The controller includes two parts: the path planning and the impedance controller. The former calculates the inverse kinematics based on the feedback information and generates the joint space commands; the latter uses the force feedback information to establish the impedance control algorithm.

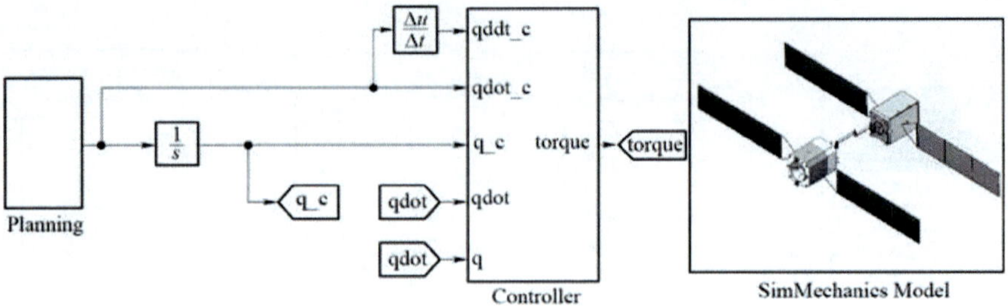

Fig. 11.5 Controller model

11.3.3 Model Configuration and Preliminary Verification

After the controller is set up, the input/output relationship between the controller and the simulation modules built by SimMechanics should be configured. In addition, the configuration of the simulation environment, the configuration of joint input/output, the construction of contact dynamics module, the configuration of visualization parameters, the addition of scope, and other post-processing work need to be completed. After the configuration is completed, the model must be verified, and the simulation can begin after the model is verified to be correct.

11.3.4 Simulation and Debugging

SimMechanics can automatically calculate kinematics and dynamics, and generates 3D visualization scenes from CAD-imported files. In the simulation process, the joint angle, angular velocity, angular acceleration, constraint force, drive force, end position, velocity, and acceleration can be measured and analyzed.

Figure 11.6 shows the contacting force (torque) curve between the manipulator end and the target satellite under one working condition when the manipulator captures the target satellite. This data can be used to evaluate the key indices such as the response speed and the force control accuracy of the manipulator, thus verifying the feasibility of the task.

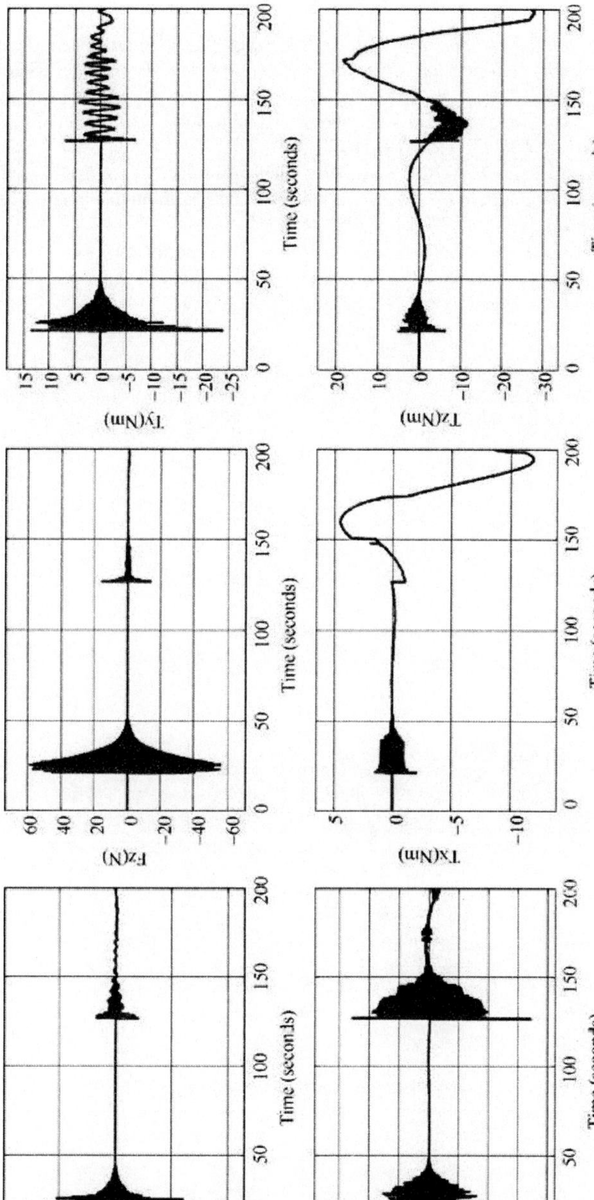

Fig. 11.6 Contacting force (torque) curve between the manipulator end and the target satellite

References

1. X. Zhang, *System Modeling and Simulation*, 2nd edn. (Tsinghua University Press, Beijing, 2015)
2. J. Liu, H. Gao, H. Luo, *Modeling and Simulation of Intelligent Robot System* (Science Press, Beijing, 2014)
3. W. Xu, B. Liang, Y. Xu, Survey of modeling, planning, and ground verification of space robotic systems. Acta Astronaut **68**(11–12), 1629–1649 (2011)
4. J.C. Piedboeuf, J. De Carufel, F. Aghili, et al., Task verification facility for the Canadian special purpose dexterous manipulator, in *Proceedings 1999 IEEE International Conference on Robotics and Automation*, Detroit, vol. 2 (1999), pp. 1077–1083
5. F. Aghili, J.-C. Piedboeuf, Contact dynamics emulation for hardware-in-loop simulation of robots interacting with environment, in *Proceedings 2002 IEEE International Conference on Robotics and Automation*, Washington, DC (2002), pp. 523–529

Chapter 12
Design Example of Large Space Manipulator

12.1 Overview

12.1.1 Engineering Background

At present, the application of large space manipulators mainly focuses on the field of manned space, such as the construction and operational support of space station. The assembly and operation support of International Space Station (ISS) is the most successful application example of large space manipulator in orbit.

The first large space manipulator used to build the ISS was the Canadarm (also known as SRMS: Shuttle Remote Manipulator System), developed by MDA (MacDonald Dettwiler and Associates Ltd.) and installed on the space shuttle. Its main tasks are to send the launched cargo modules and cargoes to the ISS through space shuttles and complete the assembly tasks. In 2001, the SSRMS (Space Station Remote Manipulator System) was installed on the ISS. It is mainly responsible for tasks such as cargo handling, capture and assisting berthing of visiting spacecraft, and assisting Extravehicular Activities (EVA) [1].

The Japanese Experiment Module (JEM) of the ISS is also equipped with a large space manipulator system—JEMRMS, which is mainly used for the operation of the test load of the Exposed Facility (EF) outside JEM. In addition, the Russian module of the ISS is proposed to deploy a large space robotic arm European Robotic Arm (ERA) developed by the European Space Agency (ESA) for various operational tasks outside the Russian module [2].

This chapter takes a space robot system similar to the above space manipulator as an example to introduce its design and verification.

Y. Wang, *Space Robotics*, Space Science and Technologies,
https://doi.org/10.1007/978-981-15-4902-1_12

12.1.2 Design Requirements

12.1.2.1 Task Requirements

The large space manipulator undertakes the construction, operational support, and subsequent expansion tasks of the space station, which can be roughly divided into the following categories [3]:

(1) Large module docking and re-docking assistance tasks;
(2) Hovering spacecraft capture tasks;
(3) Extravehicular handling and maintenance of large equipment tasks;
(4) Supporting astronauts' EVA tasks;
(5) Extravehicular state monitoring and inspection tasks.

12.1.2.2 Functional Requirements

(1) Large payload and wide range motion functions

Have the ability to move with large mass payloads at the end.

(2) Inch-worm type motion function

By connecting different target adapters, it has the ability of alternating inch-worm type motion from the shoulder to wrist.

(3) Multi-DOF motion function

Has the ability to move in multi-joint linkage mode along any path in the workspace.

(4) Multi-mode control function

Has different control modes, including position control, force control, follow-up control, etc.

(5) Multi-mode operation function

Has different operation modes, including autonomous operation, astronaut on-orbit operation, ground teleoperation, and astronaut on-orbit training mode.

(6) Capture and locking function

Has the function of capture and locking targets with end effectors.

(7) Path planning function

Has the ability of real-time path planning and offline path planning.

Table 12.1 Main technical indices of large space manipulators

Indices	Requirement
Arm length (m)	≥10
Mass (kg)	≤750
Load mass (kg)	0–25,000
DOF	7
Stopping distance at maximum load (m)	≤0.5
Position accuracy (mm)	≤45
Attitude accuracy (°)	≤1

(8) Visual measurement function

Has the ability to measure the geometric and motion characteristics of space targets.

(9) Astronaut securing function

Has the ability to secure astronauts to support their extravehicular activities.

(10) Hovering spacecraft capture function

Has the ability to capture visiting spacecraft that is in a relatively hovering state.

(11) Fault and safety function

Has the ability to ensure the safety of astronauts, robotic arms, and spacecraft in the event of malfunctions and accidents.

(12) Support on-orbit expansion function

Has the ability to expand the mechanical, electrical, and information modules on-orbit.

12.1.2.3 Performance Requirements

The main technical indices of large space manipulators are shown in Table 12.1 [4].

12.1.2.4 Interface Requirements

(1) Mechanical interface

The large space manipulator is in a furled state at the launch segment and installed outside the small column of the core module of the space station. One end of it is connected to the target adapter on the bulkhead by an end effector to maintain mechanical locking and electrical connection. The rest of the manipulators are connected to the module by the hold-down and release mechanism.

(2) Power supply interface

The large space manipulator adopts the hierarchical power supply system. The spacecraft provides a primary power supply bus and a command power supply bus for the manipulator, which are converted into a required secondary power source through the power supply module inside the manipulator.

(3) Thermal interface

The large space manipulator is connected to the module through a hold-down and release mechanism and a target adapter. The manipulator itself is adaptive to the space thermal environment. There is a thermal conductive design between the hold-down and release mechanism, the target adapter and the module to meet the temperature requirement of the mounting surface.

(4) Information interface

The large space manipulator realizes the data interaction with the spacecraft through the manipulator console in the module.

The spacecraft digital management subsystem and the GNC subsystem provide 1553B bus interface for the manipulator console, which meets the requirements of command receiving and telemetry transmission of space manipulator equipment.

The spacecraft measurement and control subsystem provides an Ethernet communication interface for manipulator console and target adapter, and video image transmission channel for the robotic arm camera.

12.1.2.5 Reliability Requirements

During the whole service life of the manipulator, its normal operational function should be ensured through maintenance and repair, and the task reliability shall not be lower than the required value.

12.1.2.6 Safety Requirements

Safety requirements for large space manipulator itself and its operators are included.

12.1.2.7 Maintainability Requirements

Requirements for large space manipulator on key component maintainability, on-orbit software injection, modification, and upgrade are included.

12.1.2.8 Testability Requirements

Requirements for large space manipulator on testability design, function, and performance ground test verification, and non-testable-item analysis and evaluation are included.

12.1.2.9 Ergonomic Requirements

In the process of use and maintenance, large space manipulator, spacecraft, and astronauts constitute a complete "human–robot-environment" system, which should meet the requirements of ergonomic design.

12.1.3 Constraints

The design constraints of the large space manipulators include the environmental conditions in launching segment and on-orbit working segment, as well as payload constraints during on-orbit operation.

12.1.3.1 Envelope Dimension Constraints

The large space manipulator is installed on the outer wall of the space station module. In the launch segment, the overall dimension is strictly limited by the size of the launch vehicle fairing; during on-orbit operation, its dynamics envelope dimension is mainly limited by the extravehicular equipment of the space station.

12.1.3.2 Mechanical Environment Constraints

Constraints on the mechanical environment are the main consideration in the launch segment.

12.1.3.3 Thermal Environment Constraints

Constraints on alternating temperature environment are the main consideration in on-orbit operation segment.

Table 12.2 On-orbit payload conditions of large space manipulator

Loads		Mass (kg)	Envelope size (mm)
Astronauts (including the human-spacesuit system)		300	Diameter $\leq$ 1000; Height $\leq$ 2000
Instruments and equipment	Medium standard payloads	$\leq$200	$\leq$600 × 600 × 500
	Small standard payloads	$\leq$100	$\leq$400 × 400 × 500
	Special payloads	$\leq$600	–
	Large cargo	$\leq$3000	Diameter $\leq$ 3000; Height $\leq$ 4000
Visiting spacecraft		$\leq$25,000	Diameter $\leq$ 4600; Height $\leq$ 20,000

12.1.3.4 Payload Constraints

There are three types of typical payloads for large space manipulators: astronauts, instruments and equipment (payloads), and visiting spacecraft. The approximate mass and dimensions of the payloads are shown in Table 12.2.

12.2 Overall System Design

12.2.1 Task Analysis

12.2.1.1 Analysis of Module Re-Docking and Docking Tasks of Large Spacecraft

The main operational object of large spacecraft module re-docking is the visiting spacecraft. Before re-docking, the main body module of the spacecraft and the visiting module are in a "straight line" configuration. After re-docking, they are in a perpendicular configuration. The special constraints on large space manipulators for large module docking and re-docking tasks include:

(1) Earth orientation of the visiting module, spacecraft, and manipulator assembly;
(2) Disturbance to the manipulator and the spacecraft attitude by the re-docking load;
(3) Operational safety and reliability of related product characteristics, such as the docking mechanisms;
(4) Operation time of the attitude change of the modules during the re-docking process.

12.2.1.2 Analysis of Capturing Visiting Module

According to the different task phases, the mission of capturing the visiting module in relative hovering state can be divided into three segments: accompanying flight segment, closing segment, and capturing segment. The manipulator performs capture operation in the capturing segment.

The special requirements for visiting module capture tasks of large space manipulators include:

(1) Collision with the spacecraft is not allowed at various phases of hovering module capture task;
(2) The manipulator shall ensure that the visiting module is in the initial position for docking after capture and before docking;
(3) The manipulator provides proper docking speed and docking force when assisting the docking process.

12.2.1.3 Analysis of Large Equipment Handling and Maintenance Tasks

In the process of spacecraft assembly and operation, the visiting module will be supplemented with a large number of instruments. Large space manipulator is not only responsible for taking out, handling, transferring various large equipment, but also responsible for the recovery of damaged equipment on the spacecraft and installation to the visiting module. After the visiting module is docked with the spacecraft, the large space manipulator (jointly with the other space manipulator on the spacecraft) performs the handling operation of large equipment.

Special requirements for large equipment handling and maintenance tasks for large space manipulators include:

(1) Combined control with the other space robotic arms in series;
(2) Meeting the requirements of smooth control during cargo handling operations;
(3) The influence of system dexterity on the operation under series connection of two arms.

12.2.1.4 Analysis of Supporting Astronauts' EVA Tasks

In order to improve the efficiency and safety of EVA, astronauts are secured on the end effectors of the large space manipulators for extravehicular activities during the normal operation of spacecraft. The spacecraft is equipped with astronaut access hatch. The astronauts go out of the module through the access hatch and are connected to the end effector of the manipulator by a special mechanism. The manipulator is operated by the astronauts in the module, and the astronauts are sent to the desired position to perform extravehicular maintenance activities.

Special requirements for large space manipulators supporting astronauts' EVA include:

(1) When the astronauts perform the EVA tasks, they should adjust the end position of the large space manipulator at any time to meet the operation and functional requirements of the extravehicular astronauts;
(2) The end effector of the large space manipulator shall provide the interface for astronaut foot restraint, which shall be convenient for astronauts to install and removal, and at the same time facilitate the astronaut's boarding/leaving the manipulator;
(3) Large space manipulators should provide safety facilities such as handrails for astronauts;
(4) Consider the maximum moving speed and acceleration that astronauts can withstand during motion planning of large space manipulators;
(5) Facilitate the autonomous and emergency transfer of astronauts under special conditions.

12.2.1.5 Analysis of Extravehicular Conditions Monitoring and Inspection Tasks

During the operation of spacecraft in orbit, it is necessary to regularly monitor and inspect the extravehicular conditions. In order to minimize the EVA of astronauts, the large space manipulator is usually used for routine monitoring and inspection activities, that is, the large space manipulator regularly performs traverse imaging to the extravehicular key parts by means of the visual cameras mounted at its end, and transmits the images to the module through the manipulator itself for observation and monitoring by astronauts. Important information is sent back to the ground through the measurement and control channel for monitoring and analysis.

When no mission is performed, the large space manipulator remains in the designated position on the spacecraft and maintains a specific normal configuration. Due to the limited length of the manipulator, it is impossible to cover all areas outside the spacecraft at the designated position. Therefore, the special requirements for extravehicular condition monitoring and inspection tasks for large space manipulators include:

(1) The large space manipulators should be able to perform alternating inch-worm type motion between the target adapters on the spacecraft;
(2) Service life and reliability requirements of related devices by frequent inch-worm type motions.

12.2.2 System Design

12.2.2.1 System Configuration and Layout

The configuration and layout of the large space manipulator system includes the Whole-arm installation layout, the arm configuration layout, and the hold-down point layout.

Fig. 12.1 Large space manipulator compressed state installation configuration

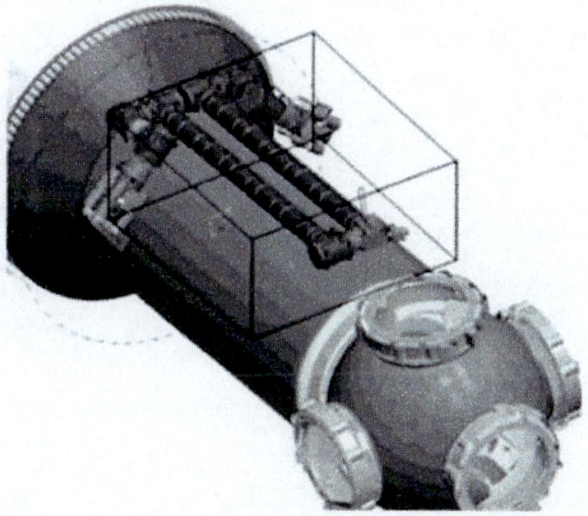

(1) Whole-arm installation layout

In order to meet the constraints of the launch vehicle envelope, the large space manipulator should reduce the height of the installation state as much as possible, and meet the requirements of the overall layout of the equipment on the spacecraft, such as avoiding the furled state envelope profile of the solar wing, avoiding the moment gyro and engine plume effects. Therefore, the furled configuration of the large space manipulator is designed to be pressed on the outer wall of the spacecraft, as shown in Fig. 12.1.

(2) Arm-body configuration layout

The large space manipulator is equipped with 7 joints, adopting the configuration of "3 + 1 + 3", that is, 3 DOFs in the shoulder (swing, yaw, pitch), 1 DOF in the elbow (pitch), and 3 DOFs in the wrist (pitch, yaw, swing). In the furled configuration, the DOFs of the shoulder and wrist of the manipulator are configured as a left-right mirror relationship. In addition, all joints adopt an offset scheme, that is, there is a certain distance between the axis of the two rods connected by the joint. This configuration relationship enables the two arm booms to be in the same plane when the large space manipulator is in the furled configuration, which is convenient for the whole arm to be held down; furthermore, the angle range of each joint is not restricted by the geometric interference of adjacent components (achievable of $\pm 180°$ or even larger range of rotation angle). The shoulder and the wrist swivel joints are connected to the end effector, so that the mechanical system of the whole manipulator is symmetrical between the shoulder and wrist, facilitating inch-worm type motion. The configuration layout of large space manipulator arm-body is shown in Fig. 12.2.

(3) Hold-down mechanism Layout

The hold-down mechanism layout of large space manipulator mainly considers its load-carrying performance in the launch segment, that is, to ensure the manipulator to be able to withstand the mechanical environment load during the launch segment

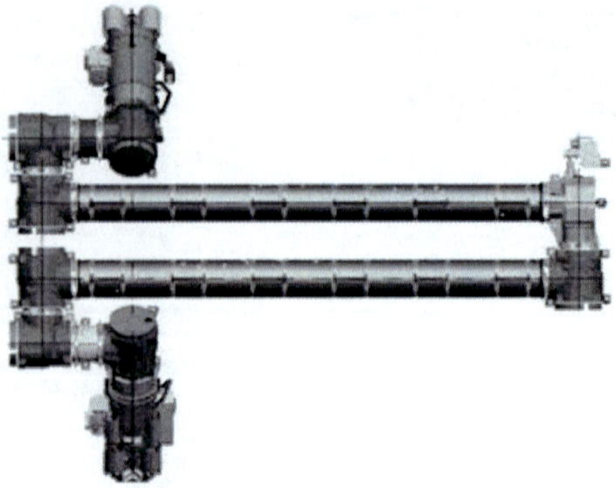

Fig. 12.2 Manipulator arm-body configuration layout

without strength damage or permanent deformation. According to the optimization results of mechanical analysis, the large-scale space manipulator adopts a hold-down and release solution of multi-point hold-down and linkage unlocking.

12.2.2.2 System Working Modes

The working modes of the large space manipulator system refer to the manipulator operating modes designed according to different working conditions. The working modes of the large space manipulator are shown in Table 12.3.

12.2.2.3 System Information Bus

The large space manipulator is installed outside the spacecraft module, and the manipulator console is installed inside the spacecraft module. The bus connection of the manipulator information system is designed as follows:

(1) The manipulator console realizes data interaction with the spacecraft digital management subsystem and the GNC subsystem through 1553B bus;
(2) The manipulator console uses 1553B bus to realize data interaction with the manipulator body connected to the base or the target adapter to control the operation of the whole arm;
(3) Various devices inside the manipulator use 1553B bus for data interaction;
(4) The video information of the camera is transmitted through the Ethernet, and transmitted to the measurement/control communication subsystem via the connected target adapters for processing;
(5) When large space manipulator and other space manipulator are combined, the combined arm uses 1553B bus for data interaction.

Table 12.3 Working modes of large space manipulator

No.	Working condition	Working mode	
1	Whole-arm standby braking state	Whole-arm standby brake mode	
2	Whole-arm servo standby state	Whole-arm servo standby mode	
3	Whole-arm follow-up motion state	Whole-arm follow-up mode	
4	Whole-arm servo control state	Whole-arm preprogramming mode	
		Whole-arm visual servo control mode	
		Whole-arm end motion mode	End parameter input linear motion
			End handle operation motion
		Single joint motion mode	Single joint parameter input motion (speed, position)
			Multi-joint linkage parameter input motion
			Single joint handle operation (speed)
		Whole-arm force control mode	

12.2.2.4 System Composition

Large space manipulator system mainly consists of the following three parts.

1. Mechanical system
   (1) Main functions: It constitutes the basic structure of large space manipulator and realizes the main configuration parameters of the system; keeps the whole arm held-down outside the spacecraft module during the launch segment, unlocks and releases it after entering the orbit; provides a mounting interface for the arm equipment and appropriate mechanical environment; receives control commands from the central controller; completes closed-loop control of joints and end effectors according to instructions, executes various motions; feeds back motion control results and status information to the central controller.
   (2) Main equipment: including the base and other structural parts, arm booms, joints, end effectors, target adapters, quick connection devices, hold-down and release mechanisms, etc.
2. Control system
   (1) Main functions: it analyzes the motion commands sent by the ground and astronauts; plans the motion trajectory of the manipulator, sets the motion

parameters, and controls the motion process of the manipulator; feeds back the status information of the manipulator to the ground and astronauts; manages system power supply and distribution, visual fusion, and various telecontrol command and telemetry information; sends terminate commands to the manipulator in case of emergency.

(2) Main equipment: manipulator console, central controller, emergency brake control device, etc.

3. Perception system

(1) Main functions: it performs visual imaging of target objects and other equipment; processes and transmits image information; identifies the position of markers on the target object, completes target position and attitude measurement; perceives end force/torque and feeds back the information; feeds back the measured information to the central controller.

(2) Main equipment: measuring cameras, visual markers, six-dimensional force/torque sensors, and Ethernet switches.

12.3 Mechanical System Design

12.3.1 Mechanical System Composition

Large space manipulator adopts 7-DOF redundant configuration scheme, as shown in Fig. 12.3. It includes 7 joints, 2 end effectors, 2 arm booms, and 1 hold-down and release mechanism. As a structural member, the housing of the central controller forms a part of the mechanical system, and the elbow camera in the visual camera system participates in the structural load in the launch segment.

The joints of large space manipulator are all rotating joints. In order to facilitate on-orbit maintenance and replacement, the joints adopt modular design and the structure is identical. Two arm booms are provided for connecting the shoulder joints, the elbow joints, the wrist joints with the housing of the central controller which is arranged at the elbow. The elbow visual camera is installed in a two-axis pan-tilt unit, which is installed on the central controller housing with a separate hold-down and release device.

12.3.2 Joint Design

The joints of large space manipulator feature high stiffness and large output moment, and its appearance is shown in Fig. 12.4. The joint integrates power supply, transmission mechanism, perception components, driving circuit, thermal control, and other functional units. The external interface includes mechanical connection interface,

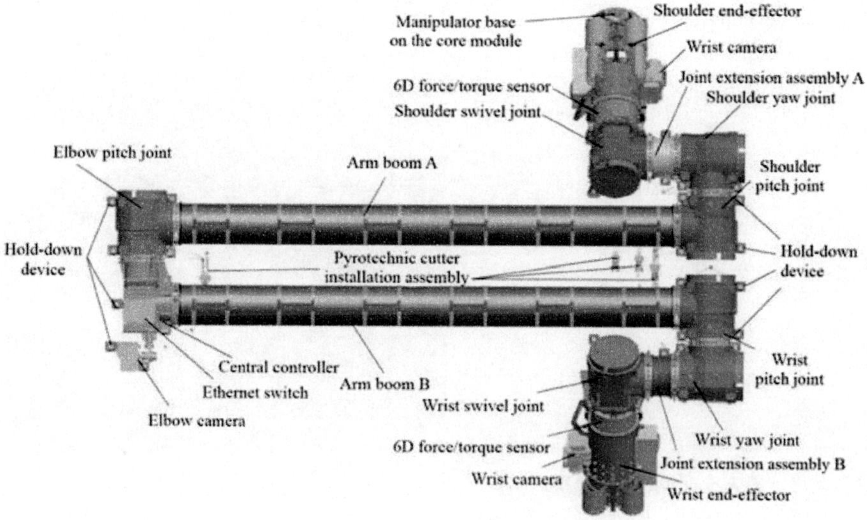

Fig. 12.3 Composition of the mechanical system of a large space manipulator

Fig. 12.4 Joint body configuration

power supply, and communication circuit interface. The joint adopts a multi-stage planetary drive system driven by two motors. Both the output shaft and the input shaft use resolvers as angle sensors. The mechanical interface uses an on-orbit repairable quick connection device to realize synchronized electrical and mechanical connection and ensure the rigidity, accuracy and reliability of the connection.

Fig. 12.5 Composition of the end effector

12.3.3 End Effector Design

The end effector of large space manipulator is mainly composed of a housing assembly, a capture module, a drag module, a locking module, an end controller, a power supply module, a six-dimensional force/torque sensor and a joint quick connection socket. The capture module and the drag module are located in the housing, the capture module is connected to the guide rail installed on the housing by a trapezoidal slider, and the drag module is fixed to the capture module. The locking module is fixed to the outer circumference of the housing. The six-dimensional force/torque sensor is mounted on the bottom of the end effector. The joint quick connection socket installed in the six-dimensional force/torque sensor realizes the quick connection function together with the joint quick connection plug mounted in the wrist/shoulder joint. The composition of the end effector is shown in Fig. 12.5.

12.3.4 Arm Boom Design

There are two sets of arm booms with basically the same structure and dimensions and completely symmetrical in configuration, the only difference is the cable installation phase. Each arm boom is composed of a boom body, a boom cable assembly, a boom thermal control assembly, and a boom quick connection plug. The boom body is composed of a housing, an end flange, an armrest connector, a lifting connector, a cable connector, and a grounding connector. The housing consists of a bare shell and a circumferential reinforcing rib, which are integrally solidified by carbon fiber. The remaining connectors are connected to the housing as a whole by the glued and threaded connection. In the boom body, the boom is the main force bearing and transmitting member, which determines the rigidity of the entire boom; the end flange provides the interface between the boom assembly and the movable joints.

12.3.5 Design of Hold-Down and Release Mechanism

There are three types of hold-down and release mechanism of large space manipulator: linkage-rod type hold-down mechanism, independent-rod type hold-down mechanism (including an end hold-down mechanism and an avoidance-type hold-down mechanism), and linkage-type mechanism. The layout of the hold-down and release device is shown in Fig. 12.6. The hold-down and release device consists of a total of 10 sets and 41 hold-down points (dotted circles in the figure).

Considering that each hold-down point of the independent-rod type hold-down mechanism adopts separate nuts to loosen respectively, in order to reduce the number of pyrotechnic devices and improve the product reliability, the adjacent products in the same plane can adopt a linkage-rod hold-down mechanism to realize interlocked release by the linkage mechanism. Each hold-down mechanism is connected with the linkage mechanism through a linkage rope, and each linkage mechanism is equipped with a pyrotechnic cutter for unlocking. As shown in Fig. 12.6, joints 2 and 3, joints 5 and 6, joint 4 and the controller are adjacent to each other, which can be linkage released by linkage-rod type hold-down mechanism.

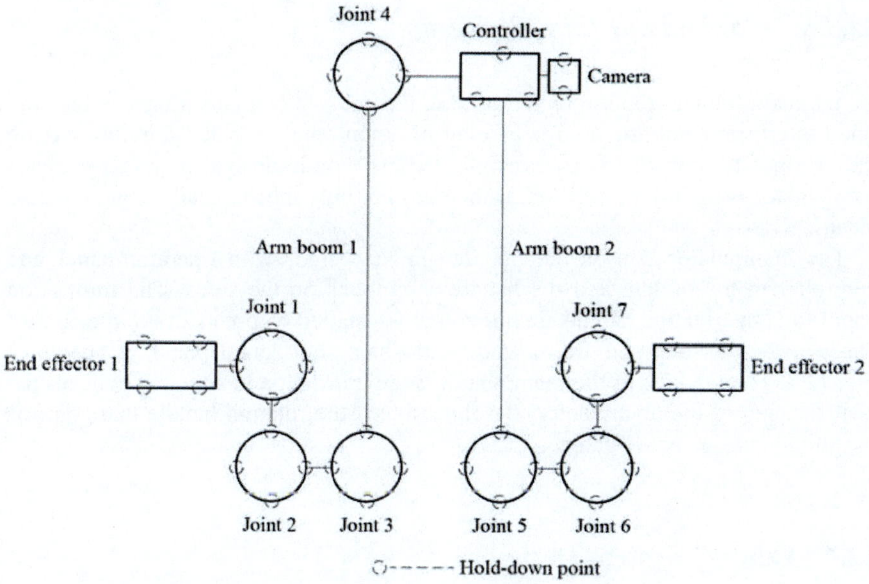

Fig. 12.6 Schematic diagram of the manipulator hold-down and release device

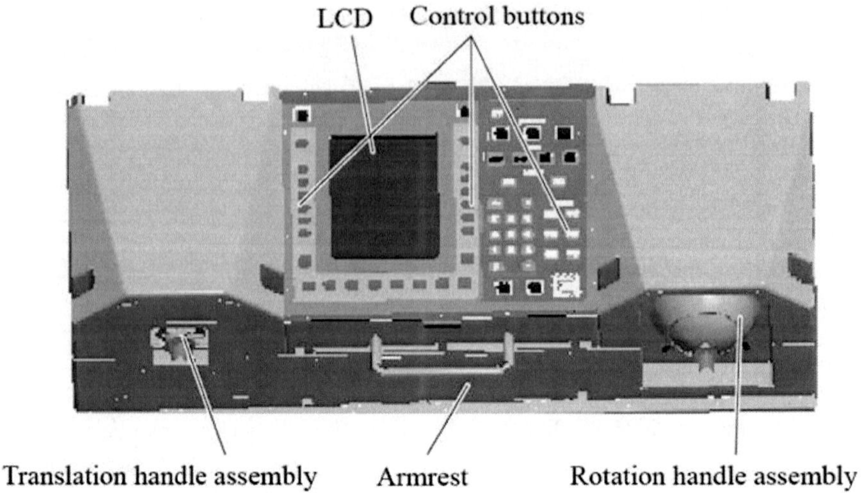

Fig. 12.7 Overall layout of the manipulator console

12.3.6 Manipulator Console Design

As the main human–robot interaction tool, the console contains a unique human–robot interface, involving a large amount of ergonomic designs. Therefore, taking the manipulator console as an example, the ergonomic design of the large space manipulator is mainly introduced. It includes not only mechanical design, but also electrical design.

The manipulator console adopts integrated design, with operator panel and controller in the middle and the handle assemblies on the sides. The translation handle and the rotation handle are horizontally installed on both sides of the control panel of the console, and are located on the same horizontal plane, as shown in Fig. 12.7. The left side of the manipulator is the translation handle that controls the displacement of the manipulator; the right side is the rotation handle that controls the attitude of the manipulator.

12.4 Control System Design

12.4.1 Composition of the Control System

The control system of large space manipulator uses distributed system architecture, including a command scheduling layer, a motion planning layer and an execution control layer. The control system consists of a manipulator console, a central

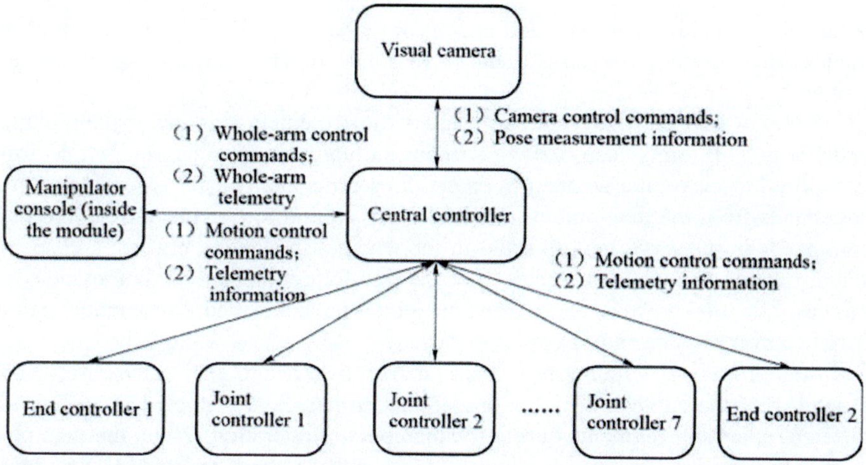

Fig. 12.8 Block diagram of the hardware composition of the manipulator control system

controller, a joint controller, an end controller, a visual camera, and an Ethernet switch, as shown in Fig. 12.8.

The manipulator control system can be operated and controlled under various on-orbit tasks. The main functions of each controller of space manipulator are as follows:

(1) Central controller: According to the status of the manipulator, camera measurement data, force sensor data, and operational control commands, the path planning algorithm is called to generate motion control commands and sends them to the joint controller; measures are taken to protect the whole manipulator in case of failure; camera pan-tilt unit control is performed; the health management for the whole manipulator is implemented.
(2) Joint controller: it controls the joint motion according to the control command of the central controller to ensure the accuracy of joint motion.
(3) End controller: it controls the motion of the end effector mechanism according to the control command of the central controller to complete capture or releasing operation of the target.
(4) Visual camera: According to the control command of the central controller, the high-quality video images of the target in the working range is obtained, and the video image is compressed and output to the instrument subsystem for display. The pose data of the target is calculated and output to the central controller.

12.4.2 System Control Strategy

The operating state of the whole-arm mainly includes the standby braking state, servo standby state, servo motion state, and follow-up state. The servo motion state is further divided into various motion control modes according to different motion modes and input sources, such as visual servo, single joint motion, multi-joint linkage motion.

After powered on, the whole-arm automatically enters standby braking state, under which all joints keep braking without motion control. When the whole-arm is required to move, the whole-arm enters the servo standby state according to the commands from the manipulator console or the ground to execute various motion commands in the servo motion state. In the whole-arm standby braking and servo standby states, the whole-arm can enter the free follow-up state by commands. In this case, the joint brake is released and the joint is not controlled, so the manipulator is free to move by the end force.

In the process of whole-arm motion control, in order to ensure the safety and reliability of the control, different operational strategies are adopted according to different operating segments during the manipulator operation. When the manipulator is performing typical operation tasks, the motion space is divided as follows and the corresponding algorithm is designed:

(1) Far-distance segment (free zone): The far-distance segment is controlled by preprogramming mode to ensure that no trajectory deviation occurs at the end of the manipulator, and the manipulator meets the requirements for position and speed at the initial point of the approaching segment as it arrives the end of the far-distance segment.
(2) Approaching segment (adjacent zone): The approaching segment is controlled according to the relative distance information of the manipulator end effector and the target. The control system judges the target position by the measuring camera and tracks the axial speed of the manipulator end to determine the relative axial speed of the end and the target. When the relative speed is lower than a certain threshold, the approaching process ends.
(3) Capturing segment (capturing zone): The capturing segment is the phase when the manipulator end effector captures the target and completes the connection. Since the manipulator end effector contacts with the target, there is a contact force between them. In this phase, the design is focused on the successful capture of the target, while considering the protection measures for the manipulator and the target under various abnormal conditions (excessive contact force, capture failure, etc.).

12.4.3 Control System Hardware Design

12.4.3.1 Central Controller

The electronic component of the central controller adopts a dual-system cold backup working mode. Each system hardware module includes a secondary power module, a CPU processor and peripheral circuits (including a memory, an algorithm control circuit, a logic control circuit, a 1553B bus interface, etc.), an A/D acquisition module,

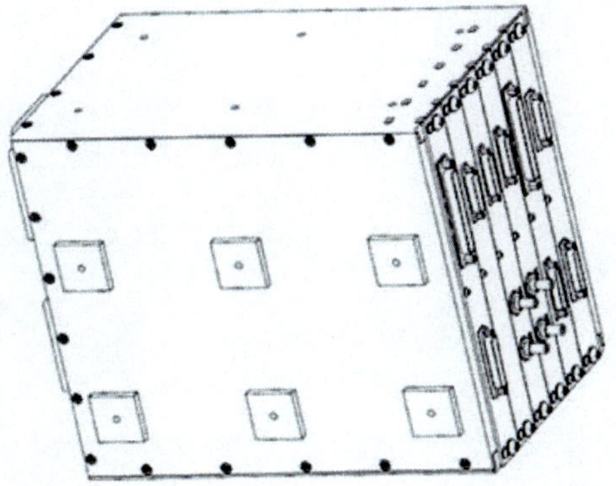

Fig. 12.9 Appearance of the central controller

a command thermal control module, a pan-tilt unit control module. The appearance of the central controller is shown in Fig. 12.9.

(1) The control board of the pan-tilt unit mainly realizes the stepper motor drive control and photoelectric switch control. The cold backup mode is adopted, and the main and backup circuits are redundant to each other. The main circuit controls the main motor, and the backup circuit controls the backup motor. The two circuits are exactly identical.
(2) The power board mainly realizes the conversion from primary power to secondary power, controls the power on/off of the main and backup systems by direct commands, and outputs telemetry status parameters of the main and backup systems of the equipment.
(3) The central controller has a main and a backup board with the completely same status. It mainly consists of a power supply circuit, a clock circuit, a reset circuit, a watchdog circuit, an internal bus drive circuit, a CPU and peripheral circuits, an FPGA module, and peripheral circuits and other circuit modules to achieve the calculation and control functions of the equipment.
(4) The A/D acquisition board mainly completes external temperature acquisition, external analog quantity acquisition, and the equipment internal state acquisition.
(5) The thermal control command board cooperates with the CPU board to control the heating circuits of the thermal control system and send OC control commands.

12.4.3.2 Joint Controller Hardware Solution

The joint controller adopts cold backup mode, which means the main and backup circuits are mutually redundant. The main circuit controls the main motor, and the backup circuit controls the backup drive motor. At any time, only one motor works, but both motor brakes need to be on or off at the same time. The switching of the main and backup circuits is realized by the remote command of the central controller. The

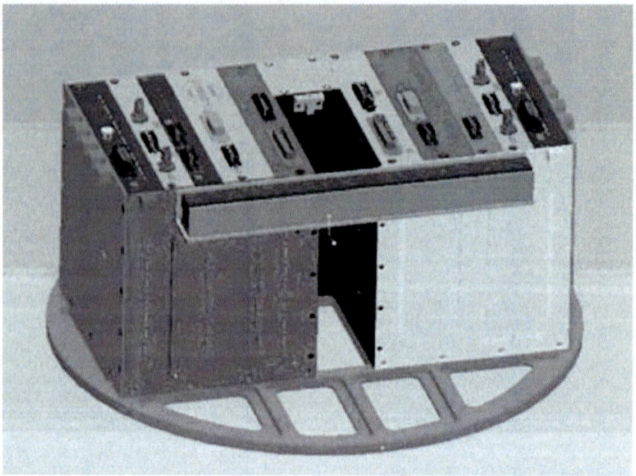

Fig. 12.10 Appearance of the joint controller

appearance of the joint controller is shown in Fig. 12.10. It mainly includes power board, CPU circuit board, FPGA circuit board, motor driver board, signal processing board, and connection structure.

(1) The power board mainly realizes the conversion from primary power to secondary power, it includes an EMI filter module, a DCDC power conversion module, a bus power protection circuit, a bus power on/off control circuit, a motor power protection circuit, a motor bus current acquisition circuit, and a thermal control processing circuit, etc.
(2) The CPU board is mainly composed of power conversion circuit, CPU circuit, PROM circuit, SRAM circuit, EEPROM circuit, and watchdog circuit, which mainly performs the management of a joint controller.
(3) The FPGA board mainly realizes the servo control of the joint motor, it includes power conversion circuit, FPGA circuit, FLASH circuit, 1553B interface circuit, and a signal level drive conversion circuit.
(4) The motor drive board is mainly used to drive the motor, including motor drive circuit, motor power supply filtering circuit, motor winding switching circuit, motor current acquisition circuit, and motor brake control circuit.
(5) The signal processing board mainly completes the solution of the resolver signal, including a power conversion circuit, a resolver excitation signal circuit, a resolver solution circuit, a revolver signal switching circuit, and a signal level drive conversion circuit.

12.4.3.3 End Controller Hardware Solution

The end controller adopts a cold backup design, and the main and backup circuits are mutually backup. It consists of the following parts: a power module control circuit, a power conversion circuit, a motor power supply control circuit, a thermal control circuit, and a band-type brake (brake control) circuit, a power drive circuit, a signal acquisition circuit, a telemetry parameter acquisition circuit, a remote command generation circuit, a FPGA circuit, the CPU and peripheral processing circuits, a 1553B bus interface circuit. The shape of the end controller is shown in Fig. 12.11. It mainly includes a power board, a motor drive board, a telemetry remote control board, a CPU board module, and the connection structure.

(1) The power board mainly realizes the conversion from primary power to secondary power, including an EMI filter module, a DCDC power conversion module, a bus power protection circuit, a bus power on/off control circuit, and a motor power protection circuit.
(2) The motor drive board mainly completes the drive of the motor, including a motor drive circuit, a motor power supply filtering circuit, a motor winding switching circuit, a motor current acquisition circuit, and a motor brake control circuit.
(3) The telemetry remote control board mainly realizes the motor control, command transmission, telemetry parameter acquisition, in-position switch state acquisition, including a switch Hall signal acquisition circuit, an in-position switch acquisition circuit, a telemetry parameter acquisition circuit, a remote command generation circuit, a FPGA circuit, etc.
(4) The CPU board mainly performs the management of the complete end controller, including a power conversion circuit, a CPU circuit, a FLASH circuit, a 1553B interface circuit, a watchdog circuit, and a signal level drive conversion circuit.

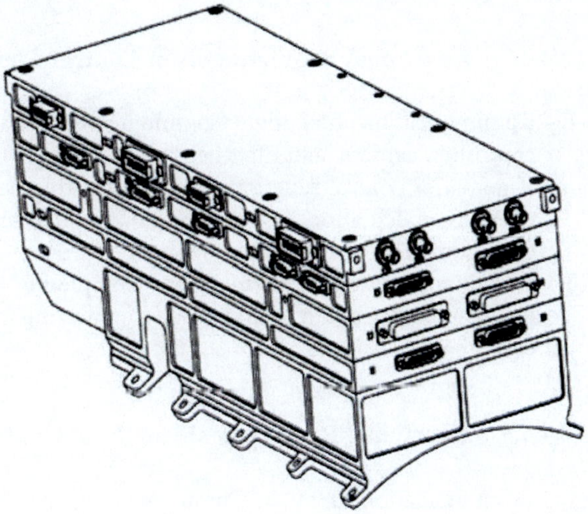

Fig. 12.11 Appearance of the end controller

12.4.4 *Control System Software Design*

The software of the manipulator control system mainly runs on the manipulator console, the central controller, the joint controller, the end controller, and the visual camera. The functional requirements for equipment software are as follows:

12.4.4.1 Functional Requirements of Manipulator Console Software

(1) Communication function: Connecting with the space station digital management system and the GNC subsystem by 1553B bus to realize data communication; communicating with the laptop computer in the module through Ethernet; communicating with the manipulator controller through 1553B bus.
(2) Human–robot interaction function: Providing the operation interface for the manipulator, including the switch button, the control panel, and the operation handle, enabling the astronaut to realize the control of the state and pose of the manipulator; the console should adopt a misoperation proof design and meet ergonomic requirements.
(3) Display function: Displaying the operation data, health status data, and video images during operation on the instrument display and providing visual feedback for the operation of the astronauts. The console shall display operational instructions and data during the operation of the astronauts.

12.4.4.2 Functional Requirements of Central Controller Software

(1) Communication function: Communicating with the joint controller, end controller, camera, and Ethernet switch through 1553B bus.
(2) Kinematics control function: Completing kinematics calculation and collision detection calculation of linear motion, visual autonomous motion, and force control motion modes.
(3) Whole-arm management function: Completing the switching and control of the whole-arm control mode, and monitoring and controlling the status of equipment.

12.4.4.3 Functional Requirements for Joint Controller Software

(1) Communication function: Communicating with the central controller through 1553B bus, and outputting the motor current, speed, and position information to the central controller.
(2) Servo control function: Controlling the joint motor according to the command of the central controller to perform closed-loop control of the position, speed, and current feedback.

(3) Position sensor solving function: Collecting the output voltage signal of the speed measuring resolver and the position measuring resolver, performing signal solving, and obtaining corresponding position information for closed-loop control of the motor and joint.

12.4.4.4 Functional Requirements of End Controller Software

(1) Communication function: Communicating with the central controller through 1553B bus, and outputting the motor current, speed, and position information to the central controller.
(2) Servo control function: Controlling the end motor according to the command of the central controller to perform closed-loop control of the speed and current.
(3) Force sensor signal solving function: Collecting the force sensor signal and solving it.

12.4.4.5 Functional Requirements of Visual Camera Software

(1) Communication function: Communicating with the central controller through 1553B bus, and outputting information such as pose measurement results to the central controller.
(2) Measurement function: Detecting, identifying, and continuously measuring the space cooperative target (target adapter visual marker), and outputting the three-dimensional pose information of the target adapter visual marker.
(3) Monitoring function: Monitoring the working area of the manipulator and outputting corresponding image information.

12.5 Perception System Design

12.5.1 Composition of Perception System

The perception system of space manipulator mainly includes an external visual measuring system, a force sensor, and internal position sensors, limiting sensors, and temperature sensors. The following will focus on the introduction of the external visual measurement system, as shown in Fig. 12.12.

(1) Wrist camera (integrated light source): It is installed in the end effector of the manipulator to realize the target perception and measurement of the manipulator end, and assists the visual servo control function.
(2) Elbow camera (integrated light source) and elbow camera pan-tilt unit (including the pitch and yaw mechanisms): the elbow camera and pan-tilt unit assembly

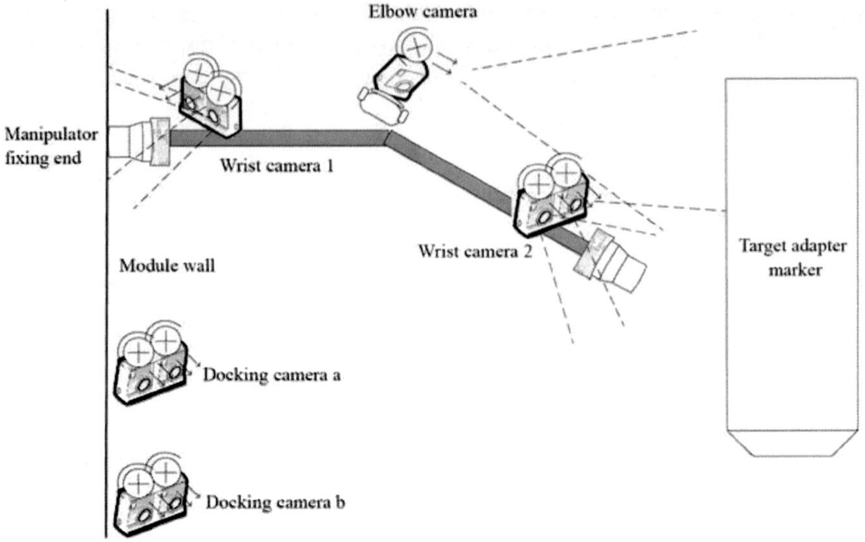

Fig. 12.12 External visual measurement system

are mounted on the central controller housing of the manipulator for monitoring the environment around the manipulator.

(3) Docking camera (integrated light source): It is installed near the spacecraft's docking port for measuring the initial docking pose of the visiting spacecraft.
(4) Visual markers: including capture visual marker and docking visual markers, installed on the target adapter and the docking port of the visiting spacecraft, respectively, to provide the target pose measurement reference for the wrist camera and the docking camera.
(5) Ethernet switch (referred to as switch): It is installed in the central controller of the manipulator for the transmission and switching of camera data.

12.5.2 Visual Measurement System Strategy

When the space manipulator is working on orbit, it will experience various stages from the long-distance target positioning to the medium-distance target approaching the close-distance precision operation on the target, and each stage requires the visual measurement subsystem. In the middle and long distances, the visual measurement subsystem only needs to monitor the working area of the manipulator; at close range, the visual measurement subsystem should be able to output high-precision three-dimensional poses of the target, and guide the end effector of the manipulator to further adjust its own pose, so as to gradually reduce the rotational and translational differences between the end effector coordinate system and the target adapter coordinate system, and finally guide the end effector to enter the capture area of the target adapter and complete the capture operation on the target adapter.

During the on-orbit operation of the manipulator, the elbow camera is far away from the module and can overlook the area under the manipulator. There is no obstacle in the observation field, and the elbow camera can completely observe the operation of the manipulator. When the visual measurement system is running, the manipulator uses the elbow camera to perform safety inspections before manipulator start, working condition monitoring during manipulator operation, and monitoring of module re-docking and secondary docking process. During the motion of the manipulator, by controlling the motion of the elbow camera pan-tilt unit, the elbow camera can maintain monitoring of the target all the time.

When the effective observation area of the target adapter enters the field of view (FOV) of the wrist camera, the manipulator uses the wrist camera to obtain high-precision, three-dimensional pose information of the target, and controls the end effector to further approach the target adapter, and completes the high-precision and high-stability operations, such as grasping of the target adapter.

12.5.3 Visual Measurement System Information Bus Design

The visual measurement system information flow interface mainly includes a 1553B bus interface and an Ethernet bus interface.

The target pose information generated by each camera is sent to the central controller through 1553B bus. The central controller performs path planning according to the target pose and sends motion control commands to each joint to form a visual closed-loop control. The central controller collects the motion information of the pan-tilt unit and obtains the target pose information and camera status information from each camera, and sends them to the manipulator console through 1553B bus, which is further sent to the digital management subsystem for processing and transmits down to the ground to be displayed through the instrument and lighting subsystem. The information flow of the visual measurement system is shown in Fig. 12.13.

12.5.4 Visual Measurement System Hardware Design

The wrist camera is composed of an optical lens (including a hood), a camera circuit box, and a lighting assembly (including a light source and a light source driving circuit box), as shown in Fig. 12.14. The arrangement sequence in the camera circuit box is (from front to back): a focal plane circuit, a signal processing circuit, a video compression circuit, an image preprocessing circuit, a pose calculation circuit, and a power control circuit.

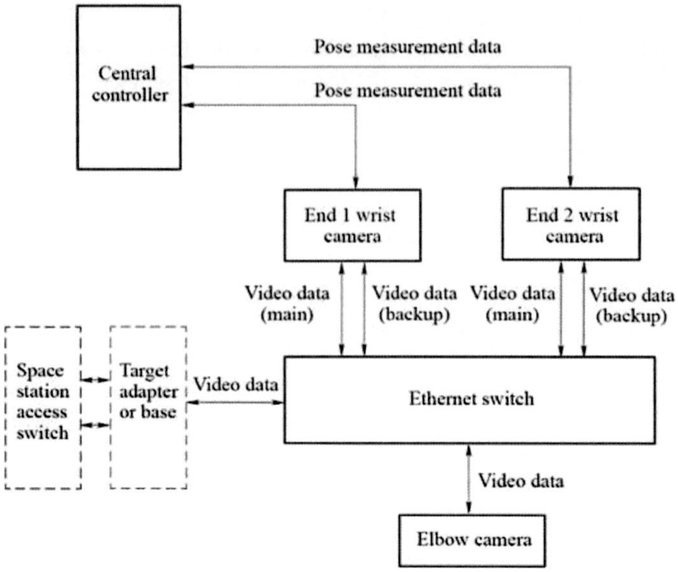

Fig. 12.13 Information flow of a visual measurement system

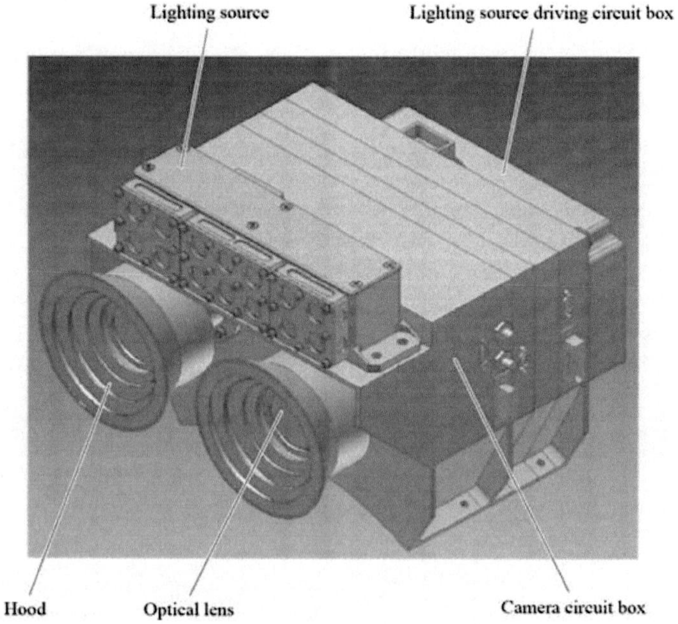

Fig. 12.14 Appearance of a wrist camera

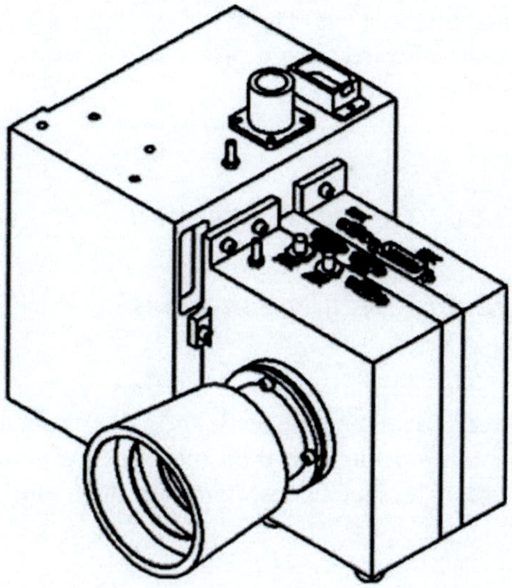

Fig. 12.15 Appearance of an elbow camera

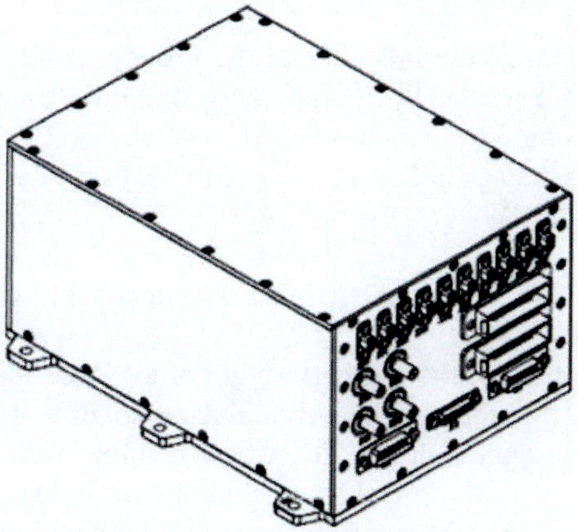

Fig. 12.16 Appearance of the Ethernet switch

The composition and interface of the elbow camera are similar to those of the wrist camera, as shown in Fig. 12.15.

The Ethernet switch adopts a four-layer plug-in structure, as shown in Fig. 12.16.

12.5.5 Visual Measurement System Software Design

The visual measurement system software mainly includes the embedded software operating in the DSP and FPGA processing chips to drive the hardware and realize system functions, and performs designated tasks and communicates with 1553B bus and Ethernet bus. These softwares include camera control software, video compression software, image processing software, pose calculation software, and switch software, etc.

12.6 Design Verification

12.6.1 Verification Items

Large space manipulator systems are required to complete the basic performance tests, basic function tests, special tests such as inch-worm type motion, deployment and re-docking, on-orbit maintenance verification tests, and task-level verification tests. The specific test items are shown in Table 12.4.

12.6.2 Verification Scheme

Due to the influence of the ground gravity environment, large space manipulators are verified by a combination of multiple methods, such as mathematical simulation, semi-physical simulation, and physical tests. Physical tests are divided into two schemes: plane motion verification and three-dimensional motion verification.

12.6.2.1 Plane Motion Physical Test Verification

Because the size and weight of the large space manipulator and its payload are very large, the microgravity simulation of its wide range of motion tests uses an air-floating support solution. The whole manipulator system is designed to be supported by 6 air bearing brackets, including: 1 bracket each under the wrist joint assembly, the elbow joint assembly, and the shoulder joint assembly, 2 brackets under arm boom 1 assembly (the higher arm boom on the air-floating platform) 1 bracket under the arm boom 2 assembly, as shown in Fig. 12.17. The air bearing equipment ensures that the manipulator can achieve the ideal microgravity environment in the horizontal plane, and support it to achieve a large range of motion in the horizontal plane. The disadvantage is that it cannot meet the requirement of three-dimensional mechanical motion verification.

Table 12.4 Large space manipulator system test items

No.	Items	Sub-items
1	Function test	Large load, wide range transfer function
		Multi-DOF motion function
		Multi-mode control function
		Multi-mode operation function
		Path planning function
		Visual servo test
		Emergency stop function
2	Performance test	Load-carrying capacity
		Working space radius
		End speed, end acceleration
		End pose accuracy test
		End force/torque test
		Stiffness test
		Braking distance test
3	Special test: inch-worm type motion	
4	Special test: three-dimensional deployment	
5	Special test: module re-docking	
6	Combined test of large and small manipulators	
7	On-orbit maintenance verification test	
8	Task-level verification test	Manipulator re-docking and docking task verification test
		Combined arm pallet handling task verification test
		Re-docking and docking semi-physical simulation verification test
		Capture of hovering spacecraft task test
		Assisting astronaut activity task test

Fig. 12.17 Physical test verification method for the plane motion of the manipulator system

12.6.2.2 Physical Test Verification of Three-Dimensional Motion

The whole system of a large space manipulator is required to test its three-dimensional motion ability in a certain range to verify that it meets the requirements of three-dimensional motion operation. Therefore, a suspended three-dimensional motion zero-gravity simulation test equipment is used to provide the simulated zero-gravity environment for the manipulator. The system has the ability of force, motion speed, and displacement measurement and compensation to ensure that the manipulator is not damaged by additional forces and moments during motion, and minimizes the effect of the zero-gravity simulation test equipment on the motion characteristics of the manipulator. The three-dimensional test system has five lifting points, including two for wrist and shoulder, and one for the elbow. Each lifting point has two-dimensional follow-up ability in the horizontal plane and a gravity compensation function in the vertical direction, thereby allowing the three-dimensional motion of the suspended portion of the manipulator. The disadvantage of this scheme is that there is motion interference between the lifting points, which does not support the wide range motion of the manipulator.

12.6.2.3 Semi-physical Simulation Verification of Robot-Assisted Docking Tasks

When the manipulator captures a module or assists the module for re-docking and docking, the capture mechanism or the docking mechanism will certainly have a contact collision during the process. Due to the complicated contact collision process, it is difficult for the contact collision model to accurately simulate the physical process, resulting in a certain error in the simulation verification; in addition, it is difficult for large space manipulator to achieve a wide range simulated zero-gravity three-dimensional motion on the ground. For this reason, in order to ensure the verification accuracy, a semi-physical simulation verification platform for large space manipulator on-orbit task has been developed to conduct semi-physical simulation verifications of the tasks such as inch-worm type motion and capture of visiting module by the manipulator.

The semi-physical simulation verification system platform for manipulator on-orbit tasks mainly includes space manipulator simulated robot, target simulated robot, end effector, and target adapter installed at the ends of the two robots, and large space manipulator system simulation software (including the manipulator model, the main body module model for manipulator installation, and the target object model). The simulation software of the manipulator system performs simulation calculation according to the motion command issued by the central controller. The real-time simulation result data is used to drive the two robot's motion, so that the end effector and the target adapter installed at the end of the robot can reproduce the actual motion state and contact collision state of the manipulator and the target object.

This system includes the mathematical models of the large space manipulator, the cabin modules and the target object, and includes physical models of the end effector and the target adapter. Since the motion of the two robots follows the simulation results, it is ensured that the motion of the end effector and the target adapter is consistent with the simulation result, and that the physical contact collision state between them is completely consistent with the state of the model calculation. The contact collision force is obtained by the force sensor and fed back into the mathematical model for the next simulation calculation. Through the continuous iteration of the above process, the semi-physical simulation verification of a large space manipulator can be realized.

References

1. B. Siciliano, O. Khatib, *Handbook of Robotics* (Springer, New York, 2007)
2. H.J. Cruijssen, M. Ellenbroek, M. Henderson et.al., The European robotic arm: a high-performance mechanism finally on its way to space, in *Proceedings of the 42nd Aerospace Mechanisms Symposium* (NASA Goddard Space Flight Center, Leiden, The Netherlands, 2014), pp. 14–16
3. J. Zhou, Chinese space station project overall vision. Manned Spaceflight **19**(2), 1–10 (2013)
4. D. Li, W. Rao, C. Hu et al., Key technology review of the research on the space station manipulator. Manned Spaceflight **20**(3), 238–242 (2014)

Chapter 13
Design Example of Planetary Exploration Mobile Robot

13.1 Overview

13.1.1 Engineering Background

Mobile robot is the main tool for wide range exploration of extraterrestrial surfaces. At present, the successful applications of mobile robot are mainly to explore the surface of lunar and Mars.

The earliest application of planetary exploration mobile robot was first seen in the 1970s, when the Soviet lunar rover Lunokhod1 and Lunokhod2 successfully landed on lunar in 1970 and 1973, which symbolized that mobile robot has entered into practical applications. On July 30, 1971, the United States manned lunar rover LRV carried by "Apollo 15" landed on lunar, which was the world's first manned mobile robot to realize space applications. In 1971, the Soviet Mars rover PROPM landed on Mars with the Soviet automatic research station explorer "Mars 3", which was the world's first mobile robot for Mars exploration [1, 2].

From 1997 to 2004, the United States successfully launched the Mars rovers "Sojourner", "Spirit", and "Opportunity". In 2012, the United States Mars rover "Curiosity" successfully landed on Mars. In 2013, the Chinese rover "Yutu" successfully landed on the surface of the moon [3].

These planetary exploration robots all adopt the wheeled movement scheme. This chapter introduces the design and verification of the planetary exploration wheeled mobile robots with an engineering example.

Y. Wang, *Space Robotics*, Space Science and Technologies,
https://doi.org/10.1007/978-981-15-4902-1_13

13.1.2 Design Requirements

13.1.2.1 Task Requirements

The main tasks of the planetary exploration mobile robot are to roam and explore the planet. In order to adapt to complex terrains of the planet surface and harsh environmental conditions, multi-mode motions should be realized according to the requirements of scientific exploration under the constraints of limited energy. The tasks of the planetary exploration mobile robot are described as follows:

(1) With the assistance of the landing platform and the transfer mechanism, the robot can safely separate itself from the lander and land on the planet surface;
(2) The robot can roam on the planet surface;
(3) The robot can adapt to various terrains such as flat surface, hill slope, rock, and dune on the planet surface. Moreover, in case of malfunctions or special conditions, the mobile performance and fault-tolerant ability of the robot can be improved by vehicle lifting, wheel lifting, and other extended functions.

13.1.2.2 Functional Requirements

(1) Hold-down function at pre-landing flight phase

The planetary exploration mobile robot is in contact with and held-down on the landing platform when in the furled launch state. The robot should have sufficient strength and rigidity to withstand various loads and environmental conditions encountered during launch, orbital transfer, and off-orbit landing.

(2) Function of configuration deployment and leaving platform after landing

The planetary exploration mobile robot changes from the furled launch state to the normal mobile state, and moves from the landing platform to the planet surface via the transfer mechanism. The robot cooperates with the transfer mechanism to safely drive off the landing platform without dangerous situations such as vehicle overturning.

(3) Wide range movement function on the planetary surface

Under the control of the ground, the planetary exploration mobile robot can move safely and smoothly on the planet surface with various steering capabilities such as in situ steering and moving steering and a certain terrain adaptability, including climbing and obstacles crossing.

(4) Function of extrication wheel from sinkage on the planetary surface

The suspension of planetary exploration mobile robot should have the ability to change the configuration. Under the control of the ground, the vehicle body and the

Table 13.1 Main technical indices of planetary exploration mobile robots

Technical index	Requirement
Mass/kg	≤40
Load-carrying capacity/kg	≥200
Mobile capability/km	5
Maximum speed/m·h^{-1}	200
Obstacle crossing capability/mm	300
Climbing capability/°	30
Body lifting height range/mm	0–500
Wheel lifting height range/mm	0–300

wheels can be lifted by suspension adjustment to improve the moving performance and meet wheel sinkage extrication requirement.

13.1.2.3 Performance Requirements

The main technical indices of planetary exploration mobile robot are shown in Table 13.1.

13.1.2.4 Interface Requirements

(1) Mechanical interface

The mobile robot is screw-connected to the base plate and side plates of the planetary rover structure, and is connected to the landing platform by a hold-down mechanism. The mechanical interface is mainly related to such parts as the main shaft, the differential, and the hold-down base.

(2) Electrical interface

The data management subsystem provides a 23–29 V power for the mobile robot with a bus. During the planet surface operation, the average power consumption of the mobile robot does not exceed 50 W, and the peak power consumption does not exceed 100 W.

(3) Thermal interface

The thermal control subsystem provides temperature measurement and heating functions for the mobile robot. Each mechanism of the mobile robot is equipped with a thermistor for temperature measurement, and a 7-way heating circuit is provided to implement active thermal control for the mobile robot mechanisms, so that the mobile robot can adapt itself to the planet surface temperature environment.

(4) Information interface

The data management subsystem provides a RS422 bus communication interface for mobile robots for system telemetry and remote control information transmission.

13.1.2.5 Reliability Requirement

The reliability of the robot throughout the life cycle is not less than the required value.

13.1.3 Constraints

The planetary exploration mobile robot should be designed under the consideration of the boundary conditions, such as the constraints of the launch vehicle on the mass and envelope dimension of the robot, and the constraints of the planetary surface temperature environment on the mechanism lubrication modes.

13.1.3.1 Envelope Dimension Constraints

The overall dimensions of the planetary exploration mobile robot are constrained by the launch vehicle, the robot configuration, and the installation mode.

13.1.3.2 Mechanical Environment Constraints

The mechanical environment constraints mainly include those in the launch phase, the descending and landing phase, and the planet surface working phase.

13.1.3.3 Thermal Environment Constraints

It should be mainly taken into account the storage temperature environment in the flight phase and the high- and low-temperature environmental constraints in the planet surface working phase.

13.1.3.4 Special Constraints of Planet Surface

(1) Atmospheric environment

There may be an atmosphere on the surface of the planet. The effects of the atmosphere on the materials and devices of the planetary exploration mobile robots should be considered. In addition, due to the presence of the atmosphere, there may be wind and dust storms on the planet surface, thus, there may be dust accumulation in local areas of the planet surface and form an extremely soft environment, which may cause the robot wheels to sink into the dust.

(2) Gravity environment

The gravity acceleration on the planet surface may be different from that on the Earth, which may cause the kinematic mechanical properties of the planetary exploration mobile robot on the planet surface different from those on the Earth ground.

(3) Planetary dust environment

When there is wind on the planet surface, the dust raised by the wind will cause wear of the mechanical devices of the planetary exploration mobile robot, and discharge between the robot and the planetary atmosphere due to the charge accumulation on the robot. There may be static electricity in the planetary dust, and electrostatic charging may occur on the robot during its movement. When the charging reaches a certain threshold, discharging may occur, which will have an influence on the sensitive components of the robot.

13.2 Overall System Design

13.2.1 Task Analysis

13.2.1.1 Driving off the Landing Platform

After landing, the planetary exploration mobile robot is located on the surface of the landing platform. It needs to be moved from the landing platform to the planet surface through the transfer mechanism, and then roams and explores on the planet surface. The robot driving off the landing platform is as shown in Fig. 13.1.

The requirements on the robot by this task are as follows:

(1) The robot autonomously changes the hold-down and furled configuration to normal moving configuration;
(2) The robot moves from the landing platform to the transfer mechanism and moves to the planet surface via the transfer mechanism;

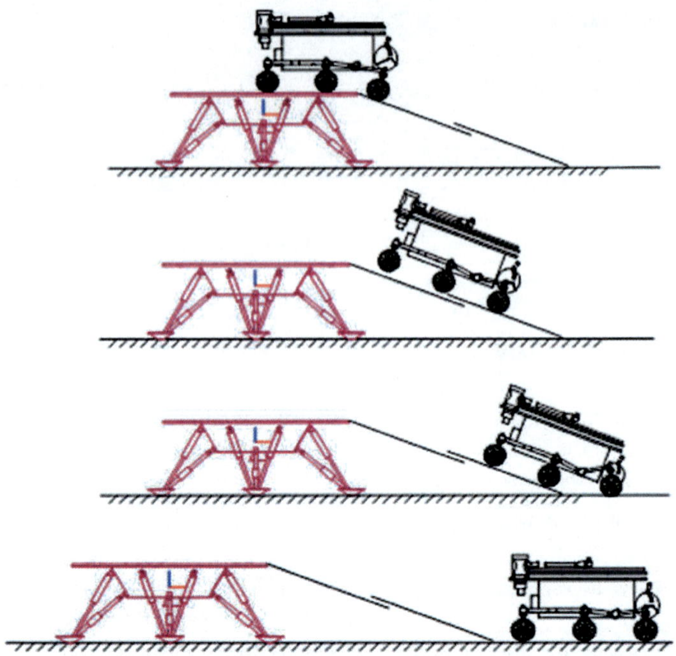

Fig. 13.1 Planetary exploration mobile robot driving off the landing platform

(3) The robot should be able to adapt to different landing conditions and safely leave the landing platform in different attitudes of the landing platform and the transfer mechanism;
(4) The coordination requirements between the robot and the landing platform and the transfer mechanism.

13.2.1.2 Planet Surface Exploration and Movement Task

After the robot moves to the planet surface, it can move autonomously according to the ground commands and safely reach the target point. It can adapt to different terrains and have different motion modes; in case of dangerous conditions such as encountering stones and deep pits, it can perform emergency obstacle avoidance; when the wheels are sunk, it can extricate from the sinkage by means of lifting the vehicle body or the wheels; it should have strong fault tolerance capability.

The typical topographical environment of the planetary surface on which the robot travels mainly includes:

(1) Flat hard terrain;
(2) Rugged gravel terrain;
(3) Soft sandy terrain;
(4) Hilly desert terrain;

(5) Crater and Gobi terrain.

The requirements on the robots by this task are as follows:

(1) To move quickly on a flat ground;
(2) To steer to avoid obstacles;
(3) To be wear and impact resistance when moving on gravel ground;
(4) With sufficient climbing ability when moving on the dune ground;
(5) To resist soil adsorption, with sufficient traction when moving on soft ground;
(6) To extricate wheels from sinkage by lifting the wheels;
(7) To move normally by the lifting the wheels in case of a wheel malfunction;
(8) With different control and motion modes;
(9) With the capability of autonomous movement and emergency obstacle avoidance under the condition of large communication delay.

13.2.2 Overall Scheme Design

The planetary exploration mobile robot overall scheme, which includes the mobile scheme, the steering scheme, the suspension scheme, the control scheme, and the perception scheme, should be analyzed and designed according to the functional requirements of the robot and the environmental characteristics of the planet surface.

13.2.2.1 Selection of Mobile Scheme

The optional mobile schemes include: wheeled, tracked, wheel-tracked, wheel-legged, planetary-wheeled schemes. Their advantages and disadvantages are detailed as follows [4, 5]:

(1) Tracked, wheel-tracked scheme: large mass and power consumption, not applicable for deep space exploration requirements on lightweight and low power consumption.
(2) Planetary-wheeled scheme: When traveling in the sand of the planet surface, there are problems such as large resistance, gravel stuck, and dust blockage.
(3) Wheeled and wheel-legged scheme: for the wheeled mobile scheme, it has been proven to have obvious advantages in terms of reliability, safety, and stability; while the wheel-legged mobile scheme is more advantageous in terrain adaptability, but its mechanism is relatively more complex and the reliability is slightly lower.

In summary, considering the technology maturity and the planetary surface environment adaptability, the wheeled mobile scheme is selected for the planetary exploration mobile robot.

13.2.2.2 Selection of Steering Scheme

Different steering modes can be selected for planetary exploration mobile robot such as differential steering, front-wheel steering, and six-wheel independent steering. The advantages and disadvantages of different steering modes are compared as follows:

(1) From the perspective of the realization of steering function, the configuration of differential steering is the simplest, but due to its unique wheel side-pushing effect, a sliding friction will be produced between the tire surface and the ground so that it will be difficult to steer on hard ground and high friction ground. On gravel roads, it may be blocked by obstacles such as stones and bumps, resulting in difficulty in steering and even failure in steering.
(2) A large turning radius is usually required for front-wheel steering. For complex terrain on the planet surface, sometimes front-wheel steering cannot meet the turning radius requirement, thus reducing the terrain adaptability of the robot.
(3) With six-wheel independent steering, not only the functions of arc and in situ steering but also the function of lateral straight movement can be realized. When all wheels are turned to 90°, the travel direction of the wheels is perpendicular to the vehicle body, thus achieving transverse walking. This steering mode can greatly enhance the terrain adaptability of the planetary exploration mobile robot.

Therefore, the six-wheel independent steering scheme is selected for the planetary exploration mobile robot.

13.2.2.3 Selection of Suspension Scheme

From the actual situation of the US Mars roaming and exploration, the wheel sinkage extrication ability and climbing ability are the most important ability of the mobile robot on celestial body surfaces. The Mars rover "Spirit" has sunk many times, and each time it takes several months to extricate. In addition, due to the influence of soft sand, the climbing ability of the planetary exploration mobile robot on the planet surface is also limited to a large extent.

In order to improve the robot sinkage extrication, obstacle crossing, and climbing capability, an active suspension is used to improve the adaptability and the survivability of the robot in the complex terrain and geological conditions of the planet surface. Compared with the passive suspension, the main improvement of the active suspension is to add an active angle adjustment mechanism between the front and rear sections of the main rocker arm of the passive suspension and a clutch at the connection between the main rocker arm and the auxiliary rocker arm. Thus, the main and auxiliary rocker arms have the ability to actively adjust the configuration (see Fig. 13.2 for details).

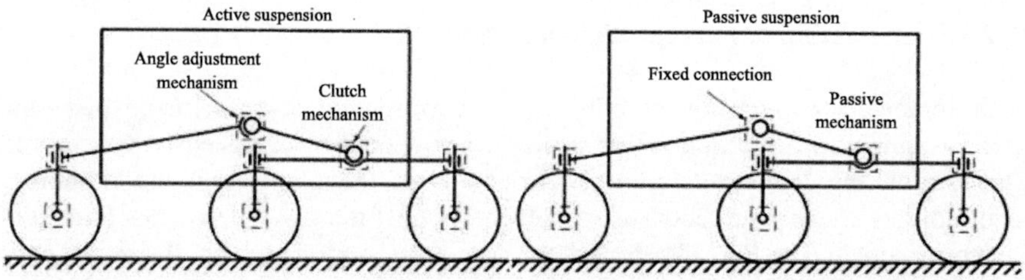

Fig. 13.2 Comparison of active suspension and passive suspension

The advantages of the active suspension are:

(1) The carriage height can be adjusted; the support structure needed for launching can be omitted; after landing, the vehicle body can be raised to normal height and can drive away from the landing platform;
(2) The robot has the function of wheeled walking, by which the ability of sinkage extrication is enhanced in case of wheel sunk;
(3) The robot can change the centroid position of the vehicle body, or lift the wheels to reduce the wheel pressure of the sunk wheel;
(4) The robot can change the centroid height of the vehicle body, thereby reducing the difference of the load and wheel driving torque between the front and rear wheels during climbing, and improving the climbing ability of the robot;
(5) In case of wheel malfunctions, the faulty wheel can be lifted. If one wheel fails, it can walk by 5 wheels; if one wheel on each side fails, it can walk by 4 wheels.

The planetary exploration mobile robots mainly adopt the active suspension scheme.

13.2.2.4 Selection of Control Scheme

The main function of the planetary exploration mobile robot control system can be classified into the motion control of the whole vehicle and the servo drive control of each mechanism. The former mainly controls the mobile robot navigation and trajectory planning by the CPU, which is equipped with the vehicle motion control software; and the later completes the servo control of each mechanism motor by FPGA. Mobile robot has 16 active mechanisms and adopts a centralized control scheme under the consideration of mechanism size and weight limitation. The CPU and FPGA are integrated into the control unit of the robot. The control unit is placed inside the carriage of the robot, which is more convenient for the temperature control and dust-proof design.

13.2.2.5 Selection of Perceptual Scheme

In the planetary exploration mobile robot, measurement devices for position and speed acquisition are required for position servo control or speed servo control. Considering the low-temperature environment of the planet and the reliability requirements, the mobile robots use revolvers for position and speed measurement of each mechanism (see Fig. 13.5 for the mechanisms). Among which, the motor ends of the drive mechanism, the steering mechanism, the angle adjustment mechanism, and the clutch mechanism are all equipped with a speed measurement resolver; in order to improve the position servo precision of the steering mechanism and the angle adjustment mechanism, a position measurement resolver is arranged at the output end of the mechanisms; in order to monitor the state of the vehicle body and the wheels during the movement of the robot, a position measurement resolver is arranged between the differential mechanism and the main and auxiliary rocker arms, respectively; the clutch is equipped with an in-position switch for interpretation of the clutch engaging or disengaging state.

The planetary exploration mobile robot must perceive the surrounding environment to complete on-orbit autonomous motion. Therefore, a navigation camera and an obstacle avoidance camera are equipped for visual navigation and autonomous obstacle avoidance control, respectively, and a sun sensor is equipped to complete the pose identification of the robot. The perception system communicates with the control system via the bus RS422.

13.2.2.6 Overall System Scheme

The planetary exploration mobile robot adopts an overall layout of an active suspension combined with a six-wheel independent steering and driving. The entire system includes a suspension (composed of 2 sets of main and auxiliary rocker arms and 1 spindle), wheels (totally 6 sets, each set contains a drive mechanism, a wheel structure, a steering mechanism, and a steering arm), angle adjustment mechanisms (2 sets), clutches (2 sets), and a differential mechanism, as shown in Fig. 13.3. The control system adopts a centralized control scheme, the control unit is placed inside the carriage. The perception system consists of revolvers, visual cameras, and sun sensors, etc.

13.2.2.7 System Configuration Parameters Design

The key dimension parameters of the planetary exploration mobile robot under nominal mobile state are shown in Fig. 13.4, and mainly include:

(1) The length of the front half of the main rocker arm l_1, length of the rear half l_2;
(2) The length of the auxiliary rocker arm l_3 (the lengths of the two halves of the auxiliary rocker arm are the same);

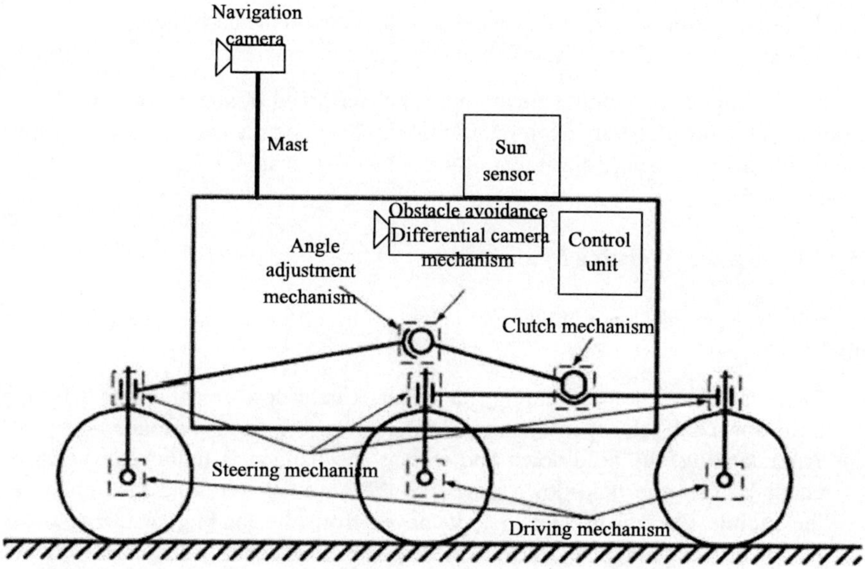

Fig. 13.3 Overall composition of the planetary exploration mobile robot

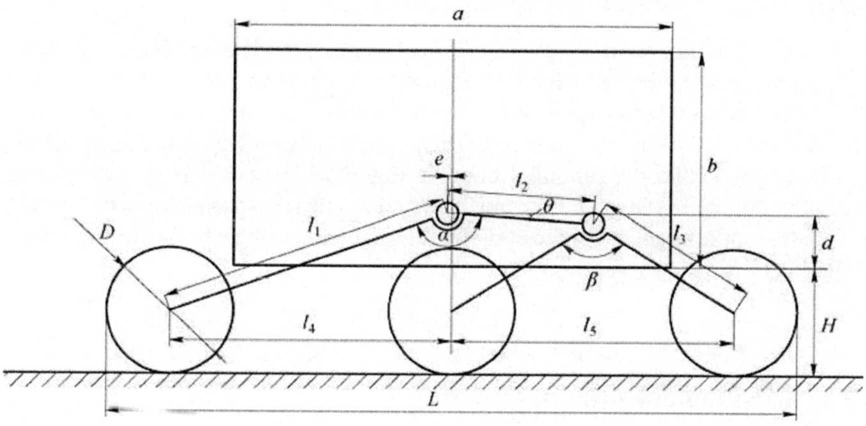

Fig. 13.4 Key dimension parameters of a planetary exploration mobile robot

(3) The angle between the main rocker arms α, the angle between the second half of the main rocker arm and the horizontal line θ;
(4) The angle between the front and rear halves of the auxiliary rocker arm β;
(5) The deviation distance e between the system centroid and the centerline;
(6) The tread between the front and middle wheels l_4, and the tread between the middle and rear wheels l_5;
(7) The height from the carriage bottom to the ground H;

(8) The installation height of the suspension spindle on the carriage *d*;
(9) Wheel diameter *D*, the distance between the front and rear wheel outer edges *L*.

The key dimension parameters are mainly determined by the overall dimension requirement of the planetary exploration mobile robot and the matching relationship of the dimensions during adjustment of the active suspension.

13.2.2.8 System Working Mode Design

The planetary exploration mobile robot will work in different modes during different mission phases:

(1) From launch to before landing, the robot is held-down on the lander by the hold-down and release mechanism, in the power-off standby mode.
(2) After landing, the hold-down and release mechanism is unlocked, the robot changes from the hold-down state into the nominal state and then enters to the mobile working mode: first, it moves from the landing platform to the planet surface, and then it moves on the planet surface according to the ground command to roam and explore.

The planetary exploration mobile robots have two working modes: the nominal motion mode and the active suspension motion mode.

(1) Nominal motion mode: This is the normal working mode of the planetary exploration mobile robot, in which it performs tasks such as moving, climbing, and obstacles negotiating when the active suspension does not involve in the work.
(2) Active suspension motion mode: This is the emergency working mode of the planetary exploration mobile robot. In this case, the robot is in a dangerous situation such as wheel failure or wheel sunk, the active suspension is in working state, supporting the robot to complete tasks such as vehicle body lifting/lowing, wheel lifting, and walking.

13.3 Mechanical System Design

The mechanical system of the planetary exploration mobile robot is shown in Fig. 13.5. According to functions, the mechanical system can be divided into four parts: the drive steering module, the active suspension module, the differential support module, and the hold-down and release module.

(1) **Drive and steering module** is mainly composed of wheels, drive mechanisms, steering mechanisms, and steering arms. It is the core component of the robot to realize walking, steering, climbing, obstacles crossing, static holding, and the only product that directly contacts with the ground for a long time. The wheel structure tread must meet the requirement of resisting rock piercing.

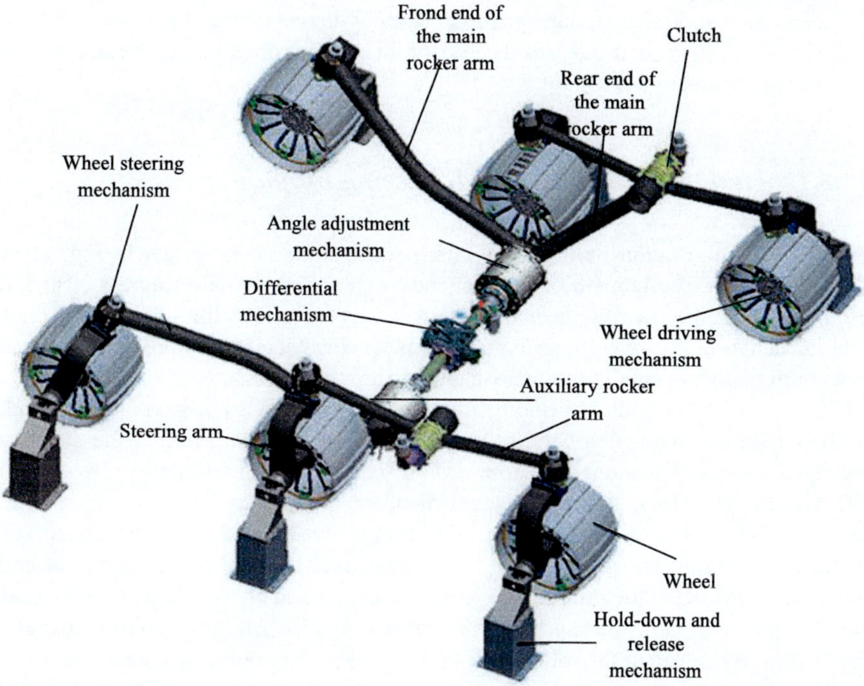

Fig. 13.5 Composition of the mechanical system of the planetary exploration mobile robot

(2) **Active suspension module** is mainly composed of the angle adjustment mechanism, the clutch, the front end of the main rocker arm, the rear end of the main rocker arm, and the auxiliary rocker arm. The functions of the active suspension module are to maintain the suspension configuration and connect with the drive steering module. During normal driving, it provides support for the carriage, and uses the clutches of the main and auxiliary rocker arms to accommodate the free bumps of the four wheels on the auxiliary rocker arm to meet the terrain adaptability requirements during driving. The active suspension module also realizes the functions such as adjustment of the carriage height, adjustment of the centroid position, and wheel lifting.

(3) **Differential support module** is mainly composed of a differential mechanism, a spindle connecting shaft and an angle adjustment mechanism adapter cylinder. It is the main connection point between the robot carriage and the suspension, with the overturning support function when the carriage centroid is deviated. When both sides of the wheels are running on uneven ground and with different speeds, it can provide a differential balance to prevent the carriage from turning or the wheels being dragged on the ground.

(4) **Hold-down and release module** is mainly composed of a hold-down and release mechanism, used to reliably lock the robot during the launch segment so that it

can withstand the load during launch without damage; after the lander lands on the planet surface, it releases the constraints on the robot, so that it can perform various types of motion.

13.3.1 Design of the Drive and Steering Module

The body configuration of the drive and steering module is shown in Fig. 13.6. The wheel drive mechanism of the planetary exploration mobile robot is arranged orthogonally to the steering mechanism and connected to it by the steering arm. The rotation center of the wheel steering mechanism is located at the middle of the wheel body with relatively small running resistance. The drive mechanism is arranged at the middle of the wheel body to avoid bending moments when driving on plain ground.

The drive mechanisms of the robot's wheel drive mechanism and the steering mechanism basically adopt the same design scheme, each consists of a Brushless DC Motor (BLDCM), a single-channel resolver, a planetary reducer, a harmonic reducer, a housing and a shafting. In order to improve the position control accuracy of the steering mechanism, a single-channel resolver is installed at the output end of the mechanism for the rotation angle measurement and closed-loop control of the steering position. Considering the temperature condition on the planet surface and the loading condition of the robot, the planetary gears, harmonic gears, and bearings all adopt a solid lubrication solution. The drive assembly adopts a contact dynamics seal design to isolate dust from entering the mechanism; a dust storage groove is

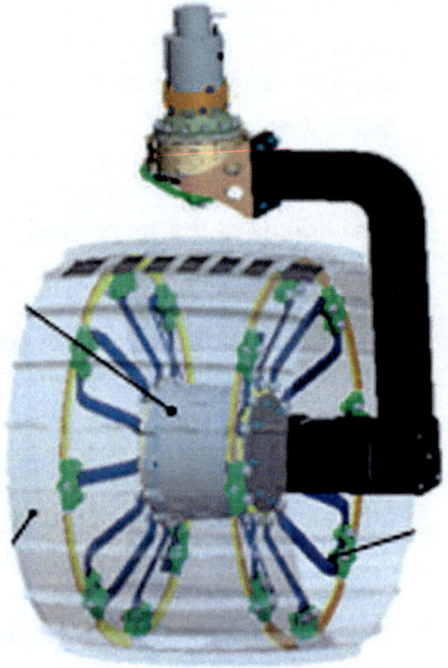

Fig. 13.6 Drive steering module

Fig. 13.7 Wheels

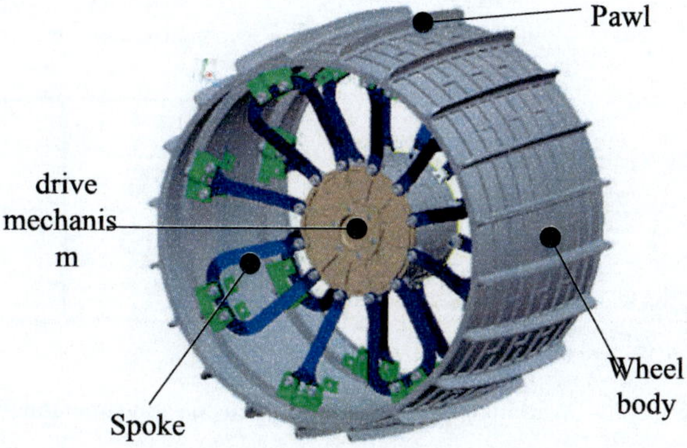

designed inside the dynamics seal to prevent the incoming dust from affecting the mechanism transmission.

The wheels take the form of drum wheels, as shown in Fig. 13.7. The wheel surface is provided with pawls to improve its gripping force and reduce the wheel-ground slip ratio. The wheel body and the drive mechanism are connected by spokes with shock absorbing effect.

13.3.2 Design of the Active Suspension Module

The body configuration of the active suspension module is shown in Fig. 13.8. The active suspension module is mainly used during vehicle height adjustment, centroid

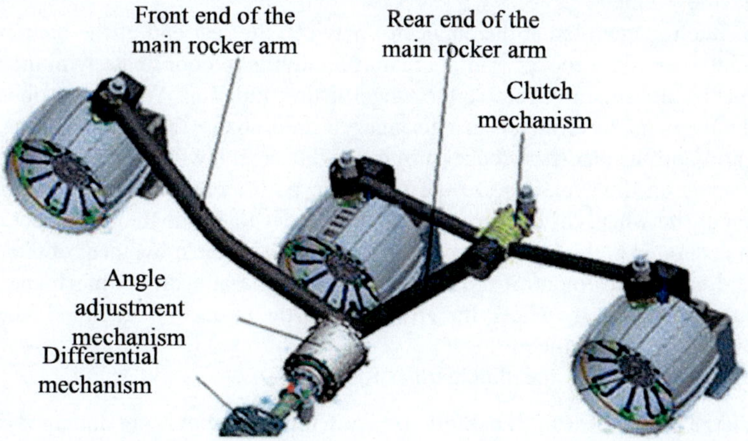

Fig. 13.8 Active suspension module

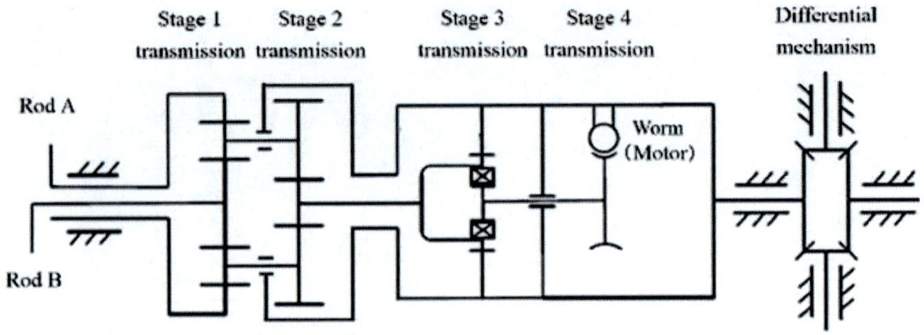

Fig. 13.9 Working principle of the suspension adjustment module

adjustment, and wheel lifting. During normal driving, the motor is powered off, the entire active suspension module does not work, only bears the static load. During operation, the suspension adjustment module adjusts the opening angle between the front and the rear ends of the main rocker arm through the angle adjustment mechanism, and coordinates with the lock or release of the clutch to realize the vehicle body lifting, centroid adjustment, and wheel lifting.

The angle adjustment mechanism includes a BLDCM, a worm gear mechanism, a harmonic reducer, a spur gear transmission, a planetary differential output mechanism, and a goniometric resolver. It realizes self-locking of the angle mechanism by the worm gear to ensure that the configuration can be maintained in case of power-off. The angle adjustment mechanism is designed in the form of single-input double-output, as shown in Fig. 13.9. The two main rocker arms are, respectively, connected to the two output shafts (A-rod and B-rod) of the angle mechanism. Therefore, the angle mechanism motor can drive the two main rocker arms at the same time, so as to ensure that the vehicle body maintains horizontal during the lifting without side rolling.

The clutch is mounted at the connector between the rear end of the main rocker arm and the auxiliary rocker arm, its main function is to coordinate with the angle adjustment mechanism to realize the wheel lifting function. When the robot is in normal movement, the clutch is in a disengaged state, so that the rear end of the main rocker arm and the auxiliary rocker arm can rotate freely, which can ensure that all three wheels on the same side of the robot can be in contact with the ground. In case one of the wheel driving function fails, in order to eliminate the impact of the friction generated between this wheel and the ground on the movement of the robot, the coordinated movement of the clutch and the angle adjustment mechanism can realize the lifting of any wheel, thereby reducing the impact of the failed wheel on the movement of the vehicle.

The requirements for the clutch are as follows:

(1) A large torque bearing capability to ensure that the robot is not damaged during wheel lifting;

(2) A self-locking function to ensure smooth engagement of the clutch under power-off conditions;
(3) An in-position signal to judge whether the clutch is engaged or disengaged.

The clutch includes a motor, a worm gear, and a ball screw. In order to determine the rotation angle between the rear end of the main rocker arm and the auxiliary rocker arm, a resolver is installed inside the clutch.

The gears and bearings in the angle adjustment mechanism and the clutch mechanism all adopt solid lubrication and contact-type dynamics sealing structure with a dust storage groove inside the dynamics sealing.

13.3.3 Design of the Differential Support Module

The differential mechanism is aimed to meet the requirements for all-terrain adaptability of the planetary exploration mobile robot and the ability to support the carriage and prevent it from turning over. The differential mechanism is connected with the main shafts of the left and right wheels. When the robot is moving on the rugged ground, the main shafts of the left and right sides can rotate relative to each other, so that the left and right wheels can all contact the ground. In this way, on one hand, all the wheels can provide the driving force for the movement of the robot, so that the driving ability of the robot does not decrease; on the other hand, all the wheels of the robot are in contact with the ground, so that the whole vehicle has a better stability, and will not fall over due to hanging of a wheel. In addition, the vehicle body of the robot and the main shafts of the wheels are connected by differential mechanisms, which requires the differential mechanism to have good strength and rigidity and small clearance so as to reduce the swing of the vehicle body during movement.

The differential mechanism uses four cross-differential gears. In order to reduce the mass and volume of the differential mechanism, two large and two small bevel gears are used; in order to improve the axial load-carrying capacity of the bevel gears, an angular contact ball bearing is used between the gear shafts and the casing; a resolver is designed between the two large bevel gears in the direction of the output shaft to measure the relative rotation angle between the left and right spindles, by which the unevenness of the road can be judged during the whole vehicle movement. The motion principle of the differential mechanism is shown in Fig. 13.10.

13.3.4 Design of the Hold-Down and Release Module

The planetary exploration mobile robot adopts a rod-type hold-down and release mechanism, which features a good rigidity and a large pre-tightening force, and is generally suitable for hold-down of large mechanisms.

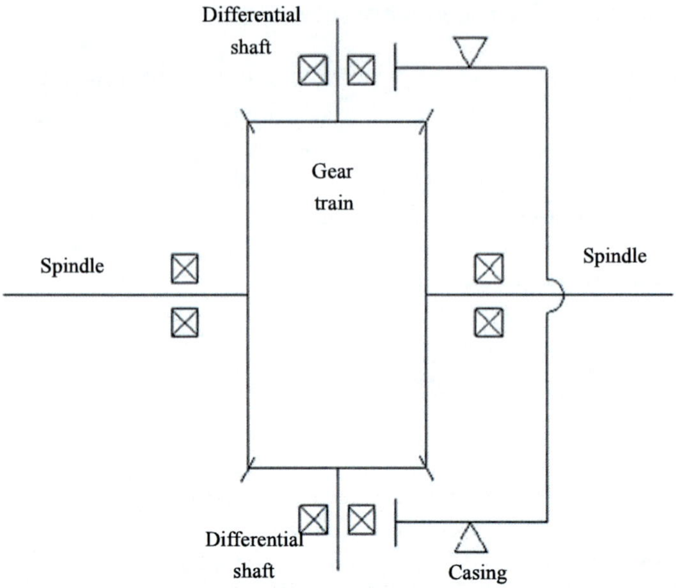

Fig. 13.10 Motion principle of the differential mechanism

The six wheels of the planetary exploration mobile robot are all held-down, as shown in Fig. 13.11. The hold-down and release mechanism consists of a hold-down base, a metal embedded part, a support base, a lock nut, a spring, an escape cap, a hold-down rod, a buffer assembly and limiting pads, as shown in Fig. 13.12.

13.4 Control System Design

13.4.1 Overall Design of the Control System

The control system of the planetary exploration mobile robot adopts a centralized architecture, and its hardware is the control unit installed inside the vehicle body. The control system is divided into two levels and contains the motion planning module of the entire robot and the drive assembly servo control module. The robot's motion planning is realized by the CPU, and the drive assembly servo control is realized by the FPGA.

(1) The robot motion planning can be divided into motion navigation and trajectory planning, including data acquisition and processing from the control system sensors, measurement information fusion, calculation of navigation control rule, and generation and transmission of the control commands.

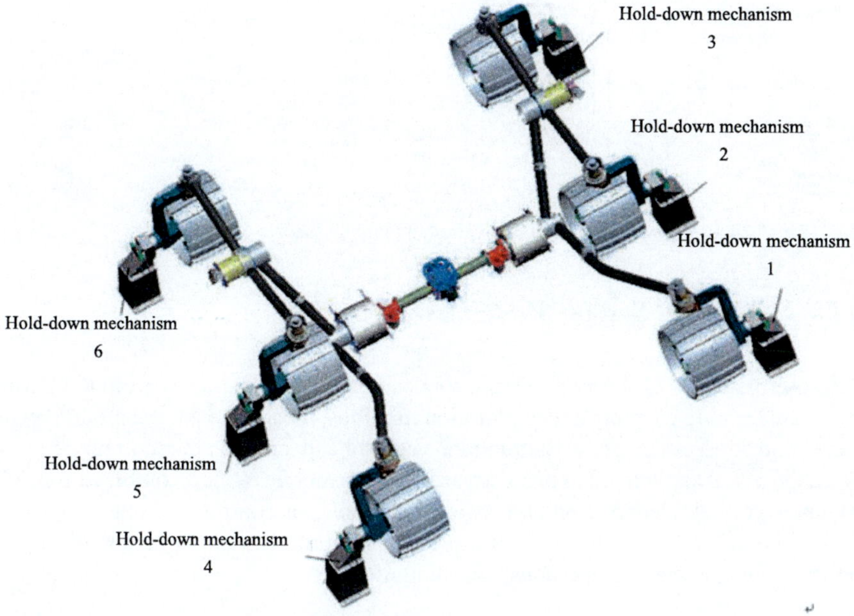

Fig. 13.11 Layout of the hold-down mechanism

Fig. 13.12 Body mechanism of the hold-down and release mechanism

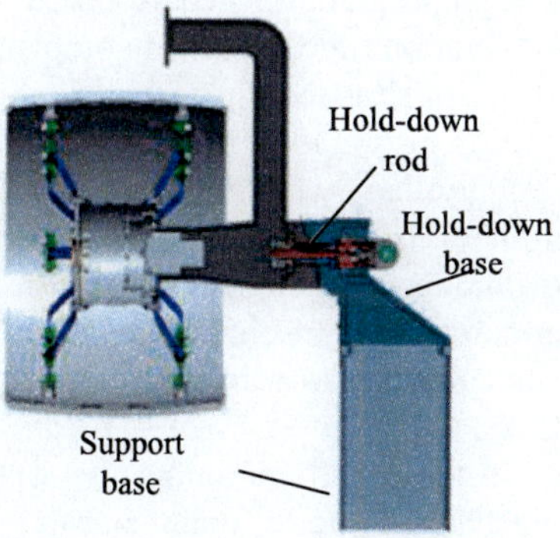

(2) The drive assembly servo control includes servo controls of the wheel steering mechanism, the drive mechanism, the angle adjustment mechanism, and the clutch mechanism.
(3) The motion planning module communicates with the drive assembly servo control module through the 1553B bus. The control flow of the planetary exploration mobile robot is shown in Fig. 13.13.

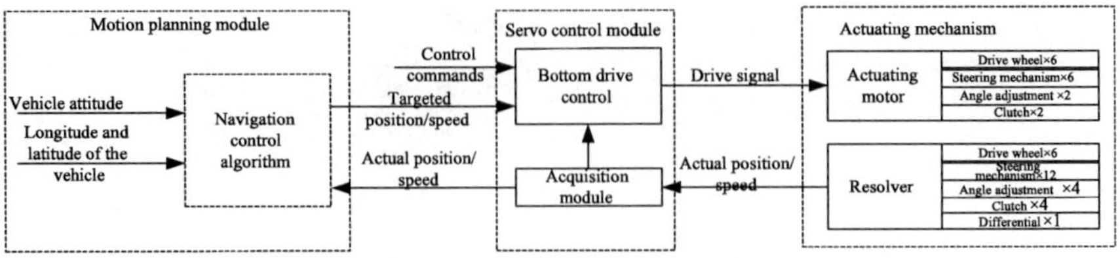

Fig. 13.13 Control flow of the planetary exploration mobile robot

13.4.2 Control System Mode Design

Because planetary exploration mobile robots are generally far away from the Earth when performing roaming and exploration missions, it cannot meet the ground real-time control requirements, so autonomous control capabilities are required.

In the actual application of planetary exploration mobile robot, a combined control mode of ground teleoperation and on-orbit autonomous control is adopted. Specifically, the control is designed with a prestorage control mode, a parameter input control mode, and an autonomous control mode.

13.4.2.1 Preprogramming Control Mode

The preprogramming control mode is that the typical motion modes of the planetary exploration mobile robot is preplanned before launch, for example, linear walking 10 m, in situ steering 180°, traveling steering in a radius of 3 m, and the planned motion information for all the mechanisms is stored in the control unit. The robot is controlled by the ground station to move in a preprogrammed mode with specific numbers by remote command. After the control unit receives the ground command, the motion planning module transmits the prestored motion commands into the drive circuit of each mechanism of the servo control module in real time according to the control cycle, so as to control the motion of each mechanism to accomplish specific tasks.

In the preprogramming mode, the motion planning module sends a prestored mechanism control command (angle, angular velocity, etc.) in one control cycle, no need to do real-time planning on orbit. In addition, the status information of each mechanism is sent to the ground monitoring and control station to monitor and analyze the entire execution process.

The preprogramming control mode is simple and reliable, but lacks dexterity, mainly suitable for situations where the environment is known and the task is fixed.

13.4.2.2 Parameter Input Control Mode

According to the status information sent by the robot telemetry and the planetary surface environment information collected by the cameras, the ground personnel decides the next motion of the robot. Firstly, the motion to be executed is verified by the simulation with ground motion support system. After the verification is passed, the parameters (motion mode, motion speed, steering radius, etc.) required for the motion of the robot is injected into the control unit through telecontrol command. The motion planning module plans the motion command of the robot in real time on orbit according to the set parameters, and sends the motion command to the drive circuit of each mechanism of the servo control module to control the motion of each mechanism to perform specific tasks.

The parameter input control mode is a man-in-the-loop control mode. The top-level task planning is mainly accomplished by the ground. Compared with the prestorage mode, it has certain dexterity. However, the robot has less autonomy, so it is more suitable for specific tasks, such as detection and sampling, in which the target location is decided by the ground station.

13.4.2.3 Autonomous Control Mode

In the autonomous control mode, the moving trajectory and the motion commands for each mechanism is planned by the motion planning module of the planetary exploration mobile robot autonomously without ground involvement according to the image information given by the navigation camera and directly sent into the servo control system without ground intervention to control the motion of each mechanism of the robots. The robot can realize the autonomous obstacle avoidance through the obstacle avoidance camera during walking.

In the autonomous control mode, the closed-loop control is realized based on the robot's own sensors such as the navigation camera and the obstacle avoidance camera without ground intervention, it is with high motion efficiency and mainly suitable for large range transfer and roaming.

13.4.3 Design of the Servo Drive Scheme

The servo control mode of the planetary exploration mobile robot is implemented based on FPGA. The control system includes the basic position loop, speed loop, and current loop control modes, the three loop control flow is as shown in Fig. 13.14. Various mechanisms of the robot select different servo control modes according to the requirements. Among them, the steering mechanism and the angle adjustment mechanism have stringent position requirements, therefore, the position servo control is selected; while the drive mechanism and the clutch mechanism have no strict requirements on the position, so speed servo control is selected.

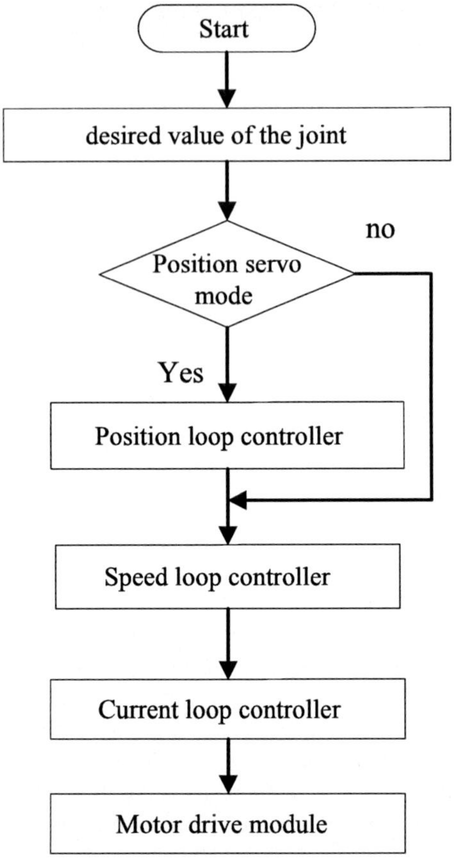

Fig. 13.14 Servo control flow of planetary exploration mobile robot

13.4.3.1 Current Loop Control Scheme

The current loop controls the motor current. The current input comes from the output of the speed loop. The current loop uses a PI (proportional integral) controller to eliminate the large delay element and obtain the input of the motor drive and controls the generation of PWM control signals.

13.4.3.2 Speed Loop Control Scheme

The speed loop controls the motor speed. It adopts a modified PID (proportional integral differential) controller and introduces a negative feedback element according to actual needs to improve the adaptability of the system. Its control object is the output shaft speed of the motor so as to reduce the impact of the nonlinear elements.

The input of the speed loop comes from the desired speed given by the vehicle motion control system and the converted equivalent motor speed of the position loop output. The motor speed feedback is obtained by the resolver's solution of the motor shaft, and the desired current of the motor is output via the controller.

13.4.3.3 Position Loop Control Scheme

The position loop controls the position of the motor. It used a P (proportional) controller to improve the stability of the system. The position loop takes the output end of the transmission system as its control object to ensure the accuracy of the closed-loop system.

The position loop input comes from the desired position issued by the vehicle motion control system. It obtains the actual position by the resolver on the output end of the transmission system and outputs the desired speed via the position controller.

13.5 Perception System Design

13.5.1 Overall Structure of the Perception System

The perception system of the planetary exploration mobile robot is mainly composed of a navigation camera, an obstacle avoidance camera, a sun sensor, and a data processing unit. The sun sensor is mounted on the vehicle body for determining the pose of the robot; the navigation camera is mounted on the mast for navigation control of the robot; the obstacle avoidance camera is mounted on the vehicle body for emergency obstacle avoidance of the robot. The perception system communicates with the control system through the bus RS422. The information obtained by the perception system measurement and processing are as follows:

(1) The long-distance panoramic images acquired by the navigation camera.

The navigation camera follows the pitch and yaw motion of the mast to image the 360° range around the robot, with an effective perceptual distance of about 10 m. The long-range panoramic images are used for global path planning.

(2) The close range detailed terrain information acquired by the obstacle avoidance camera.

The obstacle avoidance camera images the small area in front of the robot's motion and obtains detailed terrain information through image stereo matching and three-dimensional recovery. The detailed terrain information is used for local path planning and autonomous obstacle avoidance planning for the robot.

(3) The discrete information acquired by the obstacle avoidance camera and laser dot matrix.

The laser dot matrix projects light spots in the imaging area of the obstacle avoidance camera and obtains the geometric information at the light spots after imaged and visually processed by the camera. This perception method only obtains the geometric information at discrete points, it does not require binocular matching and recovery and has a fast processing speed, and is suitable for emergency obstacle avoidance in the forward close range.

When the camera FOV is in the shadow area, the camera cannot effectively image, but can image the laser spot projected by the laser dot matrix, and obtain the three-dimensional information at the laser spot for obstacle recognition.

(4) The pose information acquired by the sun sensor.

The sun sensor is mounted on the vehicle body of the planetary exploration mobile robot. It can receive the sun vector information, with which the global pose of the robot can be determined by calculation for motion planning of the robot.

13.5.2 Visual Perception System Design

Visual perception system of the planetary exploration mobile robot includes the navigation camera system and the obstacle avoidance camera system. Each system further includes a camera, a visual processing unit and real-time image information processing software, and finally connected with the control unit of the robot for communication. Considering the need for hardware backup, a binocular camera is used on the mast, and another binocular camera is installed on the left and right ends of the robot. The camera acquisition data is sent to the visual processing unit to realize the terrain reconstruction by stereo matching technology and complete the visual navigation. Besides, the image information collected by the obstacle avoidance camera is analyzed in real time during the motion of the robot, the possible obstacles are warned in time, and the travel route is re-planned through information coordination by the control unit of the robot. The image compression is realized by the hardware in the visual unit, and the compressed images are finally transmitted to the ground through the data transmission system for monitoring and analysis by ground personnel.

The visual processing unit is developed in a DSP + FPGA architecture. The three sets of visual measurement systems for navigation, left-end obstacle avoidance, and right-end obstacle avoidance have relatively independent functions. The functional division of the modules is shown in Fig. 13.15. Each visual measurement system consists of a binocular camera, a FPGA, an image buffer, an image compressor, a DSP processor, and a communication module.

The images acquired by each visual system are directly divided into two methods. In the first method, images are compressed by a special compression chip hardware

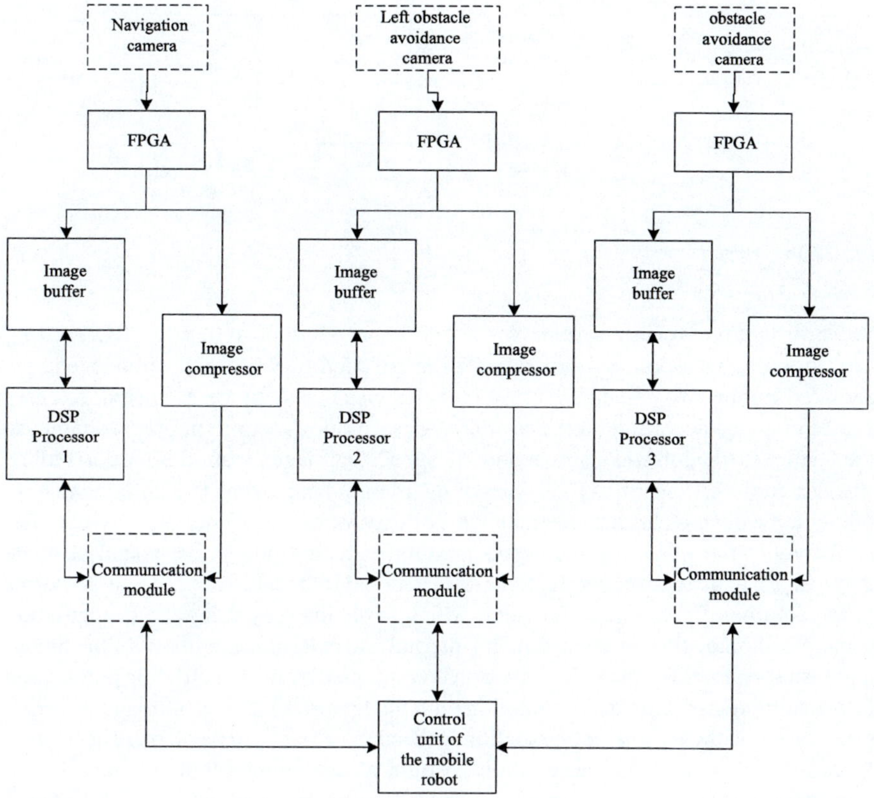

Fig. 13.15 Block diagram of hardware design scheme

and sent to the communication module, and packaged and output according to the communication protocol requirements; in the second method, images are sent to the image buffer for image feature matching, target recognition, and three-dimensional reconstruction by the DSP processor, and the resulted visual measurement is finally transmitted to the control unit via the communication module.

For terrain reconstruction, visual navigation and visual obstacle avoidance, image processing feature must be matched to obtain the relative position of the target in front of the camera by further calculation according to the basic principle of triangulation. For terrain reconstruction and visual navigation, it is necessary to match relatively dense image features to obtain dense 3D point cloud information, and then draw the terrain and topography by technical means such as point cloud registration and SLAM. For visual obstacle avoidance, only the approximate location of an obstacle or a swale should be roughly calculated to ensure that the robot can plan the path in advance and avoid the danger of obstacles or swales. The computation for visual obstacle avoidance data is less than that for terrain reconstruction or visual navigation. The image processing flow is shown in Fig. 13.16.

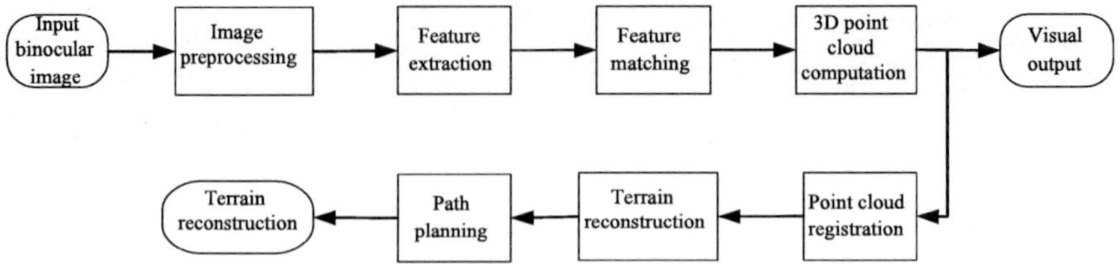

Fig. 13.16 Image processing flow

The image preprocessing refers to the filtering and Region of Interest (ROI) setting operations, and it is optional depending on actual conditions. After the image features acquired by the two cameras in the binocular visual system are extracted, features can be matched according to the similarity description function. In order to increase the feature extraction and stereo matching speed, an image pyramid is used to reduce image sampling, generally, 3–5 layers of the pyramid are taken. The initial matching points may mismatch, this mismatched point pairs can be eliminated by epipolar constraint. After obtaining the corresponding matching points, the visual obstacle avoidance can be realized by the three-dimensional information of the feature points in the common FOV of the binocular, which is obtained by using the triangulation principle of binocular stereo vision, the internal and external parameters of the binocular visual system. Terrain reconstruction is completed by further utilizing point cloud information registration in combination with the texture synthesis of image information. Based on the terrain reconstruction, a reasonable path can be planned according to the current position and destination position of the mobile robot and the position information of obstacles or swales to complete visual navigation.

13.5.3 Navigation Camera Design

The navigation camera adopts a binocular camera integrated structure design. The main technical indices are shown in Table 13.2.

With a stereovision measurement method, navigation camera can generate a set of 3D map data with an image resolution of 1032×776 and a total of 800,000 pixels. Each pixel is represented by 1 bit, which is a binary number (0 or 1).

A complete 3D map with very large data amount must be further simplified for rapid transmission. Streamline 3D map data is mainly a set of array sequences containing the position information of several obstacles (the outer envelope should not be less than 200 mm). Each element array in the sequence includes the horizontal x and vertical y coordinate components corresponding to the upper left corner vertex and the right lower corner vertex of the smallest circumscribed rectangle of the obstacle envelope. The horizontal x and the vertical y coordinate components of the upper left corner vertex are represented by 32 bits, respectively; similarly, the horizontal x and the vertical y coordinate components of the lower right corner

Table 13.2 Main technical indices of the navigation camera

No.	Technical index	Value	
1	Measuring distance	0–10 m	
2	Band range	Visible light, color imaging	
3	Maximum image resolution	1032 × 776	
4	Pixel resolution	7.75 mm at 10 m	
5	Lateral FOV angle	43° × 33°	
6	Video data output	24 bits	
7	Image transmission distance	5 m	
8	Camera installation position	Distance above the ground	1.5 m
		Camera tilt angle (pitch angle)	30°

Table 13.3 Main technical indices of obstacle avoidance cameras

Technical index	Value
FOV angle	Not smaller than 120° × 120°
Image quality (resolution)	1024 × 1024

vertex are represented by 32 bits in the sequence. The data volume of the obstacle area is totally 128 bits, therefore, the data volume of the streamlined 3D map is much smaller than the complete 3D map.

13.5.4 Obstacle Avoidance Camera Design

The obstacle avoidance cameras include the left and right obstacle avoidance cameras that can adapt to the lighting conditions of the planet surface for large angle imaging, and can clearly image the laser spots in its visible range. The main technical indices are shown in Table 13.3.

13.5.5 Sun Sensor Design

The sun sensor determines the orientation of the sun vector in the robot body coordinate system by the orientation of the measuring sun vector, thereby solving the attitude of the robot. The planetary motion mobile robots use digital sun sensors. The specific indices are shown in Table 13.4.

Table 13.4 Main technical indices of sun sensors

Technical index	Value
FOV angle	±60° × ±60°
Accuracy	0.05°

Table 13.5 Planetary exploration mobile robot test items

No.	Items	Sub-items
1	Function tests	Hold-down function in the flight phase before landing
		Function of configuration deployment and drive off platform after landing
		Function of large area movement on the planet surface
		Function of extrication from wheel sinkage on the planet surface
2	Performance tests	Load-carrying capacity
		Moving distance
		Maximum speed
		Obstacle crossing capability
		Climbing capability
		Adjustment range of vehicle height
		Wheel lifting height range

13.6 Design Verification

13.6.1 Verification Items

After the design is completed, it is necessary to test and verify whether the planetary exploration mobile robot meets the design requirements (Table 13.5). According to the characteristics of the roaming and exploration tasks of the robot, the walking ability under different terrain is the keystone of the planetary exploration mobile robot. The verification items are designed for this feature of the robot. These items include the verification of the vehicle body's climbing ability, the obstacle crossing capability, the vehicle body lifting ability, and the wheel lifting and walking ability.

13.6.2 Verification Scheme

The conventional test items can be implemented in accordance with the general test methods for spacecraft, so this book will not give a detailed description. The following descriptions apply only to special verification schemes:

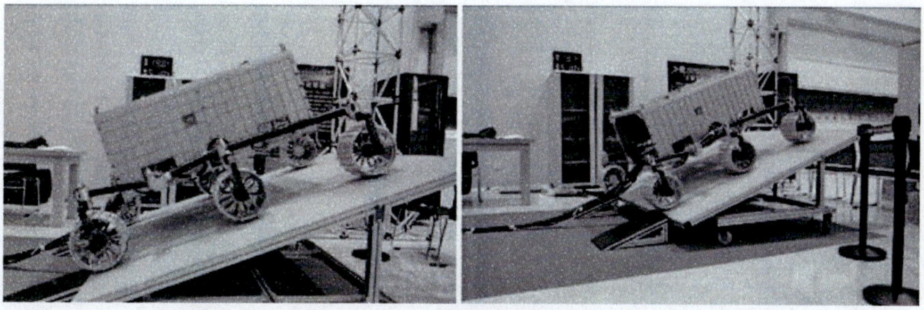

Fig. 13.17 Planetary exploration mobile robot climbing ability verification scheme

Fig. 13.18 Soft ground climbing ability verification scheme

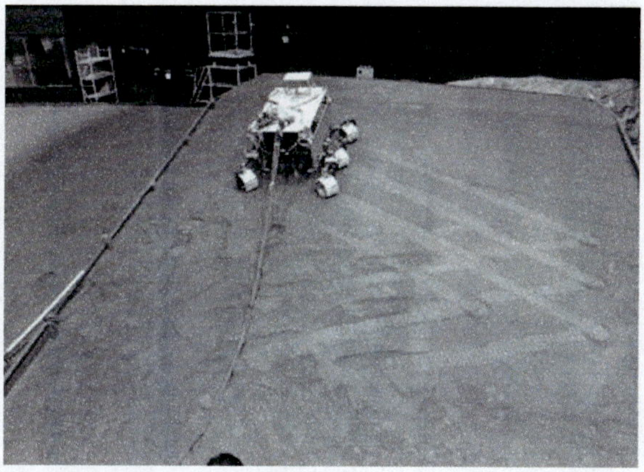

13.6.2.1 Verification of Vehicle Body Climbing Ability

When the planetary exploration mobile robot walks on the planet surface, climbing is an important working state, and its climbing performance determines its adaptability to the terrain. The verification design of the climbing ability of the robot on a simulated ramp equipment and different slope settings is as shown in Fig. 13.17.

Considering the difference between soft and hard grounds, simulated ramps made of planetary soil with different slopes are used in the test field to verify the climbing ability of the robot on soft ground, as shown in Fig. 13.18.

13.6.2.2 Verification of Vehicle Body Obstacle Crossing Capability

There are many obstacles on the planet surface such as stones. The obstacle crossing capability of the planetary exploration mobile robot is also an important technical index to measure its mobile performance. Obstacles with different heights, such as

Fig. 13.19 Verification scheme of soft ground pit crossing capability

100 and 300 mm, are simulated on the ground to test the obstacle crossing capability of the robot.

According to the verification requirements of the pit crossing capability on soft ground, pits of different diameters and depths with planetary soil are simulated in the test field to test the pit crossing capability of the robot on soft ground, as shown in Fig. 13.19.

13.6.2.3 Verification of Vehicle Body Lifting and Wheel Lifting Ability

The planetary exploration mobile robot can parallel lift/lower the vehicle body by the active suspension, thus improving its obstacle crossing capability. The robot can also achieve wheel lifting through the active suspension, including single-side single wheel lifting and both side wheels lifting, so that the robot can smoothly extricate wheels from sinkage and can still move normally in the event of wheel failure.

Hard ground and soft soil environments are constructed for verification, and the vehicle body's lifting and lowering ability from 0 to 500 mm and wheel-lifting walking ability are tested, as shown in Fig. 13.20.

Fig. 13.20 Verification scheme for parallel lifting/lowering ability of vehicle body on soft slopes

References

1. O. Witasse, F. Jansen, MARS EXPRESS ten years of European Mars exploration. Eur. Space Agency Bull. **154**, 40–51 (2013)
2. Q. Wang, D. Yu, J. Yang, Review on mission planning technologies of planetary rover. Spacecraft Eng. **23**(4), 5–12 (2014)
3. H. Li, US Manned Asteroid and Mars exploration. Aerosp. China **8**, 45–50 (2014)
4. R. Siegwart, P. Lamon, Innovative design for wheeled locomotion in rough terrain. Robot. Auton. Syst. **40**, 151–162 (2002)
5. X. Ding, K. Xu, Design and analysis on a robot with new variable structure and wheel legs. J. Central South Univ. **40**(S1), 91–101 (2009)

Chapter 14
Design Example of Planetary Surface Sampling Manipulator

14.1 Overview

14.1.1 Engineering Background

Planetary surface sampling manipulators are space robots that perform sampling tasks on the planet surface. They are usually mounted on a planetary lander or a planetary rover to perform multi-point sampling and other operations.

As we all known, both the Mars lander "Phoenix" and the Mars rover "Curiosity" are equipped with sampling manipulator. "Chang'e 5" explorer developed by China, is designed to realize China's first Moon autonomous sampling and returning mission, and the explorer is also equipped with a sampling manipulator.

In this chapter, the design and verification of a planetary surface sampling manipulator ("sampling manipulator" for short) is introduced as an engineering design example.

14.1.2 Design Requirements

14.1.2.1 Task Requirements

The sampling manipulator is aimed to realize multi-point autonomous sampling on the planet surface, and to encapsulate and transfer the collected samples.

14.1.2.2 Functional Requirements

The functional requirements of the sampling manipulator system are as follows:

(1) Multi-point sample collection function;
(2) Sample transferring, forming, and packaging functions;

Y. Wang, *Space Robotics*, Space Science and Technologies,
https://doi.org/10.1007/978-981-15-4902-1_14

Table 14.1 Technical indices of the sampling manipulator

Indices	Requirements
System mass/kg	≤40
Whole arm span/m	≥3.5
Load-carrying capacity/kg	≥5
End speed/(cm s^{-1})	≥10
End position accuracy/mm	>3
End attitude accuracy/(°)	>0.5
Single time sample acquisition capacity/cm^3	≥200
Single time sample acquisition time/min	≤5

(3) Sample container transfer (grabbing, transfer, release) function;
(4) Visual information acquisition and storage during sampling;
(5) Data uplink and downlink functions for image transmission;
(6) Functions of autonomous braking protection in case of emergency and restart after troubleshooting.

14.1.2.3 Performance Requirements

The main technical indices of the sampling manipulator are shown in Table 14.1.

14.1.2.4 Interface Requirements

(1) Mechanical interface

The sampling manipulator system is installed on the top plate of the lander with screws, and the installation mechanical interface mainly includes joints, hold-down base, and cable brackets.

(2) Electrical interface

The lander provides an unregulated bus of 23–29 V for the sampling manipulator. The average power consumption during operation doesn't exceed 200 W, and the peak power consumption doesn't exceed 350 W.

(3) Thermal interface

Adiabatic material is used between the sampling manipulator and the top plate of the lander, and the manipulator itself has the adaptability to the space thermal environment.

(4) Information interface

The lander data management system provides a 1553B bus communication interface and an LVDS interface for the sampling manipulator.

14.1.2.5 Reliability Requirements

The reliability of the sampling manipulator should meet the requirement of the mission.

14.1.3 Constraints

The sampling manipulator will experience various mission stages, such as the launch stage, the transfer stage, the braking stage, the orbit-flight stage, the landing stage, and the planet surface working stage. The main consideration in the design is the influence of the actual space environment during the whole mission stages.

14.1.3.1 Envelope Dimension Constraints

The sampling manipulator is installed on the top plate of the lander, so the size of the launch vehicle fairing has no influence on the envelope dimension of the manipulator. The main consideration is the dynamics envelope dimension constraint during the working stage.

14.1.3.2 Mechanical Environment Constraints

The constraints during the launch stage, the descending and landing stage, the planet surface working stage are mainly considered.

14.1.3.3 Thermal Environment Constraints

The storage temperature during the flight stage and the high-temperature during the planet surface working stage should be considered.

14.1.3.4 Special Constraints on Planet Surface

The sampling manipulator should consider the special requirements of the planet surface, such as dust, gravity, vacuum/atmosphere, illumination, and geological conditions.

14.2 Overall System Design

14.2.1 Task Analysis

According to the mission profile of the sampling manipulator, the design constraints are as follows:

(1) Landing. The sampling manipulator experiences five stages at a hold-down state before the explorer landing, and will be subjected to vibration, noise, impact, and other mechanical environments and thermal environment. To meet the requirements of these mission stages, geometric envelope, layout, mass, anti-mechanical environment design (vibration, impact), and the adaptability to thermal environment should be considered during the manipulator design.
(2) Sampling. After landing, the sampling manipulator is released by the hold-down mechanism, and the controller is power on. According to the instructions of the controller, the manipulator collects and transfers the sample to the container. During the stage, the control mode, sampling area, sampling form, sampling amount, sampling time, sampling force/torque, visual information acquisition capability, and the adaptability to high-temperature environment should be considered.
(3) Encapsulation and transfer. The sample container encapsulates the collected samples when the samples have reached the specified scale, then the sampling manipulator grabs the container and transfers it to the retuning-capsule at the top of the ascender. During the stage, visual measurement accuracy, workspace, accuracy, speed, path planning, load-carrying capacity, obstacle avoidance capability, fault braking capability, and the adaptability to high-temperature environment should be considered.
(4) Safe avoidance. In order to avoid the potential threat to the safe take-off of the ascender, the sampling manipulator is required to move to a predetermined safe area. The design constraints imposed by this task mainly include reachable space and the obstacle avoidance capability.

In addition to these design constraints imposed by tasks, the constraints on safety, reliability, and service life should also be considered in the design process.

14.2.2 System Design

14.2.2.1 System Configuration and Layout

The configuration and layout of sampling manipulator system mainly include three aspects: on-lander layout, layout of the sampling manipulator body, and layout of the sampling manipulator hold-down points.

Fig. 14.1 Layout of the sampling manipulator at the top of the lander

(1) On-lander layout of the sampling manipulator

As the main equipment of the sampling task, the sampling manipulator is mounted on the top plate of the lander by the hold-down device. The motion interference between the manipulator and the other devices of the lander should be considered in the layout process. The layout of the sampling manipulator at the top of the lander is shown in Fig. 14.1.

(2) Layout of the sampling manipulator body

Considering the installation space and the sampling space requirements of the sampling manipulator, the sampling manipulator body adopts "2 + 1 + 1" configuration scheme, that is, the manipulator is equipped with a shoulder yaw joint and a shoulder pitch joint, an elbow pitch joint and a wrist pitch joint. The configuration is shown in Fig. 14.2. The shoulder yaw joint is installed on the top plate of the lander, and the shoulder yaw joint and pitch joint are connected by a ring adapter; the shoulder pitch joint is connected to the elbow pitch joint by arm A; the output shaft of the elbow pitch joint is connected to arm B, the other end of arm B is connected to the wrist pitch joint; the wrist pitch joint is connected to sampler A and sampler B; a ground-touching sensor is installed outside the wrist pitch joint. In order to ensure the visibility of the sampling process, a near-view camera is installed on sampler A and sampler B, and a far view camera is installed on arm B to monitor the sampling process in a large range.

(3) Layout of sampling manipulator hold-down points

In order to ensure that the sampling manipulator can withstand the mechanical environment without damage in the launching stage, a special hold-down and release mechanism is designed. After landing on the planet surface, the hold-down and release mechanism is unlocked by the pyrotechnic device. Considering the weight distribution of the sampling manipulator, three hold-down points are designed: hold-down point A, B, and C. Point A is located near the far-view camera, point B is

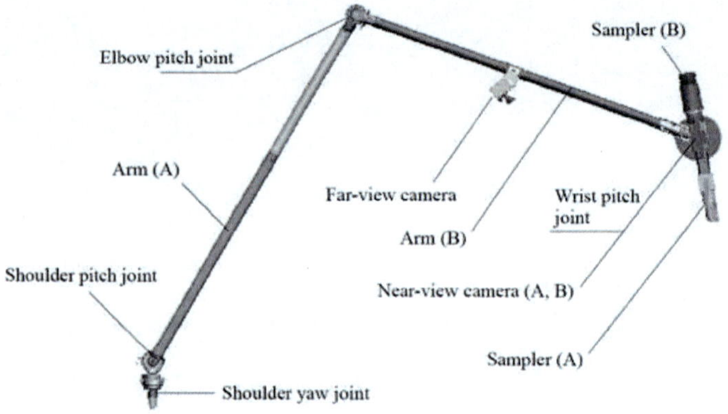

Fig. 14.2 Layout of the sampling manipulator body

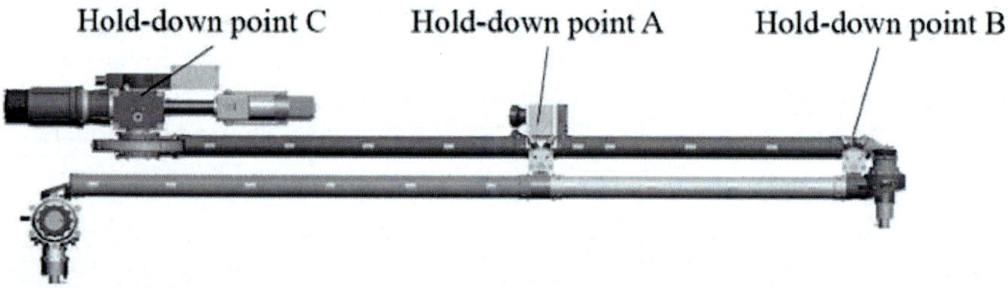

Fig. 14.3 Layout of the sampling manipulator hold-down points

located at the elbow joint, and point C is used for hold-down of the samplers and the wrist joint simultaneously. The layout of the hold-down points is shown in Fig. 14.3.

14.2.2.2 System Kinematics Parameters

In order to meet the sampling space requirements of the explorer system, considering the configuration of the lander, the height from the top surface of the lander to the planet surface, and the height from the top surface of the ascender to the top surface of the lander, the total deployed length of the sampling manipulator is design to be 3.6 m, and its kinematics parameters are determined according to the configuration and layout. The sampling manipulator coordinate system and kinematics parameters are shown in Fig. 14.4.

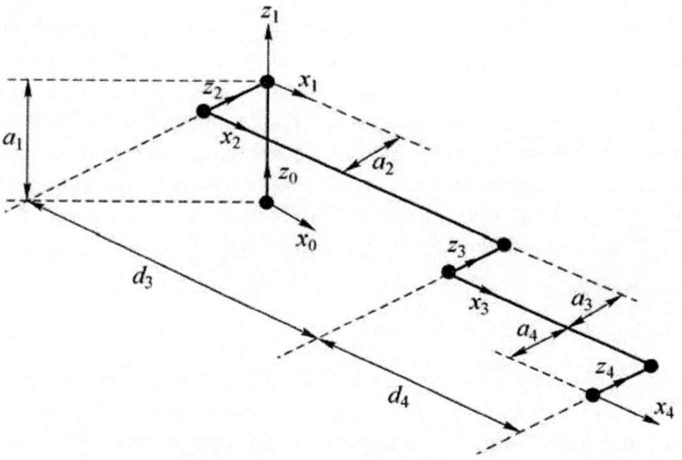

Fig. 14.4 Definition of the sampling manipulator coordinate system and parameters

14.2.2.3 System Working Mode

In order to perform autonomous sampling task on the planet surface, the sampling manipulator is required to have two working modes: preprogramming mode and teleportation mode. Teleportation mode is the default working mode, which can be subdivided into Cartesian space linear programming mode, joint space planning mode, and autonomous operation mode.

(1) Preprogramming mode

The preprogramming mode is mainly used for the sampling manipulator to realize the movement between the predetermined spatial positions according to the predetermined working path, and its main purpose is to ensure the safety, reliability, and convenience of the motion. In the preprogramming mode, the operator selects the preprogramming planning mode, sets the corresponding parameters, and inputs the preprogrammed instruction serial number (prestored in the manipulator control unit). The operation information is sent to the sampling manipulator control unit via the uplink data channel. The control unit calls the corresponding motion instruction sequence according to the instruction sequence number to realize the motion control of the sampling manipulator. The visual information provided by the camera is only used for monitoring.

During the movement, the control unit on the lander performs safety detection of the sampling manipulator, and controls the "emergency stop" of the manipulator in case of danger.

(2) Teleportation working mode—Cartesian space linear programming mode

In Cartesian space linear programming mode, the end of the manipulator can move along a linear path by coordinating the motion of the joints. In the linear programming mode, the operator selects the Cartesian space linear path planning mode and sets the corresponding parameters, inputs the target pose of the sampling manipulator. The central controller calls the path planning algorithm and generates the joint motion command according to the current and target pose information, then the motion command is sent to the sampling manipulator control unit via the uplink data channel. The visual information provided by the camera is used for monitoring purpose only.

During the movement, the control unit performs safety detection of the sampling manipulator, and controls the sampling manipulator's "emergency stop" in case of danger.

(3) Teleportation working mode—joint space planning mode

The joint space planning mode mainly enables the sampling manipulator to reach a set of desired joint angles from a set of initial joint angles, and it has two motion modes: the independent motion of a single joint in the joint space and the coordinated motion of multiple joints. In the joint space planning mode, the ground operator selects the joint space planning mode and sets the corresponding parameters, inputs the desired joint angles of the sampling manipulator. The central controller calls the path planning algorithm, and generates the joint motion command of the sampling manipulator according to the current joint angle, the target joint angle information. The motion command is sent to the sampling manipulator control unit through the uplink data channel. The visual information provided by the camera is used for monitoring purpose only.

During the movement, the control unit performs safety detection of the sampling manipulator, and controls the sampling manipulator's "emergency stop" in case of danger.

(4) Teleportation working mode—autonomous operation mode

In the autonomous operation mode, according to the 3D visual information of the sampling area provided by the lander, the planet surface environment information, the sampling target information determined by the ground operator, and the real-time target pose information provided by the camera, the sampling manipulator joint motion command is generated according to the current and target pose information by calling the path planning algorithm. The generated motion commands are sent to the sampling manipulator control unit through the uplink data channel to realize the motion control of the sampling manipulator. The visual information provided by the camera is used not only for monitoring, but also for the path planning of the whole process.

During the movement, the control unit on the lander performs safety detection of the sampling manipulator, and controls the sampling manipulator's "emergency stop" in case of danger.

14.2.2.4 Task Flow Design

Before landing, the sampling manipulator is installed in the lander and experiences several flight stages with the lander. In the planet surface working stage, the sampling manipulator is unlocked, then it performs multi-point sampling and puts the sample into the container, and encapsulates the sample container using an encapsulating device; then the sampling manipulator transfers the encapsulated sample container to the predetermined position on the ascender, after that the sampling manipulator evades to the predetermined safe position and power off. The workflow of the planet surface is shown in Fig. 14.5, and specifically as follows:

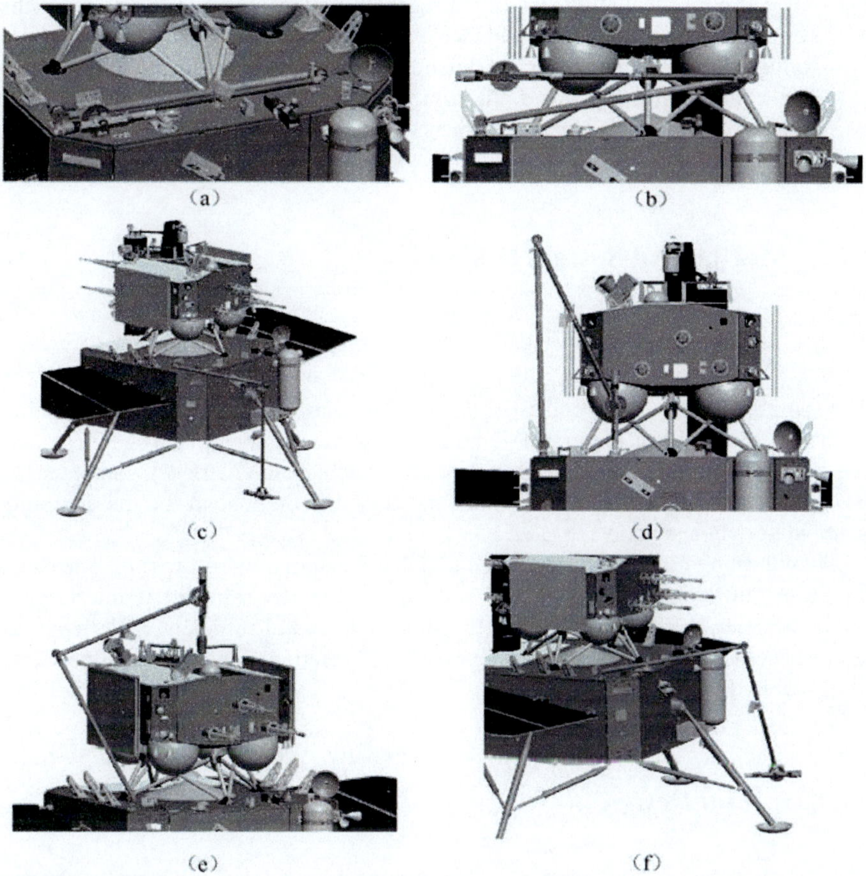

Fig. 14.5 Schematic diagram of the sampling manipulator work flow. **a** Land in hold-down state; **b** unlocked, BIT, deployment; **c** multi-point shovel collection of samples; **d** sample transfer, pouring, forming and collection; **e** container grabbing and transfer; **f** safety avoidance before take-off

(1) Accepting a releases instruction, the hold-down mechanism releases the sampling manipulator. Then the sampling manipulator power on and perform self-inspection under the instruction, and deploys according to the preprogramming mode, as shown in Fig. 14.5b.
(2) The sampling manipulator moves to the planned sampling area, and shovel samples at the designated sampling points. The far-view camera dynamically monitors the sampling process, as shown in Fig. 14.5c.
(3) The sampling manipulator transfers the collected sample to the top of the sample container (installed on top of the lander), opens the sampler and pours the sample into the container; when the sample reaches the specified scale, the sample container is encapsulated by the encapsulating device, as shown in Fig. 14.5d.
(4) After the encapsulation, the sampling manipulator autonomously grabs the sample container under the guidance of visual camera, and transfers the sample container to the sealed device at the top of the ascender, as shown in Fig. 14.5e.
(5) The sampling manipulator evades to a safe position, and the sample container is taken away from the planet by the ascender, as shown in Fig. 14.5f.

14.3 Mechanical System Design

14.3.1 System Composition

The sampling manipulator consists of a shoulder yaw joint, a shoulder pitch joint, an elbow joint, a wrist joint, arm A, arm B, a far-view camera, near-view cameras (A, B), a ground-touching sensor, hold-down mechanism (A, B, C), samplers (A, B), connector brackets, etc. The main mechanical components of the sampling manipulator are shown in Fig. 14.6.

The sampling manipulator adopts internal and external hybrid wiring. The cable bundle is laid along the arm in two ways to reduce electromagnetic interference (EMI) between the power cable and the signal cable, and sufficient cable length is reserved at the moving parts according to the joint rotation range to ensure the safety of the cable during joint rotation.

14.3.2 Joint Design

The joints are the key components of the sampling manipulator, and have the characteristics of compact structure, high-precision, wide speed range, and high reliability. The joints mainly perform the following functions:

(1) Providing mechanical, electrical, and information connections with other components, ensuring the connection accuracy and rigidity;

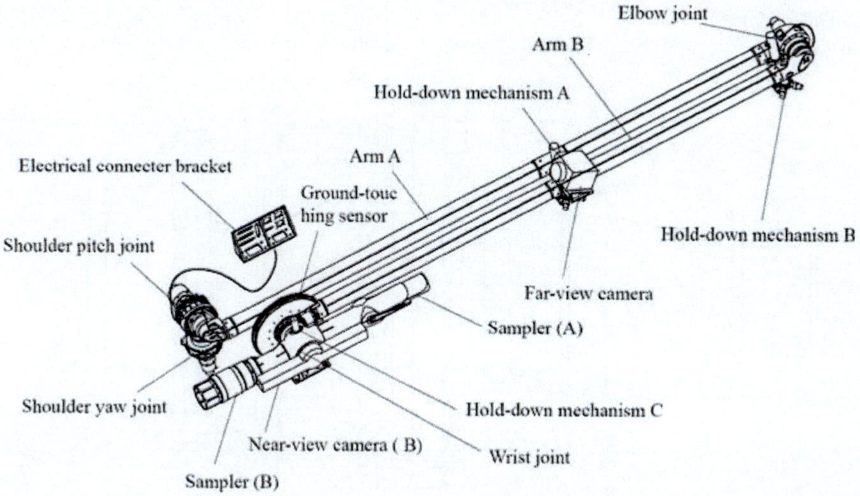

Fig. 14.6 Composition of the mechanical system of the sampling manipulator

(2) Executing the motion commands of the control unit to meet the requirements of position accuracy, speed, and torque;
(3) Braking function, the braking torque should meet the technical indices;
(4) Dust prevention function;
(5) Adaptability to various environmental requirements such as force, heat, vacuum, and radiation that may be experienced during production, integration, and test.

According to the configuration and layout, the sampling manipulator is configured with four joints: two shoulder joints (shoulder pitch, shoulder yaw), one elbow joint and one wrist joint. In order to ensure the compact structure of the sampling manipulator, based on the general consideration of the torque and speed requirements, the four joints adopt a similar structure and transmission design scheme. The joint is composed of a motor, a single-channel resolver, a planetary reducer, a harmonic reducer, a dual-channel resolver, and a joint housing, as shown in Fig. 14.7.

Considering the speed and torque stability requirements of the sampling manipulator, the sine wave BLDC motor is selected as the joint motor; considering the reliability requirement of joints, the joint motor adopts the double-winding design. In order to precisely control the joint motor, a single-channel resolver is configured at the high-speed end (motor end) of the joint as a high-precision measuring component for the absolute position and rotation speed of the motor. In addition, a dual-channel resolver with higher precision is configured at the output end of the joint shaft as a measuring component of the joint output angle and speed, which eliminates the influence of the transmission system error and achieves high-precision closed-loop control of the joint position and speed.

In order to obtain a relatively compact structure and enough output torque, planetary reducer and harmonic reducer are used in series. In order to ensure the reliable

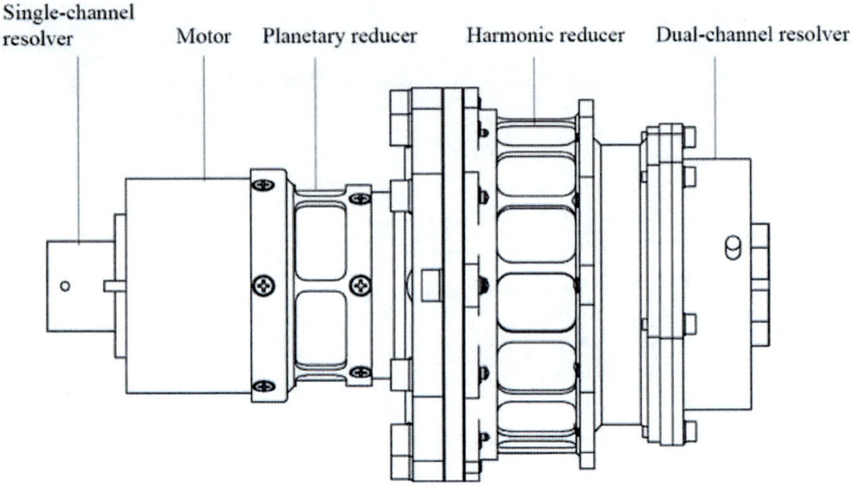

Fig. 14.7 Joint design and composition

operation of the joint transmission components (bearings, planetary reducer gears, harmonic reducer components) under vacuum and high-temperature environment, the joint adopts the hybrid lubrication scheme (solid lubrication + grease lubrication).

14.3.3 Arm Design

The arm is the main force bearing component of the sampling manipulator. Considering the constraints of strength, stiffness, mass, processing performance, and environmental adaptability, the Al/SiC matrix composites material is selected, and an equal-section hollow circular scheme is used to further reduce the mass. In order to reduce the manufacturing difficulty, arm A and B are both designed in two segments, that is, arm A and B are both assembled by two short identical segments through intermediate connectors and inner bushings, as shown in Fig. 14.8. In order to meet the installation requirements of the far-view camera, the mechanical interface for the far-view camera is designed on the arm B connector.

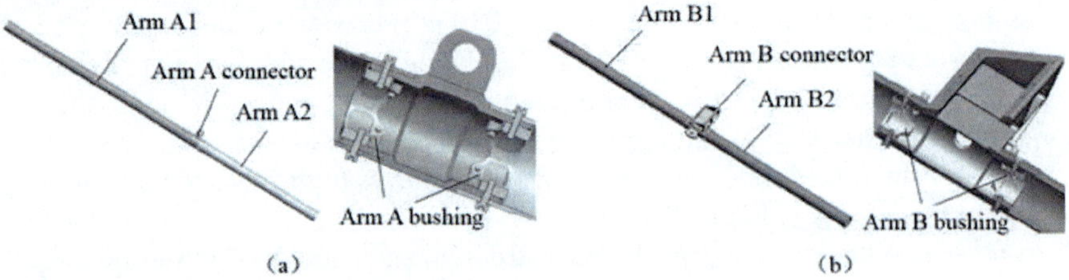

Fig. 14.8 Arm assembly of the sampling manipulator. **a** Arm A assembly; **b** arm B assembly

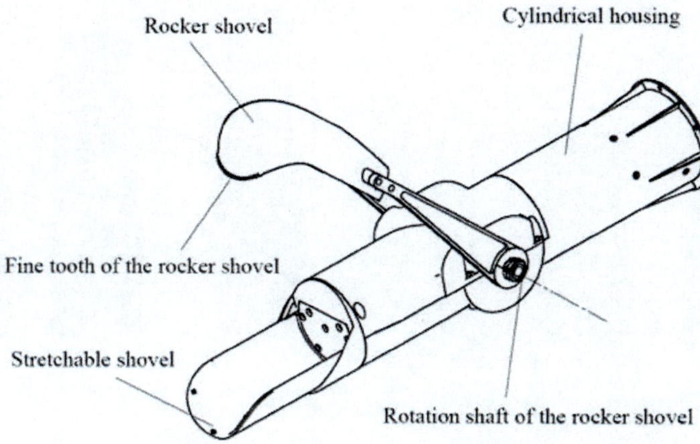

Fig. 14.9 Structure of sampler A

14.3.4 Sampler Design

The sampler is a core part of the sampling manipulator, which is used for shovel collection of samples, including sampler A and sampler B.

Sampler A is a shovel-type sampling tool, which is composed of two parts: one is a rocker shovel that can achieve rotational motion, the other is a stretchable shovel that can achieve stretching movement. The rocker shovel has a hook-shaped configuration and its edge is in a dentiform profile, which can realize excavation collection of the samples on the planet surface; the stretchable shovel can realize shovel collection of the samples by stretching movement. The rocker shovel and the stretchable shovel can also pick up and collect block samples by cooperation. The structure of sampler A is shown in Fig. 14.9.

Sampler B is a vibratory tubular sampling tool. The driving system of sampler B is located at the upper part of the sampler, and the tubular sampling head is located at the lower part of the sampler. The sampling head is mainly composed of a sampling tube and flaps. In the sampling process, the sampling tube is mainly used to store the sample, and the flaps are used to shear the samples. When the flap is fully extended, the head will close; when the flap is fully retracted into the head, the head will open. The working process of sampler B is shown in Fig. 14.10.

14.3.5 Hold-Down Mechanism Design

The sampling manipulator is designed with a total of three hold-down points (see the previous text for layout): hold-down points A, B, and C. The point A is located near the far-view camera, point B is located at the elbow joint, and point C holds

Fig. 14.10 Working process of sampler B

down the sampler and the wrist joint assembly. The hold-down mechanism is mainly composed of a hold-down base, a cushion pad, a cutter, a hold-down rod, a separation spring, and a locking nut, as shown in Fig. 14.11.

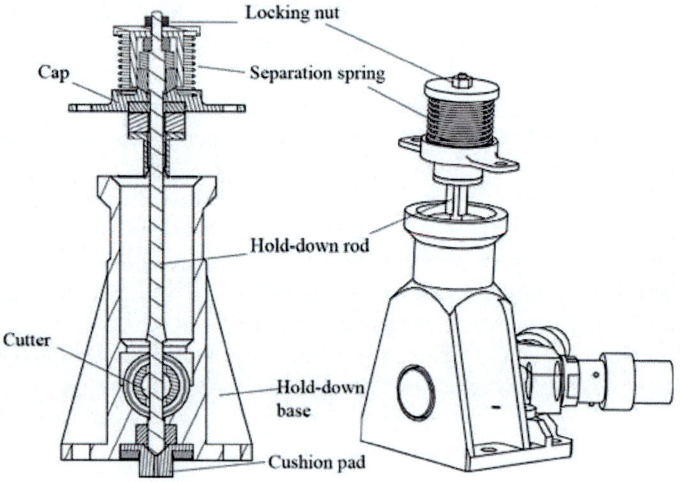

Fig. 14.11 Schematic diagram of the hold-down and release mechanism

14.4 Control System Design

14.4.1 System Architecture Design

The sampling manipulator control system is mainly composed of a task support system located on the ground and a control unit located on the lander. The sampling manipulator can be controlled with the assistance of the explorer measurement and control system and the data management system. The hardware connection relationship is shown in Fig. 14.12.

The ground mission support system is mainly composed of a central control unit, a dynamics simulation unit and a telemetry drive simulation unit. By using on-orbit state information, such as the system telemetry data, 3D visual information of the sampling area, and planetary surface environment information provided by the lander, the ground mission support system realizes the sampling manipulator working condition analysis and planning control, and generates control strategy and performs simulation verification to support the on-orbit operation of the sampling manipulator.

14.4.2 Control Unit Design

In order to realize the lightweight design, the control unit adopts centralized control mode, that is, all control hardware including joint servo control, whole arm motion planning control, and encapsulating device motion control are integrated into the control unit of the lander. The control unit adopts dual-system cold standby design and uses the 1553B bus to communicate with the lander data management system. The control unit is mainly composed of four hardware circuits:

(1) Operation control circuit: It used a FPGA+CPU operation control circuit to realize the functions of motor control, path planning, and task management. The circuit mainly includes a control FPGA circuit, a calculation CPU circuit, an AD acquisition circuit, an internal command drive circuit, an external temperature acquisition, and switch interface circuit.
(2) Motor drive circuit: It realizes the motor drive output, current, and voltage detection signal output. The circuit uses a motor drive integrated chip which mainly includes a drive chip and a current detection circuit.
(3) Image processing circuit: It realizes power supply, thermal control of the cameras, and the receiving/processing/transmission of camera image data. The circuit mainly includes a camera power supply control circuit, a camera thermal control circuit, an FPGA and DSP circuit for image data processing, and an associated interface circuit.
(4) Power supply and acquisition circuit: It mainly realizes the equipment power supply, resolver signal acquisition, bus power protection, and motor power

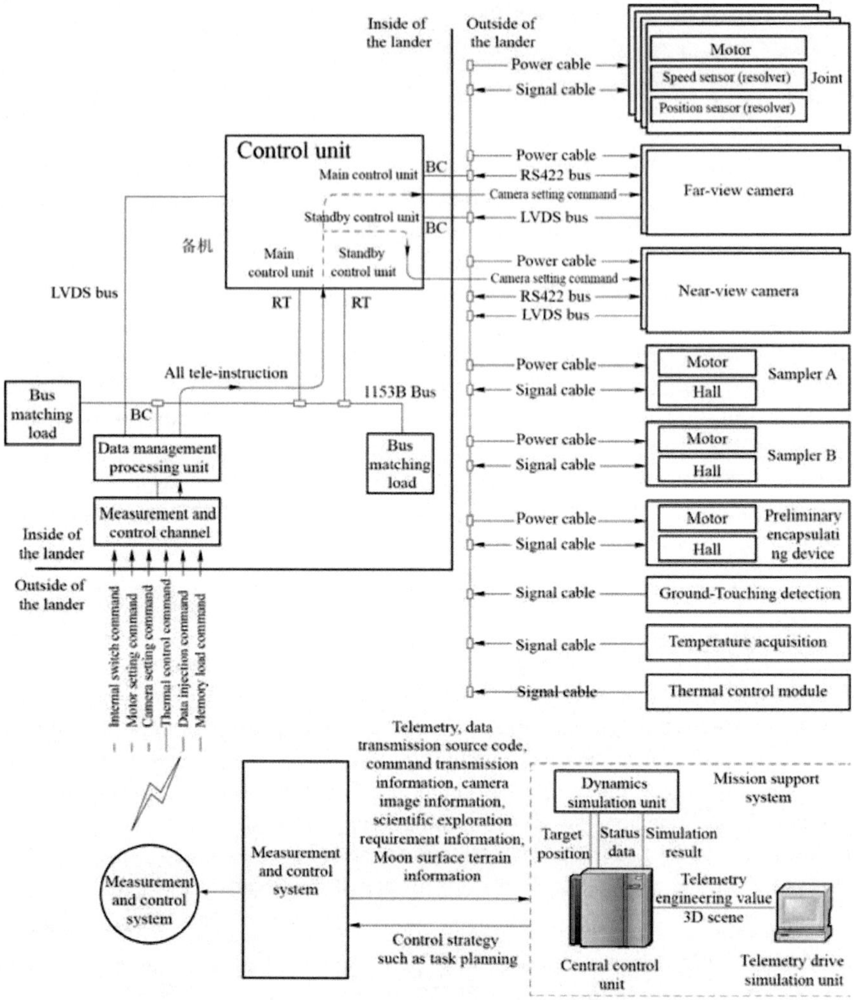

Fig. 14.12 Hardware structure of the sampling manipulator control system

on/off control. The circuit includes a power conversion circuit, a revolver signal acquisition circuit, a bus power protection, and a motor power on/off control circuit.

14.4.3 Joint Control System Design

Joint control is the core of motion control of sampling manipulator. The control unit needs to realize the control of the four joint permanent magnet motors, two DC motors of the sampler A and B, and two DC motors of the encapsulation device. The motor drive uses an integrated H-class motor-specific drive chip. The single-chip has a built-in MOSFET drive circuit and a filter circuit with a maximum switching frequency of 100 kHz, and can drive the three-phase winding of the motor.

The joint motor adopts double-winding design, and relays are arranged between the motor drive circuits of the main and standby windings, respectively, for isolation.

14.4.4 Control System Software Architecture

The control unit software adopts layered design mode, the bottom layer is the real-time operating system software, and the upper layer is the control software of the sampling manipulator. The bottom layer software is implemented by modifying a mature operating system board-level support package, which can greatly reduce the development difficulty, improve the testability and readability of software, and enhance the software safety and reliability. The control software provides functions such as task management, task scheduling, inter-task message communication, interrupt management, and memory management.

The master control software is the core of the control unit software used to control and manage the sampling tasks according to the ground commands or the prestored commands. The master control software has the following specific modes:

(1) Storage mode: The control unit enters this mode by default after power on. In this mode, it only performs normal telemetry acquisition, command uplink, and camera thermal control functions. No sampling task control of the manipulator is performed.
(2) Servo standby mode: In this mode, the control unit software responds to the motion control command, sets the parameters of the control unit, and performs forward kinematic calculation with joint position telemetry data to obtain the current end pose of the sampling manipulator.
(3) Sampling manipulator motion control mode (4 modes): In the servo standby mode, it receives mode switching command to switch to four path planning modes, performs path planning operation according to the corresponding algorithm and the motion control process requirements, and output the generated commands to the FPGA to realize the motion control of the sampling manipulator joint motors.
(4) Sampler motion control mode: it receives sampler motion control commands, performs operations according to the requirements of the motion control flow, and outputs the generated commands to the FPGA to realize the sampler motion control.
(5) Encapsulation device motion control mode: it receives related encapsulation device motion control commands, performs operations according to the requirements of the motion control flow, and outputs the generated commands to the FPGA to realize the motion control of the encapsulation device.

14.5 Perception System Design

14.5.1 Ground-Touching Sensor Design

The ground-touching sensor is located at the end of the sampling manipulator and coaxially mounted with the wrist joint. It is mainly used to obtain the accurate distance between the sampler and the planet surface when the sampling manipulator performs sampling, so as to control the sampling depth. The ground-touching sensor adopts a round-cake layout, as shown in Fig. 14.13. It is mainly composed of an inner support ring, an outer support ring, a positive copper ring, a negative copper ring, and a bonding copper ring. The positive copper ring and the negative copper ring are mounted on the inner support ring, and the bonding copper ring is mounted on the outer support ring, the inner and outer support rings are connected by several compression springs. The inner and outer support rings are made of aluminum alloy material, and the copper ring is made of bronze. When the ground-touching sensor is grounded, the outer support ring will be pressed and the compression spring between the inner and outer rings will be pressed, so that the bonding copper ring will contact with the positive and negative copper rings. Once the positive and negative copper rings are connected, a "ground-touching" signal will be sent out; when the ground-touching sensor keeps away from the ground, the outer support ring will be reset by the compression spring, and the bonding copper ring will be separated from the positive and negative copper rings, and a "no ground-touching" signal will be sent out.

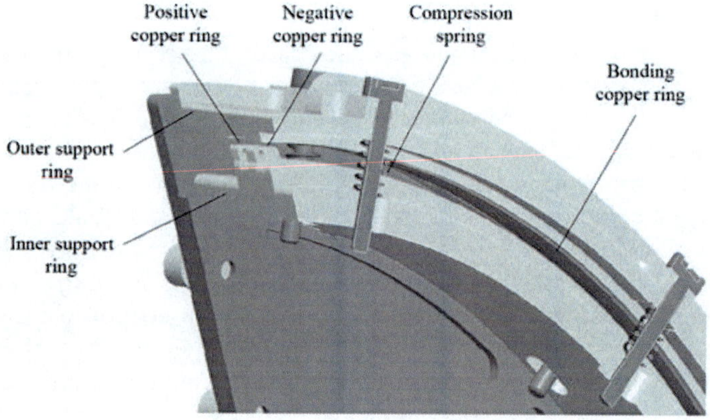

Fig. 14.13 Ground-touching sensor

14.5.2 Visual System Design

To meet the visual monitoring and visual guidance requirements in the sampling task, the sampling manipulator is designed with a far-view camera and two near-view cameras, all of which realize the image information transmission by the LVDS bus protocol.

The far-view camera is installed on the arm B connector of the sampling manipulator, mainly used for visual monitoring of the tasks, such as manipulator sampling, sample transfer and pouring, sample encapsulation, sample container grabbing and transfer, and sample container release.

The far-view camera mainly composes of an optical lens and a focal plane box. The lightweight structure design requires the integration of optical, mechanical, electronic, and thermal components to achieve the purpose of compact structure and weight reduction. In addition, the structural design should fully consider the thermal deformation effects of different materials in the thermal environment. The structure of the far-view camera is shown in Fig. 14.14.

The two near-view cameras adopt a cold standby and an integrated design by sharing the mechanical and electronic parts, that is, cameras A and B share a common structural base, and share the electrical interfaces such as the control bus interface, image data transmission interface. The integrated design can effectively reduce the mass of the near-view cameras and the number of cables. The composition of the single near-view camera is shown in Fig. 14.15.

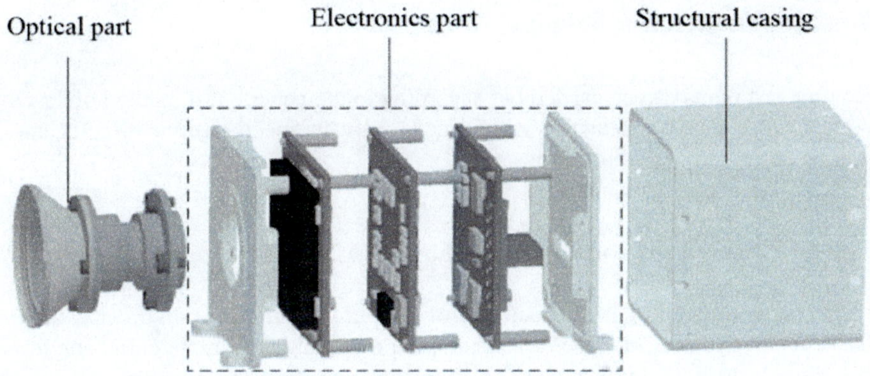

Fig. 14.14 Structure of the far-view camera

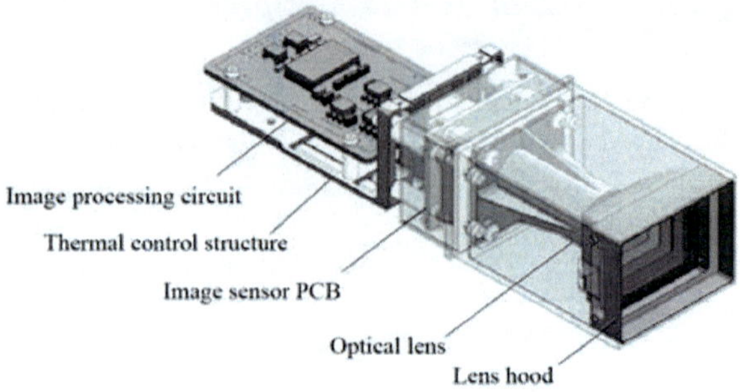

Fig. 14.15 Schematic diagram of a single near-view camera

14.6 Design Verification

14.6.1 Verification Items

As the key equipment of sampling and returning mission, in addition to the necessary environmental tests (Table 14.2), the sampling manipulator also needs to carry out special tests which have an important impact on the success of the mission, such as ground anti-dust test, high- and low-temperature adaptability test and adaptability test of landing slope.

14.6.2 Verification Scheme

Routine test items can be carried out according to the general test method of spacecraft, so no detailed introduction will be made, only the specific verification schemes will be described.

14.6.2.1 Sampling Function and Performance Test

Planet surface sampling is the main task of sampling manipulator. In order to verify the function and performance of the sampling manipulator, sample collection tests are carried out by using a variety of simulated samples and setting multiple sets of operating parameters. The specific method is: the sampling manipulator performs sample collection, and the parameters such as the position, attitude, and motion speed of the end sampler relative to the sampling area are controlled to simulate the actual sampling condition and test its various functions; in this process, the performance parameters such as the load-carrying capacity, motion speed, motion accuracy, and sampling amount of the sampling manipulator should be measured.

Table 14.2 Planetary surface sampling manipulator system test items

No.	Test item	Remark
1	Function tests	Multi-point sampling function
		Sample transfer, forming and packaging function
		Sample container transfer (grabbing, transfer, release) function
2	Performance tests	Load-carrying capacity
		End speed
		End position accuracy
		End attitude accuracy
		Single sampling capability
		Single sampling time
3	Function and performance test of Component	Component performance tests, such as joint, sampler, preliminary encapsulation device
4	Environment test of Component	Component environmental adaptability test, such as joints, samplers
5	Interface matching test	Matching of mechanical and electrical interfaces between internal devices of the test system and between the system and other devices of the explorer
6	Failure mode test	Troubleshooting exercises for typical failure modes to verify the effectiveness of the troubleshooting measures
7	Environment test	Space environment tests, such as mechanics, thermal, vacuum, and EMC
8	Service life and reliability test	Service life and reliability tests

14.6.2.2 Anti-thermal Environment Test

The sampling manipulator needs to continuous sampling. In order to verify the performance of the sampling manipulator and its components in vacuum/atmosphere and high-/low-temperature environment, the anti-thermal-vacuum environment test is required during the product development stage. Use vacuum tank and infrared heater are used to simulate the thermal vacuum environment, as shown in Fig 14.16.

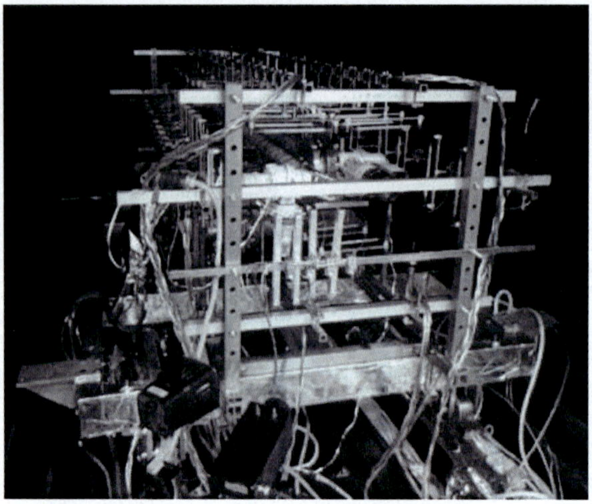

Fig. 14.16 Anti-thermal environment tests for the sampling manipulator

14.6.2.3 Anti-dust Test

The sampling manipulator system may be affected by floating dust during operation. If the dust enters into the moving components, it may affect the motion performance of the moving components. Therefore, the sampling manipulator system requires performing anti-dust tests. The test is carried out on a special dust simulation equipment, which can control the parameters such as dust particle size, dust concentration, dusting speed, temperature, and test time to simulate the actual working conditions on the planet surface.

14.6.2.4 Motion Performance Test Under High Temperature

The high-temperature environment of the planet may have an adverse effect on the motion performance of the sampling manipulator. In order to verify the influence of high-temperature environment on the positioning accuracy and other performance parameters, it is necessary to carry out high-temperature motion performance test. In order to simulate the gravity and thermal environment of the planet simultaneously, an inclined floating platform and an infrared heating equipment is used, as shown in Fig. 14.17.

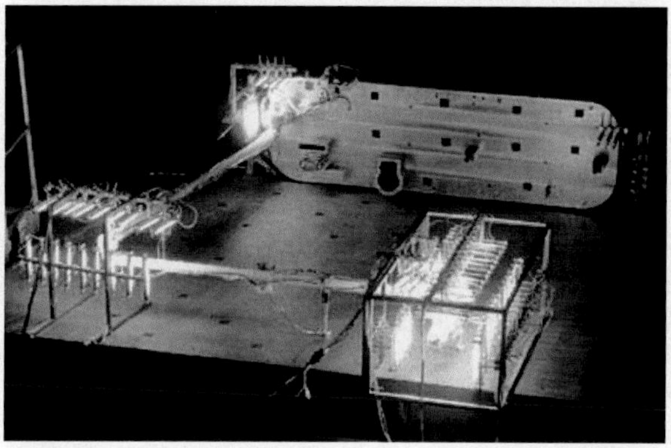

Fig. 14.17 Motion performance test under high temperature

Chapter 15
Current State of Space Robots

15.1 Evolution of Space Robots [1, 2]

In 1981, the Shuttle Remote Manipulator System (SRMS), developed by Canada, was put into orbit by the space shuttle Columbia, becoming the world's first on-orbit operating robot for space applications [1]. After that, Germany developed the technical demonstration robot ROTEX, which verified the robot's ground teleoperation technology. In 1997, a free-flying robot (ETS-VII) developed by Japan entered orbit and completed technical verification of on-orbit operations such as space target capture and ORU replacement [3]. In 2007, the United States completed a similar verification with "Orbit Express" project [4]. The construction and operational needs of the International Space Station (ISS) strongly promote the development of space robotic technology. In 2001, the Space Station Remote Manipulator System (SSRMS), was launched into orbit; in 2008, the Special Purpose Dexterous Manipulator (SPDM) and the Japanese Module Remote Manipulator System (JMRMS) entered the International Space Station [5–8]. These space robots have been successfully applied in the construction, maintenance of the International Space Station, and extravehicular activity. In 2011, the US humanoid space robot Robonaut2 entered the International Space Station and successfully carried out the technical demonstration of various smart operations, proving that there is a great potential for space robots to replace astronauts in space operations [9]. Up to now, countries around the world are constantly proposing new concepts of space robot applications. For example, the United States has proposed the Phoenix Program and the Robotic Servicing of Geosynchronous Satellites (RSGS) program, Germany has proposed the German Orbital Servicing Mission (DEOS) project [10].

The application of planetary exploration robots first appeared in the 1970s. Russian lunar rover Lunokhod1 and Lunokhod2 successfully landed on the moon in 1970 and 1973, marking the practical application of planetary exploration robot. On July 30, 1971, the US manned lunar rover LRV carried by Apollo 15 landed on the moon, becoming the world's first manned planetary exploration robot for space applications. In 1971, Russian Mars rover PROP-M landed on Mars with the

Y. Wang, *Space Robotics*, Space Science and Technologies,
https://doi.org/10.1007/978-981-15-4902-1_15

Soviet automatic research station Mars 3, becoming the world's first Mars exploration robot (unfortunately, the deployment of the Prop-M rover failed). From 1997 to 2004, the US "Sojourner", "Spirit" and "Opportunity" Mars rovers successively landed on Mars, pushing the Mars exploration to a climax [11]. In 2012, the US "Curiosity" Mars rover successfully landed on Mars, becoming the most powerful Mars exploration robot to date. It should be noted that the later US Mars rovers were all equipped with a manipulator as the main tool for exploration tasks, so that the planetary exploration robots have the ability to operate on the planetary surface. Since 2005, in order to meet the demands of the United States for comprehensive and in-depth exploration of the moon, Jet Propulsion Laboratory has led the development of a large lunar surface mobile robot—the All-Terrain Hex-Legged Extra-Terrestrial Explorer (ATHLETE) [12]. Its main feature is to use the legged locomotion and can change to wheeled locomotion by changing the configuration. It is a new type of planetary exploration robot with good adaptability to the planetary surface.

As the most powerful tool for humans to explore and utilize space, space robotics will be widely used with the deepening of human space activities.

15.2 Current Researches on Space Robotics

So far, space robotics has been widely used in engineering tasks such as manned space stations and planetary exploration. Moreover, spacecraft on-orbit service based on space robotics technology has received increased attention from countries all over the world, and several on-orbit experiments have been performed. With the development of robot products for the above applications, the types of products have become increasingly rich. According to its application, space robotics can be divided into three states: engineering application state, including the completed or ongoing engineering application; technical demonstration state, including the completed or ongoing flight demonstration state; and underdevelopment state, which means it is currently in the state of product development and ground verification. The following will introduce the state-of-the-art applications of typical products of space robots.

15.2.1 On-Orbit Operation Robots

Since the United States began to develop space robots in the 1970s, the development of space robots has experienced a history of nearly 50 years. During this period, a number of space robot products have been developed and applied. Table 15.1 summarizes the worldwide developments of on-orbit operation robots that have been made public.

Table 15.1 On-orbit operation robots

No.	Product/mission	Organization/country	Application	Application state
1	Shuttle remote manipulator system (SRMS)	Canada Space Agency (CSA), Canada	Space shuttle state inspection; assembly and construction of ISS; satellite capture and release; module capture and docking assistance	Engineering application: first launched in 1981, and retired in 2011. A total of 5 sets were developed
2	Space station remote manipulator system (SSRMS)	Canada Space Agency (CSA), Canada	ISS module assembly and service; visiting spacecraft docking assistance; equipment handling; assisting astronaut EVA	Engineering application: launched in 2001, still in service
3	Special purpose dexterous manipulator (SPDM)	Canada Space Agency (CSA), Canada	ISS small equipment operation; Technical verification for satellite refueling; module replacement; ISS maintenance services	Engineering application: launched in 2008, still in service
4	Japanese experimental module remote manipulator system (JEMRMS)	Japan Aerospace Exploration Agency (JAXA), Japan	Payload handling, installation operations on Exposed Facility of Japanese Experimental Module of the ISS	Engineering application: launched in 2008, still in service
5	European robotic arm (ERA)	Europe Space Agency (ESA)	Inspection and maintenance of the Russian module of the ISS; Extravehicular payloads handling; Equipment replacement, etc.	Under development
6	ROTEX manipulator	Deutsches Zentrum für Luft-und Raumfahrt (DLR), Germany	Truss structure assembling; Electrical plug connecting/disconnecting; ORU exchange; Floating object grasping; Teleoperation	Technical demonstration: launched in 1993, completed technical demonstration
7	Orbit express manipulator	Defense Advanced Research Projects Agency (DARPA), USA	On-orbit cooperative target capture; ORU replacement; On-orbit refueling	Technical demonstration: launched in 2009, completed technical demonstration

(continued)

Table 15.1 (continued)

No.	Product/mission	Organization/country	Application	Application state
8	Engineering Test Satellite 7 (ETS-VII) robot arm	National Space Development Agency (NASDA), Japan	On-orbit cooperative target capture; ORU and truss structures handling; Antenna assembly	Technical demonstration: launched in 2007, completed technical demonstration
9	Robonaut2 humanoid robot	National Aeronautics and Space Administration (NASA), USA	Instrument operation in ISS; inspection and replacement of extravehicular equipment; human-robot cooperation	Technical demonstration: launched in 2011, completed technical demonstration
10	TianGong-2 space lab robot	Harbin Institute of Technology (HIT), China	In-cabin smart operation test; on-orbit maintenance test using tools	Technical demonstration: Launched in 2016, completed technical demonstration
11	Phoenix program robot arms	Defense Advanced Research Projects Agency (DARPA), USA	Non-cooperative target capture; Retired satellite payload reusing; load reassembly and reconstruction	Under development
12	German Orbital Servicing Mission (DEOS) robot arm	Deutsches Zentrum für Luft-und Raumfahrt (DLR)	Non-cooperative target capture	Under development
13	Skyworker on-orbit servicing robot	Carnegie Melon University, USA	Autonomous assembly, monitor and maintenance of large space structures	Under development
14	Lemur IIb mobile robot	National Aeronautics and Space Administration (NASA), USA	Spacecraft assembly and maintenance	Under development

15.2.2 Planetary Exploration Robot

To date, there are two types of planetary exploration robots. One is the operation robots (or manipulator) configured on an explorer (lander) to perform sampling and manipulating functions, such as the sampling manipulator on the US "Phoenix" Mars explorer; the other is the mobile robots with locomotion and exploration capability (generally called a rover), such as the US "Curiosity" Mars rover. The rovers can be further divided into manned rovers and unmanned rovers. Some rovers are also equipped with manipulators with operation capabilities, such as the three generations of Mars exploration rovers developed in the United States ("Sojourner", "Spirit and Opportunity", and "Curiosity").

From the perspective of the current application, mobile robots mainly use wheeled mobile systems, including the four-wheel, six-wheel, and eight-wheel types. In particular, the six-wheel mobile system with rocker-bogie configuration occupies the absolute dominant position, which is the result of optimization of overall considerations about mobile performance, resource occupation, and compact layout, and also is the most widely used configuration worldwide. In recent years, a new type of mobile robot has emerged: leg mobile robot is more flexible and stable, which can walk on various terrains (such as plain, sand, and mountains) and cross large obstacles (such as ditch, ridge, and steps), and can replace human to perform various exploration and operation tasks in complex and high-risk environments; the disadvantage is that the moving speed is relatively slow; the leg-wheel hybrid exploration robot is a new type of mobile robot developed by combining the advantages of the legged and wheeled mobile robots, with more advantages in both terrain adaptability and moving speed; the humanoid walking robot is a human-like robot, it adopts a two-leg configuration and imitates human walking to realize the body support and continuous coordinated movement by the executive device instead of muscles. Compared with the traditional wheeled mobile systems, there are still some difficulties in the engineering application of the humanoid two-legged mobile system.

The main countries and regions that carry out development and technology research of planetary exploration robot products include the United States, Russia, Europe, and China. The typical products are shown in Table 15.2.

15.3 Summary

Space robots are widely used in various space fields, play an important supporting role in human space activities such as manned space flight and deep space exploration. For future tasks such as on-orbit assembly of large facilities, on-orbit maintenance of faulty spacecraft, and the development and utilization of space resources, robotics research institutions around the world are constantly putting forward new space robot concepts, and are developing various prototypes to carry out ground demonstrations of space missions, all of which indicate that space robotics will play an irreplaceable role in the future human exploration of the unlimited universe. However, compared with the task requirements of future robots, the current space robots still have the following shortcomings:

(1) The tasks that can be supported are relatively simple.

From the perspective of the applications that have been implemented, the tasks performed by space robots are relatively simple: for the on-orbit operation robots, they can mainly perform simple tasks such as the capture of hovering targets or low-speed cooperative targets, module transfer, and equipment handling, known as intra-

Table 15.2 Planetary exploration robots

No.	Product/mission	Organization/country	Application	Application state
1	Lunokhod1 rover	Russia	Lunar surface exploration	Engineering application: Launched in November 1970, mission ended in September 1971
2	Lunokhod2 rover	Russia	Lunar surface exploration	Engineering application: Launched in January 1973, mission ended in June 1973
3	Apollo15 manned rover	NASA, US	Manned lunar surface exploration	Engineering application: launched in July 1971, mission ended in August 971
4	"Sojourner" Mars rover	NASA, US	Mars surface exploration	Engineering application: Launched in December 1996, mission ended in September 1997
5	"Spirit and Opportunity" Mars rover	NASA, US	Mars surface exploration; sampling and scientific experiment	Engineering application: the "Spirit" was launched in June 2003, the "Opportunity" was launched in July 2003, and all landed on Mars in January 2004. The "Spirit" ended its mission in March 2010, and the "Opportunity" ended its mission in February 2019
6	"Curiosity" rover	NASA, US	Mars surface exploration, scientific research	Engineering application: Launched in November 2011, landed on Mars in August 2012, still in service
7	"Yutu-1" Lunar rover	China	Lunar surface exploration	Engineering application: Launched in November 2013, still in service
8	"Yutu-2" Lunar rover	China	Lunar surface exploration	Engineering application: Launched in November 2018, still in service
9	Marsokhod rover	Russia	Mars surface exploration	Under development
10	ATHLETE wheel-legged mobile robot	NASA, US	Lunar multi-terrain exploration; Transportation	Under development
11	Lemur IIb mobile robot	NASA, US	Planetary surface exploration (adaptive to steep terrain)	Under development

(continued)

Table 15.2 (continued)

No.	Product/mission	Organization/country	Application	Application state
12	Exomars rover	EADS	Mars surface exploration; Sampling	Under development
13	Mars 2020 Rover	NASA, US	Mars surface exploration; Scientific experiments; Sampling	Under development
14	Space Exploration Vehicle (SEV) planetary exploration version	NASA, US	The planetary exploration version SEV is used for manned exploration on planetary surface (SEV's sealed module can be used separately for manned space flight)	Under development

and extra-vehicular equipment operation, while the tasks such as high-speed target acquisition, precision assembly, multi-arm coordination or multi-robot joint operations are still in the ground test stage, their on-orbit application still takes time; for planetary exploration robots, the current robots can perform the exploration missions on relatively flat terrain, the 30° slope or above is still a challenge for the robots. Sampling operation is also limited by geological conditions, and sampling in a hard and deep layer is still difficult.

(2) Insufficient space environment adaptability.

At present, space robots can basically meet the requirements of the low-load long-life lubrication, the load-carrying capacity considering the degradation of material performance and the electronic components reliability through design methods, but the problems of long-life transmission and lubrication under high-precision, high-load conditions, the adaptability of materials and components in extremely high- or low-temperature environments, and the energy supply capacity under long-term non-lighting conditions are yet to be resolved.

(3) Relatively low autonomy and intelligence.

There are mainly two operational control modes of current space robots: preprogrammed automatic control and teleoperation control modes, both of which have natural defects. The former requires the robot to perform actions according to the preprogrammed control program, which greatly limits the types of tasks and execution efficiency of the robots; the latter requires humans to perform continuous control of the robot, this human-in-the loop control mode greatly increases the human workload, and there is also a delay between human and robot, which results in low operation efficiency and reduced operation safety as the latency increases. In both control modes, the robot has almost no autonomy. At present, the robot's autonomous target capture task under visual guidance has completed the on-orbit test and will subsequently be put into application gradually to improve the autonomy of the space robots. However, compared with the urgency and high requirements of the missions, there are still clear gaps to be filled.

References

1. B. Siciliano, O. Khatib, *Handbook of Robotics* (Springer, New York, 2007)
2. K. Yoshida, Achievements in space robotics. IEEE Robot. Autom. Mag. **16**(4), 20–28 (2009)
3. M. Oda, Space robot experiments of NASDA's ETS-VII satellite—an overview of the project and experiment results. J. Space Technol. Sci. Publ. Jpn. Rocket Soc. **14**(1), 146–150 (2013)
4. R.B. Friend, Orbital express program summary and mission overview. SPIE Sens. Syst. Space Appl. **II**, 1–11 (2008)
5. G. Gibbs, S. Sachdev, Canada and the international space station program: overview and status. Acta Astronaut. **51**(1–9), 591–600 (2002)
6. A. Abramovici, The special purpose dexterous manipulator (SPDM) systems engineering effort. J. Reducing Space Mission Cost **1**, 177–200 (1998)
7. E. Coleshill, L. Oshinowo, R. Rembala et al., Dextre: improving maintenance operations on the international space station. Acta Astronaut. **64**(9–10), 869–874 (2009)
8. N. Sato, Y. Wakabayashi, JEMRMS design features and topics from testing, in *Proceeding of the 6th International Symposium on Artificial Intelligence and Robotics and Automation in Space: i-SAIRAS 2001* (Canadian Space Agency, St-Hubert, Quebec, Canada, 2001)
9. T. Ahlstrom, A. Curtis, M. Diftler, et al., Robonaut 2 on the international space station: status update and preparations for IVA mobility, in *AIAA SPACE 2013 Conference and Exposition* (2013), pp. 211–230
10. D. Reintsema, J. Thaeter, A. Rathke, et al., DEOS—The german robotics approach to secure and de-orbit malfunctioned satellites from low earth orbits, in *Proceedings of the 10th International Symposium on Artificial Intelligence, Robotics and Automation in Space: i-SAIRAS 2010* (Sapporo, Japan, 2010), pp. 244–251
11. L. Pedersen, D. Kortenkamp, D. Wettergreen, et al., A survey of space robotics, in *Proceedings of the 7th International Symposium on Artificial Intelligent, Robotics and Automation in Space: i-SAIRAS 2003* (Nara, Japan, 2003)
12. B.H. Wilcox, T. Litwin, J. Biesiadecki et al., ATHLETE: a cargo handling and manipulation robot for the moon. J. Field Robot. **24**(5), 421–434 (2007)

Chapter 16
Future Prospects of Space Robots

Space robots have appeared and developed along with mission requirements and has been continuously upgraded and iterated with the propulsion of related technologies. It is foreseeable that the future development of space robotics will be closely related to the tasks of manned spaceflight, deep space exploration, and on-orbit services. Driven by the mission demands, space robotics will integrate the latest achievements in the development of science and technology, and will be constantly improved in form, function, and performance to meet the needs of space missions.

16.1 Space Robot Products

The development of space missions first raises a new morphological demand for space robot. Traditionally, space robots responsible for module/equipment transfer and handling, test load picking and placing tasks are mainly arm-type operation robots. Under the complicated and precise requirements on space operation tasks, two-arm coordinated operation robots and humanoid robots have appeared successively. To meet the requirements of narrow space operation, space application of flexible robots and soft robots have been gradually explored. In the celestial exploration such as moon, Mars, the traditional mobile robots have adopted the form of steering and suspension mechanisms. With the diversification and complexity of the terrain to be explored and the increased requirement for exploration efficiency, the legged, jumping, and flight movement are expected to become new choices for planetary exploration robots. In the future, with the emergence of new mission requirements and the development of science and technology, more and more new robots will enter space applications.

In addition, the rapid development of technologies such as the Internet of Things (IOT), cloud computing, and Artificial Intelligence (AI) has also promoted the profound changes in space robots technology. On the one hand, due to the expanding scope of human exploration of the universe, and on the other hand, due to the demand

Y. Wang, *Space Robotics*, Space Science and Technologies,
https://doi.org/10.1007/978-981-15-4902-1_16

to make more in-depth use of space resources, sharing, coordination, and intelligence have become the inevitable trend of the future development of space robotics. For example, if the dispersed robots, such as the on-orbit operation robotic system that performs service tasks (i.e., spacecraft assembly and maintenance) and the space mobile robotic system that performs research tasks on the planetary surface, are rationally constructed into a networked robotic system, their space task execution capability, information perception and comprehensive utilization capability, self-determination, planning and decision-making ability, as well as the risk prevention and reliable working ability will be greatly improved, at the same time, the resource consumption of the whole system will be reduced to a certain extent.

16.1.1 Soft Robot

Conventional robots typically consist of a number of rigid structure and motion pairs that connect the structure with a high stiffness together. This traditional "rigid" robot has the advantages of fast motion, accurate pointing, good repeatability, and easy control. However, there are also some deficiencies such as the minimum use space is limited by the size of the robot structure, and there may be a large contact force when contacting with an unexpected environment. Therefore, in recent years, the concept of bionic soft robots has been proposed, which takes the advantages of low rigidity of soft robot structures, reduces the contact force during operation, and enables it to work in a narrow and complex environment without considering the risk of collisions.

Unlike traditional robots based on rigid materials, soft robots originated from natural mollusks, which are designed to imitate the biological soft-bodied structures in nature, and are made of soft elastic materials that can produce large strains [1, 2]. Therefore, theoretically, they can have more DOFs and continuous deformation ability, and they can show the softness that is closer to living creatures. They are a new type of bionic continuum robots.

Based on materials science, mechanology, and control science, soft robots can move and operate by utilizing the physical properties of soft materials. Soft robot research and development is an emerging and rapidly developing field. In recent years, bionic soft robots that imitate octopus, elephant trunk, inchworm, and other animals have been successively and successfully developed and have attracted attention from more and more researchers and engineers.

The European Commission's Framework Program 7 (FP7) started to implement the OCTOPUS Integrating Project in 2009. The participating organizations include world leading research institutions such as the Scuola Superiore Sant'Anna (Italy), the Italian Institute of Technology, the University of Reading (UK), the University of Zurich (Switzerland), and the Weizmann Institute of Israel. In this project, the octopus' muscle structure and nerve system has been analyzed extensively and deeply in biology, by which the robot design is guided. Harvard's George Whitesides team has built a four-legged soft robot with four feet in an "X" configuration using a soft and flexible polymer [3]. It can perform bending motion through air charging and discharging, and complete "inchworm-type motion" by combining simple gait planning. The soft robot can pass through obstacles that may block most rigid metal robots, highlighting the unique advantages of soft robots in narrow space environments. Professor Walker et al. of Clemson University, USA, have undertaken intensive research on flexible elephant trunk robots, using artificial pneumatic muscle as a driving device to develop the OctArm series of flexible robots. The octopus soft robot Octopus Gripper developed by FESTO in cooperation with Beihang University adopts a soft silicone structure. After connected with compressed air, its tentacle can bend inward driven by the pneumatic muscles and can firmly grab objects by the suckers installed on its inner side. Because of its soft materials and special way of gripping, the Octopus Gripper can gently and safely grab objects of different sizes and shapes, and will play an important role in the field of unstructured environments such as space debris capture. Xi'an Jiaotong University, China University of Science and Technology, Tianjin University, Tongji University, Zhejiang University, Huazhong University of Science and Technology, and Shanghai Jiaotong University have also started soft robot research in recent years.

Soft robots mainly made of elastic matrix materials, and the movement is realized by continuous changes its shape in space. The main driving modes are as follows:

(1) Shape Memory Alloy (SMA) drive. SMA is a kind of alloy material that can completely eliminate its deformation at a lower temperature and recover to its original shape after heating. It has a high mass stress ratio and is widely used as a flexible driver material. However, the stress generated by SMA is greatly impacted by temperature changes, which is difficult to reliably control in different environments.
(2) Pneumatic/hydraulic drive. The pneumatic/hydraulic drive is a mature technology with fast reaction speed and high-power density. The disadvantage is that, due to the limitation of the auxiliary system, it requires air and liquid circulation systems, and the driving equipment is large.
(3) Electroactive polymer (EAP) drive. EAP is a new type of polymer functional material that can produce large changes in size or shape under electric field excitation. When the external electrical excitation is canceled, it can return to its original shape and volume, but it requires a higher driving voltage.
(4) Wire drive. As a traditional under-actuation method, it realizes the variable curvature control of each segment of the flexible unit by different wire groupings.

The soft robot is a brand-new type of robot with lightweight, good adaptability to the target shape, and small collision force with the environment. It is naturally low risky in the use process, especially suitable for use in unmanned space conditions. Potential applications of soft robots include:

(1) Adaptive capture of target. For space targets such as space debris and failed satellites, especially space rotation targets, the soft robots for target capture can greatly reduce the requirements for target measurement accuracy and the impact load during the capturing, so as to meet the operational safety requirements.
(2) Narrow space operation. In the spacecraft internal equipment maintenance and space station module operation, there are constraints such as narrow operating

space and large equipment interference, which may easily lead to equipment damage. These risks can be avoided with soft robots.

(3) Long-distance handling of space facilities. Space is a micro-gravity environment, in which the robot generally only needs to overcome the inertial force when performing object handling tasks. Therefore, the use of large-sized soft robots for long-distance handling of objects has a great advantage in cost-effectiveness.

16.1.2 Flying Robot

The application exploration of free-flying robots in space orbits has been well known, but the idea of using them for planetary surface explorations has not yet been implemented. In the planetary surface exploration missions, the most challenging task for the designer is to improve the adaptability to the terrain in addition to overcoming the extreme environment. The topography and geological conditions of the planet's surface are full of uncertainty, which poses a great threat to the safety and reliability of mobile robots. Therefore, it is always a dream for designers to effectively reduce or avoid the influence of planetary surface topographic uncertainty during exploration. In recent years, with the rapid development of UAV technology, it is more attractive to use flying robots to perform exploration tasks on planets with atmosphere environment, and relevant in-depth research is also in progress.

As early as 2005, NASA's Ames Research Center proposed the idea of a micro-flying robot and supported the development of a prototype. In 2017, NASA's New Frontiers Program proposed a rotorcraft robot to be used for Titan exploration. This rotorcraft robot developed by the Applied Physics Laboratory of Johns Hopkins University is called Dragonfly, which uses Titan's low-gravity and dense atmosphere to conduct flight exploration missions.

In recent years, bionic flying robots (flapping-wing flying robots) have been developed rapidly [4]. Such robots can integrate lifting, hovering, and propulsion functions into a flapping-wing system, which is characterized by small size, good maneuverability, and strong anti-interference ability, and more suitable for the planetary surface environment where there is atmosphere but the airflow is unstable. Bat Bot B2 developed by the California Institute of Technology is a representative bionic flying robot (a bat-like robot). During the flight, the bat robot can visually perceive the information of the surrounding environment, while monitoring its own flight state in real time through an inertial measurement device and five encoders located on the joints. At present, the bat robot can perform autonomous flying and cruise, and can make various maneuvers such as oblique flight and steep dive.

The greatest constraint on rotorcraft and flapping-wing robots comes from the need for the atmosphere. Some planets have no atmosphere, or the atmosphere is very thin, where the flying robots cannot work. Of course, there are flying robots that can realize flying by using the same method as spacecraft propulsion, but such flights will cause fuel consumption, which will reduce the operational life and flight efficiency of the robot.

There is no doubt that the flying robot has a great application potential in the planetary surface exploration mission. Its moving efficiency and exploration range are better than the wheeled or legged mobile robot, but the flying robots also have obvious disadvantages. In addition to relying on the atmosphere, the flying height also causes an increase in the distance between the robot and the target, so it is not suitable for detailed scientific survey tasks. Therefore, in the future, the flying robot may be organically combined with the wheeled or legged robots, that is, a large area general survey is realized by flying robot, and local fixed-point detailed investigation by wheeled or legged mobile robots.

16.1.3 Space Cloud Robot

With the continuous development of robots in space applications, the demand for space robots has also evolved from single robot, single application to multi-robot, multi-task support. Multi-robot knowledge sharing and semantic information exchange, information fusion and processing, multi-robot coordinated control and operation, system upgrade and task function expansion are all issues that need to be urgently solved. Moreover, the rapidly developing cloud computing science provides new solutions to these problems [5].

Cloud robotics is the result of a combination of cloud computing technology and robotics. Cloud robotics have two advantages: First, the powerful computing and storage capabilities of cloud computing provides a more intelligent "brain" for robots; second, robots distributed in different locations and with different capabilities can overcome space constraints by information sharing and multi-robot collaboration to complete larger scale and more complex tasks. In addition, the internet and cloud computing can effectively enhance the self-learning ability of robots, shorten the robot development time, and overcome the limitations of single robot self-learning.

On the Humanoids conference 2010, Dr. Kufner of Carnegie Mellon University proposed the concept of cloud robotics for the first time, which evoked extensive discussions. According to Kufner's idea, cloud robotics are a combination of cloud computing and robotics. Like other network terminals, the robot itself does not need to store all data information nor have superior computing capability. They only need to connect to relevant servers and get needed information when required.

Space cloud robotics is a new subject combining cloud computing and space robot. The space "cloud" stores all the information, with superb computing power. The space robot connects to the space "cloud" server and obtains the required information. Potential applications for space cloud robotics include:

(1) All terrain exploration on the moon and Mars. With further deepening of planetary exploration activities, robot should not only expand their exploration range,

but also explore various types of terrain, such as deep pits, caves, and cracks. In this case, a single robot is insufficient in the breadth and depth of exploration, and multi-robot joint exploration becomes a necessity. Moreover, in future planetary exploration missions, such as on moon and Mars, the use of multiple low-configuration robots has more advantages in cost-effectiveness than the use of a small number of high-cost robots. In conclusion, cloud robot technology can better meet the above requirements.

(2) On-orbit assembly and maintenance of large spacecraft. In the future, the size of a super-large spacecraft can reach a kilometer, and its assembly and construction will mainly rely on space robots. The assembly involves multi-robot and a wide range of precise coordination. The use of cloud robot can better coordinate tasks and actions between the robots and improve work efficiency.

16.1.4 Space Multi-robot System

Space multi-robot system refers to a system consisted of a number of robots working in collaboration in an extraterrestrial space to perform a space task. The multi-robot system has the advantages of performing complicated operations that are difficult for a single robot; improving the efficiency of the robot system during the operation; enhancing the environment adaptability of the robot system; and more options for a space task.

Helsinki University of Technology, Luleå University of Technology, and the University of Tokyo have jointly researched multi-robot collaboration and reconstruction strategies and the system adaptability to complex environments in the project Society of Multiple Robots for Space (SMURFS). In the NASA's project Robotic Construction Crew (RCC), a multi-robot system collaboration technology has been studied for the construction of planetary surface living modules, which focuses on multi-robot collaboration in the operation of large objects, including assembly of components, cooperative transportation of objects, etc. [6].

Potential applications for space multi-robot systems include:

(1) On-orbit construction of large space facilities. With the increasing complexity of space missions, the future spacecraft will develop toward large size, such as space solar power stations, large space science platforms and on-orbit service platforms. Due to the rigid constraints on the geometry envelope dimensions and weight of the launch vehicle, this type of spacecraft usually cannot enter orbit in the traditional way of one launch and on-orbit deployment, but need multiple launches and on-orbit assembly/manufacture to complete the construction step by step. For a large spacecraft, the robot is required not only to complete the capture and handling of functional modules, but also to assemble and adjust them with high precision. In this case, it is hard for a single robot to meet the task requirements and must be achieved through multi-robot collaboration.

(2) Exploration of extraterrestrial surface complex terrain. As human exploration of extraterrestrial body progresses, extreme terrains such as caves, deep pits,

and cracks are more attractive to researchers. There are often many dangers and unknown factors in such terrains that can easily cause damage or failure of the robot. The multi-robot collaborative exploration can avoid serious damage to the entire system by mutual rescue in the event of danger, or can achieve a certain degree of recovery of the damaged robot through mutual maintenance, thus greatly improving the system's risk prevention capability. Therefore, multi-robot systems have higher application value in the future extreme terrain explorations of extraterrestrial bodies.

16.1.5 Artificial Intelligence Space Robot

As one of the important extensions of Artificial Intelligence (AI), an inevitable trend in the future development of space robot is the deep integration and application of AI, such as computer vision, speech recognition, natural language processing and machine learning, to form "AI + space robotics" intelligent space robot products.

At present, space robot has a limited level of intelligence and insufficient autonomy and collaboration capabilities. Space robots are mainly used in known and structured environments, and their observation objects are mostly cooperative targets; space robot and human are usually in a master-slave relationship, it acts as a tool to passively receive instructions, perform tasks, and feedback necessary data.

With the gradual integration of AI and space robot, space robot will make a breakthrough in the following aspects:

(1) Environmental perception.

Space robot will fuse multi-modality perception data such as vision, hearing, force, tactile and slippage, and extract multi-dimensional and multi-level information from it, to establish a consistent interpretation and complete description model of the space working environment and the space operation target, thereby improving the reliability, robustness, credibility, and accuracy of space robot's perception and providing a more comprehensive reference and a more reliable basis for motion planning and behavioral decision of space robots.

(2) Human–robot interaction.

In the process of human–robot collaboration of on-orbit operation or planetary exploration, the space robot should understand the intentions expressed by human natural communication means such as gaze, expression, gesture, speech, and body movements, thereby greatly improving the efficiency and reliability of human–robot interaction, and enabling space robot to truly act as human partners or assistants to achieve integration and collaboration.

(3) Autonomous planning.

Based on the analysis and judgment of the working environment and the understanding of human intention, the space robot should autonomously plan working procedures, make decisions and execute motion instructions, that is, it can intelligently form a complete space mobile and operation plan and implement it.

(4) Self-learning.

On the basis of realizing dynamics update and intelligent reasoning of existing knowledge and adaptively establishing optimal control strategy, space robot can continuously acquire new skills and new knowledge through self-learning and mutual learning, and steadily improve task analysis and task planning ability, and eventually, get rid of the dependence on humans to intelligently and autonomously complete mobile and operational tasks.

AI will greatly improve the autonomy and collaboration capabilities of space robot, promote the overall breakthrough of space robot, and continuously improve the perception, cognition, and control capabilities of space robot in complex working environments, as well as the ability to handle complex, emergency, and extreme situation.

16.2 Space Robot Technologies

16.2.1 Dynamic Modeling Technology

With the rapid development of the space industry, the application requirements for space robot are increasingly diversified and complicated, such as the need for space robot to complete on-orbit assembly of large space structure, on-orbit maintenance of faulty spacecraft, and exploration of complex terrains. Corresponding to these diverse missions, the shape of the space robot is also developing in a diversified direction. New robot such as flexible robot and soft robot has put forward new challenges to traditional robot kinematics, especially dynamics theory. Driven by these new space robot applications, future space robot dynamics need to solve the following two problems:

(1) Nonlinear flexible multi-body dynamics. Cable-driven robot and soft robot will be used more and more in future field of space robot because of their lightweight, good dexterity, and ease of large-scale. This kind of space robot, especially soft robot, theoretically has infinite DOFs, its geometric nonlinearity and material nonlinearity cannot be ignored, and the multi-disciplinary coupling effects of mechanical, electrical, liquidity, and chemical sciences, which has dramatically increased the difficulty of robot modeling, must be considered in modeling [7]. Therefore, it is necessary to consider and study the modeling methods by the inherent characteristics of such robots and simplify the modeling on this basis.

(2) Efficient dynamic modeling and solution methods. The on-orbit assembly accuracy of the large spacecraft optical system has reached micron order, which puts forward a higher control precision requirement for the space robot that performs operational tasks. The dynamic model of the robot itself and the operating object for feed forward control is introduced into in the space robot motion control algorithm, which can effectively reduce the influence of the shortcomings such as too much flexibility, low fundamental frequency, and slow response of the assembled system and the operating object, and improve the operational accuracy. Due to the real-time requirements of robot control, the system model should be able to fully reflect the system characteristics on the one hand, and complete the numerical solution in the control cycle of the robot on the other hand.

16.2.2 Motion Planning and Control Technology

Traditional space robot tasks are relatively simple, and its motion is planned mainly according to the robot's own motion performance. With the increase of task complexity and object diversity, the motion of space robot should be planned under the consideration of more environmental constraints, real-time task requirement, multi-robot collaboration, and the requirement for intelligent and autonomous operation.

1. Motion planning under the consideration of environment and target characteristics

For on-orbit operation robot, the current motion planning mainly focused on the kinematics level, and the planning algorithms do not fully consider the dynamics characteristics of the robot, the spacecraft (base), and the operating object. For large flexible space target operation, high-speed target capture and other tasks, considering in advance the dynamics characteristics of related objects in motion planning can help to improve the control characteristics of space robots, but will also lead to the complexity of motion planning algorithms.

2. Autonomous Real-Time planning

With the increasing requirements of space task on operation accuracy and response speed of space robot, the relatively mature ground offline preplanning method is no longer applicable. Space robot should real-time plan path based on mission objectives, operating environment and operational object information, so as to meet the task requirements of on-orbit operation robots for on-orbit capture of noncooperative targets, and of planetary exploration robot in dynamics and dangerous environments on the planetary surface.

3. Intelligent planning and control of multi-robot coordinated motion

In the multi-robot system, robots can effectively complement each other in terms of environment perception, control decision-making, and task execution, so that the whole system has higher adaptability and robustness. On this basis, by using the research results of AI technology, collaborative planning and control between multiple robots can be achieved on the basis of the comprehensive information of "environment-robot-target" perceived by multi-robot, which can realize the optimal motion of a single robot and the objective of coordinated operation of multiple robots at the same time, and will become an important development trend for multi-robot systems in the future.

16.2.3 Force Control Technology

Force control of robot has been developed for nearly 40 years, from stiffness control and damping control to hybrid force-position control, further to impedance control and various adaptive and intelligent algorithms, but there is no unified and mature theory that is universally applicable in engineering practice. Since force is not a state quantity in kinematics equations, force control does not belong to the scope of state-based modern control theory. However, the research on robot force control will promote the development of modern control theory in turn. With the increased application of space robot, it is expected that space robot force control technology will achieve great development in the following aspects:

(1) Force control based on kinematics models. Due to the constraints of realization complexity, computing capacity, cost, etc., the existing force control of robot is mostly developed on the basis of PID control. With the increasing number of robot contacting operation tasks, the higher requirements, the development and maturity of high-speed bus and high-speed processors, model-based force control will gradually replace simple force control methods in engineering practice, while robust adaptive methods will play an increasingly important role in ensuring model adaptability and robustness.
(2) Force control based on multi-sensor fusion. The current force control is mainly realized by three methods: current open-loop control, joint force sensor, and end force sensor. However, compared with human beings, this ability of force perception processing is far from enough. In order to make the robot have faster, more accurate, and comprehensive tactile response capability, a multi-source tactile information fusion system based on pressure-sensitive electronic skin and artificial hair cell sensor will surely become an important development trend of force control, and bring the possibility of a more efficient and intelligent robot tactile control.
(3) Intelligent force control. Space robots are faced with problems such as complex and changing environments and limited human intervention in their tasks. With

increasingly frequent and complex space contacting tasks, space robot must be provided with certain autonomous operating capabilities to complete the decision-making and planning process independently on certain occasions with known and unknown conditions. Up to now, methods such as deep reinforced learning have been applied in force control of robots. Compared with traditional force control methods, the use of machine learning to realize the intelligent force control of space robot and improve the space operation ability, has potential advantages in force control performance, environmental adaptability, and energy efficiency.

16.3 Summary

With the continuous expansion of the human space exploration, the tasks for space robot are becoming more and more complex, and the forms of space robot are increasingly diversified. Corresponding with the application requirements and mission characteristics of space robot, space robot products and technical systems are also constantly enriched and improved. However, due to various limitations of human's ability to reach in the vast space, unmanned operation and exploration will still be the best way for human to explore the universe in the far future. Therefore, the ultimate goal of researchers and engineers is to improve the dexterity, reliability, environmental adaptability, and intelligence of space robot. Taking full advantage of the latest scientific and technological achievements in machinery, perception, control, materials, mechanics, and artificial intelligence, the rapid development of space robotics theory and technology will be promoted, so that space robot will become a powerful tool for human to explore the boundless universe and use space resources.

References

1. C. Lee, N. Kim, Y.J. Kim et al., Soft robot review. Int. J. Control Autom. Syst. **15**(1), 3–15 (2017)
2. Y. Cao, J. Shang, K. Liang et al., Review of soft-bodied robots. J. Mech. Eng. **48**(3), 25–33 (2012)
3. R.F. Shepherd, F. Ilievski, W. Choi et al., Multigait soft robot. Proc. Natl. Acad. Sci. **108**(51), 20400–20403 (2011)
4. G. Wang, D. Chen, Z. Zhang et al., The current research status and development strategy on biomimetic robot. J. Mech. Eng. **51**(13), 27–44 (2015)
5. G. Tian, Y. Xu, Cloud robotics: concept, architectures and key technologies. J. Shandong Univ. (Eng. Sci.) **44**(6), 47–54 (2014)
6. A. Stroupe, T. Huntsberger, B. Kennedy et al., Heterogeneous robotic systems for assembly and servicing, in *Proceedings of The 8th International Symposium on Artificial Intelligence, Robotics and Automation in Space: i-SAIRAS 2005*, Munich, Germany (2005)
7. T. Li, G. Li, Y. Liang et al., Review of materials and structures in soft robotics. Chin. J. Theor. Appl. Mech. **48**(4), 756–766 (2016)